A Survey of Basic Mathematics

A Survey of Basic Mathematics

Fred W. Sparks
FORMERLY TEXAS TECH UNIVERSITY

Charles Sparks Rees
UNIVERSITY OF NEW ORLEANS

Fourth Edition

McGRAW-HILL BOOK COMPANY

NEW YORK   ST. LOUIS   SAN FRANCISCO   AUCKLAND   BOGOTÁ   DÜSSELDORF
JOHANNESBURG   LONDON   MADRID   MEXICO   MONTREAL   NEW DELHI
PANAMA   PARIS   SÃO PAULO   SINGAPORE   SYDNEY   TOKYO   TORONTO

This book was set in Times Roman by Progressive Typographers.
The editors were Carol Napier and Stephen Wagley;
the designer was Albert M. Cetta; the production supervisor was Leroy A. Young.
Webcrafters, Inc., was printer and binder.

A SURVEY OF BASIC MATHEMATICS

Copyright © 1979, 1971, 1965, 1960 by McGraw-Hill, Inc. All rights reserved.
Printed in the United States of America. No part of this publication
may be reproduced, stored in a retrieval system, or transmitted, in any
form or by any means, electronic, mechanical, photocopying, recording, or
otherwise, without the prior written permission of the publisher.
2 3 4 5 6 7 8 9 0 WCWC 7 8 3 2 1 0 9

Library of Congress Cataloging in Publication Data

Sparks, Fred Winchell, date
   A survey of basic mathematics.

   1. Mathematics—1961–   I. Rees, Charles Sparks,
joint author. II. Title.
QA39.2.S67   1979   513'.123   78-17075
ISBN 0-07-059902-5

Contents

	Preface	**xi**
Chapter 1	**Review of Arithmetic**	**1**
1.1	Graphical Representation of Positive Integers	1
1.2	The Hindu-Arabic Number System	1
1.3	Addition	2
1.4	Subtraction	3
	Exercise 1.1	5
1.5	Multiplication	9
1.6	Division	12
1.7	Zero in Division	14
	Exercise 1.2	15
1.8	Fractions	19
	Exercise 1.3	23
1.9	Addition and Subtraction of Fractions	27
	Exercise 1.4	29
1.10	Multiplication and Division of Fractions	31
	Exercise 1.5	33
1.11	Percentage	35
	Exercise 1.6	37
1.12	Scientific Notation and Approximations	41

1.13	The Metric System	42
1.14	Denominate Numbers	43
	Exercise 1.7	45
1.15	Calculators	49
1.16	Sets	50
	Exercise 1.8	53
1.17	Chapter Summary	55
	Exercise 1.9, Review	57
	Exercise 1.10, Chapter Test	61

Chapter 2 Polynomials — 63

2.1	Introduction	63
2.2	Definitions	64
2.3	Axioms of Addition	64
2.4	Law of Signs for Addition	66
2.5	Addition and Subtraction of Monomials and Polynomials	67
2.6	The Relation of Equality	68
	Exercise 2.1	71
2.7	Axioms and Theorems of Multiplication	75
2.8	Law of Signs for Multiplication	76
2.9	Exponents in Multiplication	77
2.10	Products Involving Polynomials	77
2.11	Symbols of Grouping	78
	Exercise 2.2	81
2.12	Division	85
2.13	Monomial Divisors	85
2.14	Quotient of Two Polynomials	87
	Exercise 2.3	89
2.15	Numbers to Various Bases	93
	Exercise 2.4	97
2.16	Chapter Summary	101
	Exercise 2.5, Review	103
	Exercise 2.6, Chapter Test	105

Chapter 3 Special Products and Factoring — 107

3.1	The Product of Two Similar Binomials	107
3.2	The Square of a Binomial	108
	Exercise 3.1	109
3.3	The Product of the Sum and Difference of the Same Two Numbers	111
3.4	Product of Selected Polynomials	111
	Exercise 3.2	113
3.5	Common Factors	115
	Exercise 3.3	117
3.6	The Difference of Two Squares	119
3.7	The Sum or Difference of Two Cubes	119
	Exercise 3.4	121
3.8	Quadratic Trinomials	123
	Exercise 3.5	125
3.9	Chapter Summary	127
	Exercise 3.6, Review	129
	Exercise 3.7, Chapter Test	131

Chapter 4 Algebraic Fractions — 133

4.1	Definitions and Fundamental Principles	133
4.2	Reduction to Lowest Terms	135
4.3	Multiplication of Fractions	135
	Exercise 4.1	137

4.4	Lowest Common Denominator	143
4.5	Addition of Fractions	143
	Exercise 4.2	145
4.6	Complex Fractions	151
	Exercise 4.3	153
4.7	Chapter Summary	157
	Exercise 4.4, Review	159
	Exercise 4.5, Chapter Test	161

Chapter 5 Relations, Functions, and Graphs 163

5.1	Formulas and Charts	163
5.2	Numerical Geometry	164
	Exercise 5.1	167
5.3	Relations	175
5.4	Functions	175
	Exercise 5.2	177
5.5	The Rectangular Coordinate System	179
5.6	The Graph of a Function and a Relation	179
	Exercise 5.3	183
5.7	Chapter Summary	185
	Exercise 5.4, Review	187
	Exercise 5.5, Chapter Test	189

Chapter 6 Exponents and Radicals 191

6.1	Positive Integral Exponents	191
	Exercise 6.1	195
6.2	Negative Exponents	199
	Exercise 6.2	203
6.3	Roots of Numbers	207
6.4	Fractional Exponents	207
	Exercise 6.3	209
6.5	Simplification of Radicals	213
6.6	Rationalizing Monomial Denominators	213
	Exercise 6.4	215
6.7	Changing the Order of a Radical	219
6.8	Products of Polynomials Whose Terms Contain Radicals	219
6.9	Rationalizing Binomial Denominators	220
6.10	Addition of Radicals	220
	Exercise 6.5	223
6.11	Complex Numbers	227
	Exercise 6.6	229
6.12	Chapter Summary	231
	Exercise 6.7, Review	233
	Exercise 6.8, Chapter Test	237

Chapter 7 Linear Equations and Inequalities 239

7.1	Open Sentences	239
7.2	Equations	240
7.3	Equivalent Equations	241
	Exercise 7.1	243
7.4	Linear Equations in One Variable	247
	Exercise 7.2	249
7.5	Linear Inequalities	253
	Exercise 7.3	255
7.6	Solution of Stated Problems	257
	Exercise 7.4	261

7.7	Chapter Summary	269
	Exercise 7.5, Review	271
	Exercise 7.6, Chapter Test	273

Chapter 8 Fractional Equations   275

8.1	Fractional Equations	275
	Exercise 8.1	277
8.2	Fractional Inequalities	279
	Exercise 8.2	281
8.3	Chapter Summary	283
	Exercise 8.3, Review	285
	Exercise 8.4, Chapter Test	287

Chapter 9 Ratio, Proportion, and Variation   289

9.1	Ratios	289
	Exercise 9.1	291
9.2	Proportion	295
	Exercise 9.2	297
9.3	Variation	301
	Exercise 9.3	303
9.4	Chapter Summary	307
	Exercise 9.4, Review	309
	Exercise 9.5, Chapter Test	313

Chapter 10 Systems of Linear Equations and Determinants 315

10.1	Linear Equations in Two Variables	315
10.2	Graphs of Linear Equations in Two Variables	316
10.3	Systems of Two Linear Equations; Graphical Method	317
10.4	Independent, Inconsistent, and Dependent Equations	318
	Exercise 10.1	321
10.5	Elimination by Addition or Subtraction	323
10.6	Elimination by Substitution	324
	Exercise 10.2	325
10.7	Stated Problems	331
	Exercise 10.3	333
10.8	Determinants of the Second Order	339
	Exercise 10.4	341
10.9	Systems of Three Linear Equations	345
10.10	Determinants of the Third Order	345
10.11	Cramer's Rule	347
	Exercise 10.5	349
10.12	Chapter Summary	357
	Exercise 10.6, Review	359
	Exercise 10.7, Chapter Test	365

Chapter 11 Quadratic Equations   367

11.1	Introduction	367
11.2	Solution by Factoring	368
	Exercise 11.1	371
11.3	Solution by Completing the Square	375
	Exercise 11.2	377
11.4	The Quadratic Formula	381
	Exercise 11.3	383
11.5	Radical Equations	389
	Exercise 11.4	391
11.6	Solution of Formulas	395
11.7	Equations in Quadratic Form	395

11.8	The Pythagorean Theorem	395
	Exercise 11.5	397
11.9	Quadratic Inequalities	403
	Exercise 11.6	405
11.10	Chapter Summary	409
	Exercise 11.7, Review	411
	Exercise 11.8, Chapter Test	417

Chapter 12   Logarithms   419

12.1	Introduction	419
	Exercise 12.1	423
12.2	Properties of Logarithms	425
	Exercise 12.2	427
12.3	Common, or Briggs, Logarithms	429
12.4	Scientific Notation	429
12.5	The Characteristic and Mantissa	430
	Exercise 12.3	431
12.6	Use of Tables to Obtain Mantissa	433
	Exercise 12.4	437
12.7	Given Log N, to Find N	439
	Exercise 12.5	441
12.8	Logarithmic Computation	443
	Exercise 12.6	445
12.9	Chapter Summary	451
	Exercise 12.7, Review	453
	Exercise 12.8, Chapter Test	459

Chapter 13   Numerical Trigonometry   461

13.1	Introduction	461
13.2	Trigonometric Ratios of an Acute Angle	462
13.3	Use of Tables and Calculators	463
	Exercise 13.1	469
13.4	Solution of Right Triangles	473
13.5	Some Applications	475
	Exercise 13.2	479
13.6	Trigonometric Ratios for Angles between 90 and 180°	483
13.7	The Oblique Triangle	484
13.8	The Law of Sines	484
13.9	The Ambiguous Case	485
	Exercise 13.3	487
13.10	The Law of Cosines	493
	Exercise 13.4	497
13.11	Chapter Summary	501
	Exercise 13.5, Review	503
	Exercise 13.6, Chapter Test	507

Course Test   **509**

Answers   **517**

Index   **539**

Preface

In this, the fourth edition of "A Survey of Basic Mathematics," we have endeavored to present the basic subject matter of preparatory mathematics in such a way that the student will not only have an opportunity to master the basic skills but also to develop the habit of logical, rigorous thinking which is a necessity for the pursuit of more advanced courses. The book includes (1) a comprehensive review of arithmetic; (2) some algebra, including polynomials, fractions, functions, exponents and radicals, linear and fractional equations, ratio, proportion, variation, quadratic equations, inequalities, and logarithms; and (3) numerical trigonometry. The work on inequalities occurs in several chapters. With only a few exceptions, the problems in this edition are new.

The overall content of the text and the type of exercises are the same as in the third edition. The order of some of the chapters has been changed, and some material has been shifted from one chapter to another. A few topics have been added, notably the metric system, inequalities, and an optional use of calculators. We have also added a chapter summary, a lengthy review exercise, and a short chapter test at the end of each chapter. In addition, we have included a course test at the end of the book. This may be used as a pretest, as an end-of-the-course test, or even as both so as to measure actual progress. We hope that this will enable the instructor to see whether the chapter can be omitted and help students to determine whether they are satisfied with their knowledge of the material treated in the chapter.

We have used boldface throughout the text to emphasize key points and to make it easier to locate specific material.

Most of the sections have been rewritten, and others have been carefully revised. The commutative, associative, and distributive axioms are introduced in the first chapter and used in the development of the processes of arithmetic and of algebra. These axioms are repeated in Chapter 2 on polynomials so that a teacher who wishes to omit Chapter 1 or to give only a brief treatment of it can do so without omitting essential material.

We give exercises a normal lesson apart and in groups of four similar ones of increasing difficulty. From time to time, we make the statement that certain problems in the next exercise can now be worked. We hope that this will be helpful to instructors who may want to spend more than one day on an exercise. There are 59 regular and 13 review exercises in the book besides 13 chapter tests. These exercises are made up of some 3300 problems. Answers are given in the book to three-fourths of the problems and answers to the others are available in a pamphlet. Answers to all problems in the review exercises and chapter tests are in the book.

We think that the book contains enough material to allow the instructor considerable choice in deciding on the topics to cover in a three-semester-hour course. There are enough problems so that only one-fourth of them are needed in a semester for an average class.

The authors are happy to have received many useful suggestions from the reviewers. For their help in this regard, we want to thank Professor Jerald T. Ball, Chabot College–Valley Campus; Professor Neale Fadden, Belleville Area College; Professor Mark Saks, Community College of Philadelphia; Professor Ann Miller, Southern Illinois University at Carbondale. We also thank the staff of McGraw-Hill for their valuable advice and help, especially Carol Napier and Stephen Wagley.

The original author, Fred W. Sparks, has been joined in this edition by his namesake, Charles Sparks Rees.

A note to the student

As a student in a mathematics course, you must do some hard work including a great deal of drilling. Both the "how" and the "why" aspects of mathematics are important. Logical structure, which forms the basis of mathematics, rests on a few undefined terms, definitions, and axioms. These form a basis for proving theorems about the numbers, expressions, and operations we are to use. *Ask questions in class* about the portions of the assignment that you do not understand. This will help you learn some of the things that you do not know and will please your instructor since all of us like to have thoughtful questions from our students. You acquire skill in mathematical operations through practice and drill, and you get these by working the problems in the exercises. The space provided in the text for working the problems cannot and does not allow room for experimentation. *You should work the problem on scratch paper first, then record the essential steps and results in the allotted space.*

Remember that learning mathematics is similar to climbing a ladder. It is not hard if you take the steps one by one as you come to them. If you miss a step, there is danger of not being able to continue the climb. It is essential to do the job a step at a time if you are to get to the top.

Fred W. Sparks
Charles Sparks Rees

A
Survey
of
Basic
Mathematics

Review of Arithmetic

We all use numbers in our daily operations. We may count the number of pieces of bacon and divide by the number of people present to see what our fair share is. We ordinarily need to count, add, subtract, multiply, and divide numbers during a day to see how much we owe and how much change we are due. We may even divide cost by the number of units purchased to find the unit cost, and we may want to express that unit cost in decimal form for use in making comparisons.

1.1 GRAPHICAL REPRESENTATION OF POSITIVE INTEGERS

We shall use the term **positive integer** to indicate a number that is used in counting objects. The symbols 1, 2, 3, 4, 5, . . . are used to represent the positive integers. We can represent them along a half line by marking the beginning point with 0, marking off segments of equal length on the line, and using 1, 2, 3, 4, . . . to designate the right-hand ends of these segments, as indicated in Fig. 1.1. We use a half line rather than a line segment since there is no last positive integer.

1.2 THE HINDU-ARABIC NUMBER SYSTEM

The number system which most of the world uses is based on 10 and powers thereof and is called the Hindu-Arabic number system. It was devised in India about A.D. 500 and spread to the cities of the

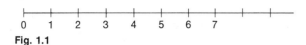

Fig. 1.1

eastern shores of the Mediterranean and to North Africa by A.D. 800, after being incorporated in arithmetic by an Arabian scholar. It was not in general use in Europe and the British Isles until around 1600. The symbols 0, 1, 2, 3, 4, 5, 6, 7, 8, and 9, called **digits,** are used in writing numbers; but the value of a symbol other than 0 depends in part on its position. In 2725, the symbol 5 represents five units, the 2 next to the 5 represents 2 tens or $2(10) = 20$, the 7 represents $7 \times 10^2 = 7(100) = 700$, and the 2 on the left represents $2 \times 10^3 = 2(1000) = 2000$; hence, 2725 represents two thousand seven hundred twenty-five. A dot is placed to the right of the units digit to separate the whole number from the fraction, as shown in Fig. 1.2, and is called a **decimal point.** The decimal point in 807,634.2591 is between the 4 and 2, and the number is

$8(100,000) + 0(10,000) + 7(1000) + 6(100)$
$\quad + 3(10) + 4(1) + 2(0.1) + 5(0.01)$
$\quad\quad\quad\quad\quad\quad + 9(0.001) + 1(0.0001)$

which is read "eight hundred seven thousand six hundred thirty-four and two thousand five hundred ninety-one ten-thousandths." This can also be read "807,634 point two five nine one."

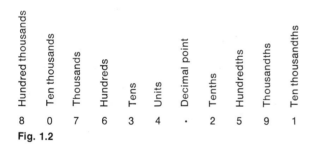

Fig. 1.2

1.3 ADDITION

In addition problems, the numbers added are called the **addends,** and the result of the addition is called the **sum.** In $4 + 3 = 7$, the addends are 4 and 3 and the sum is 7.

In this section, we shall assume that the reader knows the sum of each pair of counting numbers, or integers, from 0 to 9.

We can get the sum of 3 and 4 by going to the right 3 units from 0 and then going 4 units from there or by going 4 units from zero and then 3 units from there. We thus find that $3 + 4 = 4 + 3 = 7$.

We can get the sum of 2, 3, and 6 by adding 2 and 3 and then adding 6 to that sum. Thus, $2 + 3 = 5$ and $5 + 6 = 11$. We can indicate this order of procedure by writing $2 + 3 + 6 = (2 + 3) + 6 = 5 + 6 = 11$. A little experimenting with the number line will show us that $3 + 2$ and $2 + 3$ are equal and that $(2 + 3) + 6 = 2 + (3 + 6) = 2 + (6 + 3) = (2 + 6) + 3$. We are thus led to two of the basic assumptions of arithmetic.

If A, B, and C represent numbers, these assumptions are

$$A + B = B + A \tag{1.1}$$

which is called the **commutative law for addition** and

$$(A + B) + C = A + (B + C) \tag{1.2}$$

which is called the **associative law for addition.**

We shall further illustrate the use of the commutative and associative laws with examples that involve four numbers.

EXAMPLE 1 Using the commutative and associative laws, prove that $[(2 + 3) + 4] + 5 = 3 + [5 + (4 + 2)]$.

Solution

$[(2 + 3) + 4] + 5 = [(3 + 2) + 4] + 5$
$\quad\quad\quad\quad\quad\quad$ by commutative law
$\quad\quad\quad\quad\quad = [3 + (2 + 4)] + 5$
$\quad\quad\quad\quad\quad\quad$ by associative law
$\quad\quad\quad\quad\quad = 3 + [(2 + 4) + 5]$
$\quad\quad\quad\quad\quad\quad$ treating $(2 + 4)$ as
$\quad\quad\quad\quad\quad\quad$ a single number and
$\quad\quad\quad\quad\quad\quad$ using associative law
$\quad\quad\quad\quad\quad = 3 + [5 + (2 + 4)]$
$\quad\quad\quad\quad\quad\quad$ by commutative law
$\quad\quad\quad\quad\quad = 3 + [5 + (4 + 2)]$
$\quad\quad\quad\quad\quad\quad$ by commutative law

The example above illustrates that *we can arrange the addends in any addition problem in any order we please and that we can combine them in any order that is convenient.*

In order to obtain the sum of two or more numbers we arrange them with the units digits in a column, the tens digits in a separate column, the hundreds digits in a third column, etc., and then draw a line below the bottom number. We then add the digits in each column and use the principle of **carrying** to get the sum. Carrying is based on the fact that 10 units is 1 ten, 10 tens is 1 hundred, 10 hundreds is 1 thousand, ...

In Example 2 we show the steps in the addition process that are to be performed mentally at the left of the column and indicate with flow lines the place where the result of each operation is recorded.

EXAMPLE 2 Find the sum of 625, 36, 453, 247, and 519.

Solution We write the addends in a column as shown below. The digits 1 and 3 appearing above the hundreds and tens columns are the digits that have been carried. The steps in the addition that are to be performed mentally are shown at the left, and the flow lines and arrows indicate the place where the result of each operation is to be recorded.

1. The sum of the digits in the units column is 30, or 3 tens (digits carried) and 0 units (sum)

2. The sum of the digits in the tens column, including the 3 that is carried, is 18. This is 18 tens, or 1 hundred and 8 tens

3. The sum of the digits in the hundreds column, including the 1 that is carried, is 18, or 18 hundreds, which we write as 1 thousand and 8 hundreds

```
  1 3
  6 2 5
    3 6
  4 5 3
  2 4 7
  5 1 9
1 8 8 0
```

If the addends in an addition problem involve decimal fractions, we write the numbers so that *the decimal points form a vertical column*. Then the columns are added.

EXAMPLE 3 Find the sum of 361.52, 5278.41, 23.12, 0.37, and 7.64.

Solution We write the above numbers in the form shown at the right with the decimal points in one column and then perform the addition as in Example 2.

```
 361.52
5278.41
  23.12
   0.37
   7.64
5671.06
```

Probably the most often used **check for addition** is to add the columns upward and then downward. If no mistake is made, the same sum will be obtained in the two operations. This, however, is not an infallible check, since it does not guarantee that an error was not made in both operations.

ZERO IN ADDITION Another assumption about numbers we shall make is that for every number A

$$A + 0 = 0 + A = A \qquad (1.3)$$

Thus we have $3 + 0 = 3$ and $0 + 15 = 15$. The number 0 is called the **additive identity.**

1.4 SUBTRACTION

Associated with the concept of zero is the concept of an **additive inverse.** Thus for any number S we assume that another number $-S$ exists such that

$$S + (-S) = (-S) + S = 0$$

Hence $3 + (-3) = 0$ and $(-14) + 14 = 0$. We then define **subtraction** by saying that

$$M - S \quad \text{means} \quad M + (-S)$$

If one number is subtracted from another, the result is called the **difference;** the number being subtracted is the **subtrahend,** and the other is the **minuend.** By use of the number line or ray, we begin at M and go S units toward 0 and arrive at R, as in Fig. 1.3. It may help to realize that writing $M - S = R$ really means finding the number R that must be added to S to produce M; that is, $R + S = M$. Thus $9 - 5 = 4$ since 4 is the number that must be added to 5 to produce 9.

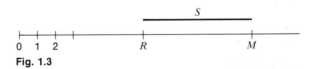

Fig. 1.3

EXAMPLE 1 Subtract 1852 from 3647.

Solution We begin by writing the subtrahend under the minuend, as below, and perform the operations.

3647	minuend
−1852	subtrahend
1795	remainder or difference

We subtract the units digit 2 of the subtrahend from the 7 and get the 5 that is placed in the unit position. Since the tens digit 5 is greater than the 4, we borrow 1 hundred from the 6 hundred and have (4 + 10), or 14 tens, and take 5 tens from the 14, thereby obtaining the 9 that is in the tens digit position of the difference. Before subtracting the hundreds digits, we must recall that we have only 5 (not 6) hundred in the minuend since we borrowed 1 hundred; furthermore, we realize that 8 is greater than 5 and therefore borrow 1 thousand or 10 hundred. Now we subtract 8 from 6 − 1 + 10 = 15 and get 7. Finally, we subtract 1 from 3 − 1 and get 1; hence, the remainder is 1795. This can be checked without any additional writing by seeing whether the sum of the remainder (bottom number) and subtrahend (middle number) is the minuend (top number).

If there is a decimal point in the minuend or subtrahend, we align the decimal points before subtracting.

EXAMPLE 2 Subtract 271.346 from 654.27.

Solution We write the numbers in the usual form and have

654.270   subtrahend
271.346   minuend
382.924   remainder

This can be checked as usual by seeing whether the sum of the remainder and minuend is the subtrahend.

EXERCISE 1.1 ADDITION AND SUBTRACTION

0 Fill in the blanks in the diagram given below in such a way that each entry is the sum of the number in the first column to the left of the blank and the number in the first row above the blank. Check your table against the one given in the answers. Study each one you miss and then form another chart. Repeat as desirable.

+	1	4	7	10	6	12	3	0	9	5	8	11	2
3													
7													
8													
2													
10													
1													
0													
12													
5													
4													
9													
11													
6													

Chapter 1: Review of Arithmetic

Give the name of the law or laws used in justifying each of the following.

1. $3 + 5 = 5 + 3$
2. $4 + 7 = 7 + 4$
3. $2 + (3 + 5) = (2 + 3) + 5$
4. $(1 + 6) + 3 = 1 + (6 + 3)$
5. $1 + (3 + 7) = 7 + (1 + 3)$
6. $(2 + 5) + 8 = (5 + 8) + 2$
7. $(4 + 2) + 9 = (9 + 2) + 4$
8. $(3 + 1) + 6 = (6 + 1) + 3$

Find the sum indicated; perform the additions rapidly and repeat each one that you miss.

9. $2 + 5 =$
10. $7 + 6 =$
11. $9 + 4 =$
12. $5 + 4 =$
13. $2 + 3 + 8 =$
14. $3 + 5 + 4 =$
15. $4 + 5 + 6 =$
16. $3 + 6 + 7 =$
17. $1 + 2 + 3 + 7 =$
18. $2 + 3 + 4 + 8 =$
19. $3 + 1 + 4 + 6 =$
20. $5 + 2 + 2 + 4 =$

Find the sums. Check at least the first eight by adding from top to bottom and from bottom to top.

21. 15
    22
    34
    76

22. 36
    21
    47
    53

23. 45
    21
    37
    86

24. 74
    36
    52
    47

25	261 107 489 764	**26**	593 184 237 645
27	428 357 595 674	**28**	534 257 463 349
29	372 5814 29 446	**30**	813 3742 38 8349
31	28 357 8143 264	**32**	7683 254 79 387
33	791.36 84.721 37.87 159.45	**34**	4971.3 368.42 71.86 235.6
35	8.76 87.6 876.54 8765.4	**36**	2345.67 765.4 543.21 76.5

Subtract the second number from the first.

37	59 − 34	**38**	76 − 42
39	84 − 23	**40**	97 − 85
41	83 − 47	**42**	75 − 48

8 | Chapter 1: Review of Arithmetic

43   94
     − 56

44   32
     − 13

45   597
     − 264

46   873
     − 251

47   438
     − 125

48   634
     − 214

49   738
     − 257

50   893
     − 276

51   726
     − 358

52   435
     − 376

53   789.36
     − 287.15

54   805.79
     − 694.88

55   57.423
     − 13.541

56   2.8793
     − 1.7886

57   426.34
     − 174.8

58   57.46
     − 29.376

59   405.28
     − 397.32

60   38.4
     − 29.637

Show that the following arrays have the same sum for each row, column, and main diagonal. Such arrays are called **magic squares.**

61 $\begin{bmatrix} 4 & 8 & 3 \\ 4 & 5 & 6 \\ 7 & 2 & 6 \end{bmatrix}$

62 $\begin{bmatrix} 2 & 9 & 4 \\ 7 & 5 & 3 \\ 6 & 1 & 8 \end{bmatrix}$

63 $\begin{bmatrix} 9 & 23 & 10 \\ 15 & 14 & 13 \\ 18 & 5 & 19 \end{bmatrix}$

64 $\begin{bmatrix} 7 & 12 & 14 \\ 18 & 11 & 4 \\ 8 & 10 & 15 \end{bmatrix}$

1.5 MULTIPLICATION

If one number is multiplied by another, the two numbers are called the **factors,** and the result of the multiplication is called the **product.** Thus in 7 × 6 = 42 the factors are 7 and 6, and the product is 42. This product can be obtained by adding 7 sixes (or 6 sevens), but this and other products of numbers from 1 to 10 should be memorized if you do not already know them.

Multiplication is indicated by the times sign × and also by parentheses

3(10) = 3 × 10

Use of a center point, another way of indicating multiplication, is generally limited to expressions like $a \cdot a \cdot a \cdot a$, which means the same as $a \times a \times a \times a$.

Table 1.1 indicates the product of any two integers from 1 to 10. To find the product of two numbers, we locate one of them in the left-hand

Table 1.1

×	1	2	3	4	5	6	7	8	9	10
1	1	2	3	4	5	6	7	8	9	10
2	2	4	6	8	10	12	14	16	18	20
3	3	6	9	12	15	18	21	24	27	30
4	4	8	12	16	20	24	28	32	36	40
5	5	10	15	20	25	30	35	40	45	50
6	6	12	18	24	30	36	42	48	54	60
7	7	14	21	28	35	42	49	56	63	70
8	8	16	24	32	40	48	56	64	72	80
9	9	18	27	36	45	54	63	72	81	90
10	10	20	30	40	50	60	70	80	90	100

column and the other in the top row; then their product is the entry in the row of the first number and the column of the second.

EXAMPLE 1 Find the product of 7 and 9 and that of 9 and 7.

Solution To find 7×9, we locate 7 in the left-hand column and 9 in the top row, then note that 63 is the entry in the row that 7 is in and the column that contains 9 at its top. We find 9×7 in the same manner and see that it also is 63.

This example illustrates one of the laws of multiplication.

For any two numbers A and B

$$A \times B = B \times A \quad (1.4)$$

This is known as the **commutative law of multiplication.**

We can see that $2 \times (3 \times 5) = (2 \times 3) \times 5 = 30$ by performing the indicated multiplications. This illustrates another of the laws of multiplication.

For any three numbers A, B, and C

$$A \times (B \times C) = (A \times B) \times C \quad (1.5)$$

This is known as the **associative law of multiplication.**

EXAMPLE 2 Prove that $2 \times (3 \times 7) = (7 \times 2) \times 3$ by use of the associative and commutative laws.

Solution

$$\begin{aligned}
2 \times (3 \times 7) &= 2 \times (7 \times 3) &\text{by commutative law} \\
&= (2 \times 7) \times 3 &\text{by associative law} \\
&= (7 \times 2) \times 3 &\text{by commutative law}
\end{aligned}$$

A third law deals with situations of the type $4 \times (2 + 5)$, which involves both addition and multiplication. Here, the sum of the number in the parentheses is 7 and $4 \times 7 = 28$. If, however, we multiply each of the numbers in the parentheses by 4 and add the products, we get $8 + 20 = 28$. Hence, $4 \times (2 + 5) = (4 \times 2) + (4 \times 5)$. We assume that this property holds for any situation in which the sum of two numbers is to be multiplied by a third number, and state this assumption in this way:

$$A \times (B + C) = (A \times B) + (A \times C)$$
$$\text{distributive law} \quad (1.6)$$

This law is called the **distributive law** with respect to multiplication. Its usefulness is appreciated in algebra, where the quantities enclosed in parentheses cannot be readily combined as 2 and 5 can in $4 \times (2 + 5)$. This law states that we can distribute the multiplication without adding first.

The distributive law can be extended to cover situations in which more than two numbers appear in the parentheses.

EXAMPLE 3

$$\begin{aligned}
7 \times (2 + 4 + 3 + 5) \\
= (7 \times 2) + (7 \times 4) + (7 \times 3) + (7 \times 5) \\
= 14 + 28 + 21 + 35 = 98
\end{aligned}$$

which is true since $7 \times (2 + 4 + 3 + 5) = 7 \times 14 = 98$.

The numbers 1 and 0 play special roles in multiplication. We define the product of any number A and 1 to be A, or, stated in symbols,

$$A \times 1 = 1 \times A = A \quad (1.7)$$

1 is the **multiplicative identity.**

MULTIPLICATION BY ZERO In the next chapter [see (2.23)], we shall show that if A stands for any number, then

$$A \times 0 = 0 \times A = 0 \quad (1.8)$$

For the present we shall accept statement (1.8) as a fact.

We shall next show by means of several examples how the distributive law and the principle of place value are used to obtain the product of two numbers where at least one of them contains two or more digits.

EXAMPLE 4 Find the product of 2 and 143.

Solution

$$\begin{aligned}
2 \times 143 &= 2 \times (1 \text{ hundred} + 4 \text{ tens} + 3 \text{ units}) \\
&\qquad\text{by principle of place value} \\
&= (2 \times 1 \text{ hundred}) + (2 \times 4 \text{ tens}) \\
&\quad + (2 \times 3 \text{ units}) \quad \text{by distributive law} \\
&= 2 \text{ hundreds} + 8 \text{ tens} + 6 \text{ units} \\
&\qquad\text{by multiplication table} \\
&= 286 \quad \text{by principle of place value}
\end{aligned}$$

EXAMPLE 5 Find the product of 4 and 348.

Solution We shall show the steps in the solution. The reader should supply the reason for each step.

$4 \times 348 = 4 \times (3 \text{ hundreds} + 4 \text{ tens} + 8 \text{ units})$
$= 12 \text{ hundreds} + 16 \text{ tens} + 32 \text{ units}$
$= (10 + 2) \text{ hundreds} + (10 + 6) \text{ tens}$
$\quad + (30 + 2) \text{ units}$
$= 10 \text{ hundreds} + 2 \text{ hundreds} + 10 \text{ tens}$
$\quad + 6 \text{ tens} + 30 \text{ units} + 2 \text{ units}$
$= 1 \text{ thousand} + 2 \text{ hundreds} + 1 \text{ hundred}$
$\quad + 6 \text{ tens} + 3 \text{ tens} + 2 \text{ units}$
$= 1 \text{ thousand} + 3 \text{ hundreds} + 9 \text{ tens}$
$\quad + 2 \text{ units}$
$= 1392$

The usual procedure for obtaining the product 4×348 of Example 5 will now be discussed. We begin by writing

$$\begin{array}{r} 348 \\ \underline{4} \\ 1392 \end{array}$$

We then multiply 4 by 8 and get $32 = (3 \text{ tens} + 2 \text{ units})$. We put the 2 in the units position and carry the 3 tens by adding them to the 16 tens obtained by multiplying 4 by 4 tens. That gives $3 + 16 = 19$ tens $= 9$ tens $+ 10$ tens. We then write the 9 tens in the tens position and carry the 10 tens $= 1$ hundred, by adding them to the $4 \times 3 = 12$ hundreds. This gives 13 hundreds $= 1$ thousand $+ 3$ hundreds, which we put in position.

EXAMPLE 6 Find the product of 372 and 4628.

Solution

$372 \times 4628 = 4628 \times 372 \quad$ by commutative law
$\qquad = 4628 \times (300 + 70 + 2)$
$\qquad \qquad$ by principle of place value
$\qquad = 4628 \times (2 + 70 + 300)$
$\qquad \qquad$ by commutative law
$\qquad = (2 \times 4628) + (70 \times 4628) + (300 \times 4628)$
$\qquad \qquad$ by distributive and commutative laws

Each of the products in the parentheses above is called a **partial product.** Each can be computed by methods previously discussed, and the complete product can then be obtained by addition. The usual method for computing this product is given below.

$$\begin{array}{rr} & 4\,6\,2\,8 \\ & 3\,7\,2 \\ \hline 4628 \times 2 = & 9\,2\,5\,6 \\ 4628 \times 70 = & 3\,2\,3\,9\,6\,\emptyset \\ 4628 \times 300 = & 1\,3\,8\,8\,4\,\emptyset\,\emptyset \\ \hline \text{Sum of the partial products} = & 1{,}7\,2\,1{,}6\,1\,6 \end{array}$$

The zeros, shown as $\emptyset$, with a line through them, are not usually written in the partial products since they contribute nothing to the sum in their columns.

EXAMPLE 7 Find the product of 37.2 and 4.628.

Solution The sequence of digits in each of these numbers is the same as in one of the numbers in Example 6. Consequently, the sequence of digits in this product is the same as in Example 6, and we need only decide on the proper position of the decimal point. There are two methods for doing this. One is the **commonsense method.** If it is used, we notice that the product is about 40 by about 5; hence, the product should be near 200. The only position for the decimal point in the sequence of digits 1,721,616 (as obtained in Example 6) giving a number near 200 is between the 2 and the 1. Therefore, $37.2 \times 4.628 = 172.1616$.

The other method for locating the decimal is by a rule, but, like many rules, it can be justified. In order to help arrive at the rule, consider

$P = 2.7(0.43) = (2 + 0.7)(0.4 + 0.03)$
$\qquad = 2(0.4) + 2(0.03) + 0.7(0.4) + 0.7(0.03)$

This product includes some (7) tenths times some (3) hundredths; hence it involves some thousandths. This conclusion could have been reached by adding the number of digits to the right of the decimal in 2.7 and the number to the right of the decimal point in 0.43.

LOCATING THE DECIMAL POINT This type of argument can be used to arrive at the conclusion that *the number of digits to the right of the decimal point in a product is equal to the sum of the numbers of digits to the right of the decimals in the two factors.*

EXAMPLE 8 Find the product of 31.4 and 4.732.

Solution The sequence of digits in this product is readily seen to be 1485848. The sum of the number of digits to the right of the decimal in the factors is $1 + 3 = 4$. Therefore, the product is 148.5848.

If an integer bigger than 1 has no integral factors besides itself and 1, it is called a **prime number.** Thus, 2, 3, 5, 7, 11, 13, and 17 are prime numbers, but $6 = 3(2) = 2(3)$ is not a prime; it is called a **composite number.** The **fundamental theorem of arithmetic** states that any positive integer can be written as a product of prime numbers. This can be done in only one way except for the order of the factors. For example, $105 = 7(5)(3)$ and $44 = 2(2)(11)$. It is believed (Goldbach conjecture) that any even integer can be written as the sum of two prime numbers. For example $8 = 5 + 3$ and $16 = 13 + 3 = 11 + 5$.

Problems 1 to 28, 45 to 52, and 77 to 84 in Exercise 1.2 may be done now.

1.6 DIVISION

Division is the inverse of multiplication and is defined by the statement

$a \div b = x$ if and only if $b \times x = a$ where $b \neq 0$

The number a is called the **dividend** or **numerator**, the number b is called the **divisor** or **denominator**, and x is the **quotient**. In terms of them

Dividend $\div$ divisor = quotient

EXAMPLE 1 $12 \div 4 = 3$ since $4 \times 3 = 12$. Just as multiplication is a shorthand method of addition, we can think of division as a shorthand method of subtraction. If a and b are integers, with a being a multiple of b, then $a \div b$ is the number of times b must be subtracted from a to obtain zero. Thus, $12 - 3 - 3 - 3 - 3 = 0$; hence $12 \div 3 = 4$. Further

$$16 \div 7 = \frac{16}{7} = \frac{14 + 2}{7} = \frac{14}{7} + \frac{2}{7} = 2 + \frac{2}{7} = 2\frac{2}{7}$$

EXAMPLE 2   Find the quotient if 1974 is divided by 7.

Solution

$$\frac{1974}{7} = \frac{1900 + 70 + 4}{7}$$
$$= \frac{1400 + 500 + 70 + 4}{7}$$
$$= \frac{1400}{7} + \frac{570 + 4}{7}$$
$$= 200 + \frac{560 + 10 + 4}{7}$$
$$= 200 + \frac{560}{7} + \frac{10 + 4}{7}$$
$$= 200 + 80 + \frac{14}{7}$$
$$= 200 + 80 + 2 = 282$$

The usual process of short division is a condensed version of the solution just given for Example 2. We shall begin by dividing 7 into 19.

$$7 \overline{|1\ 9\ 7\ 4}$$
$$2\ 8\ 2$$

That gives a quotient of 2, which is written below the 9, and a remainder of 5. This is 5 hundreds = 50 tens and along with the 7 tens in 1974, we have 57 tens. We divide 57 by 7 and get a quotient of 8 and a remainder of 1 ten, or 10. Along with the 4 in 1974 this gives 14 units, which we divide by 7 and get 2. Thus the quotient is 282, as written below 1974. As a check $7(282) = 1974$.

EXAMPLE 3   Find the quotient if 95,228 is divided by 7.

Solution   We shall give the solution without explanation:

$$7 \overline{|9\ 5{,}2\ 2\ 8}$$
$$1\ 3{,}6\ 0\ 4$$

If the dividend and divisor contain two or more digits, we use the long division process, which is a condensation of the ancient method of obtaining a quotient by repeated subtractions. We shall explain the process by showing how it would be used to divide 150,475 by 325. The process would consist of the following steps.

1. Subtract 325 from 150,475.
2. Subtract 325 from the remainder obtained in the first step.
3. Subtract 325 from the second remainder.
4. Continue this process until a remainder less than 325 is obtained.

The number of subtractions made in the process is the required quotient. This long process can be abbreviated in the following way. First, we notice that 150,475 is between $400 \times 325 = 130{,}000$ and $500 \times 325 = 162{,}500$. Hence, we subtract 130,000 from 150,475 and get 20,475. The latter is the remainder after making 400 successive subtractions. Now, 20,475 is between $60 \times 325 = 19{,}500$ and $70 \times 325 = 22{,}750$, so we subtract 19,500 from 20,475 and get 975, which is the remainder after 460 subtractions. Obviously, if we subtract 3×325 from 975 we get 0. So the quotient is $400 + 60 + 3 = 463$. A study of the usual long division process given in Example 4 will reveal that it is just a condensation of the above process.

EXAMPLE 4 Divide 150,475 by 325.

Solution We write the dividend in the position indicated below and then proceed as follows:

1. Start with the first digit to the left in the dividend and move to the right until a number greater than the divisor is obtained.

$$
\begin{array}{r}
4\,6\,3 \\
325\overline{)1\,5\,0\,4\,7\,5} \\
1\,3\,0\,0 \\
\hline
2\,0\,4\,7 \\
1\,9\,5\,0 \\
\hline
9\,7\,5 \\
9\,7\,5 \\
\hline
0\,0\,0
\end{array}
\qquad \text{quotient} \atop \text{dividend}
$$

This is the first partial dividend, and here it is 1504. Divide 1504 by 325 and get 4 with a remainder. Write 4 as the first digit in the quotient, placing it above the last digit in the first partial dividend. The remainder is obtained by subtracting the product $4 \times 325 = 1300$ from 1504 and thus obtain 204.

2. Bring 7 down from dividend and obtain 2047. $2047 \div 325 = 6$ with remainder. We write 6 as second digit in quotient and obtain remainder as in step 1: $6 \times 325 = 1950$. Then $2047 - 1950 = 97$.

3. Bring 5 down from dividend to obtain 975. Then $975 \div 325 = 3$. Write 3 as third digit in quotient and proceed as in steps 1 and 2, or $3 \times 325 = 975$.

$$
\begin{array}{r}
4\,6\,3 \\
325\overline{)1\,5\,0\,4\,7\,5} \\
\rightarrow 1\,3\,0\,0 \\
\hline
\rightarrow 2\,0\,4\,7 \\
1\,9\,5\,0 \\
\hline
\rightarrow 9\,7\,5 \\
\rightarrow 9\,7\,5 \\
\hline
0\,0\,0
\end{array}
\qquad \text{quotient} \atop \text{dividend}
$$

The remainder is a number less than the divisor such that

Dividend = (divisor × quotient) + remainder

If the dividend is not a multiple of the divisor, a remainder will be obtained in the final step in the division process. For example, in the division below, the final step requires the division of 22 by 3. This yields the quotient 7 and the remainder 1.

$$3\overline{)5^29^22}$$
$$1\,9\,7 \qquad \text{remainder, } 1$$

We shall now consider the position of the decimal point in a quotient. We have seen that the number of digits to the right of the decimal in a product is equal to the sum of the numbers of digits to the right of the decimals in the factors. This is true regardless of the names of the factors and product. In particular it is true in

Divisor × quotient = dividend

if the number of digits to the right of the decimal is finite in all three numbers. Thus, the (finite) number of digits to the right of the decimal in the dividend is equal to the sum of the numbers of digits to the right of the decimal points in the divisor and quotient.

LOCATING THE DECIMAL POINT IN A QUOTIENT Consequently, if finite, *the number of digits to the right of the decimal point in the quotient of two numbers is equal to the number in the dividend minus the number in the divisor.*

EXAMPLE 5 Find the number of digits to the right of the decimal point if 9.415 is divided by 3.5.

Solution Since there are three digits to the right of the decimal point in the dividend and one in the divisor, it follows that there are $3 - 1 = 2$ in the quotient. The quotient is 2.69.

If the dividend and divisor contain decimals, the quotient and remainder are obtained just as if there were no decimals but the decimal must be properly placed in the quotient and remainder.

EXAMPLE 6 Find the quotient and remainder if 3.8621 is divided by 0.83.

Solution We shall find the sequences of digits in the quotient and remainder in the usual manner and then locate the decimal points.

```
            4 6 5
    _____
0.83│3.8 6 2 1
     3 3 2
     ─────
       5 4 2
       4 9 8
       ─────
         4 4 1
         4 1 5
         ─────
             2 6
```

Hence, the quotient is 4.65, and the remainder is 0.0026.

If the dividend has fewer digits to the right of the decimal than there are in the divisor, we add zeros on the right of the dividend until it has at least as many decimal places as there are in the divisor.

EXAMPLE 7 Divide 28.3 by 1.863 and give the result to two decimal places.

Solution We must add enough zeros to have two more decimal places in the dividend than in the divisor; hence, we add 4 zeros before dividing and have $28.30000 \div 1.863$. Performing the division in the usual manner, we find the quotient to be 15.19.

EXAMPLE 8 Show that if we divide 28.4 by 1.32, we obtain the quotient $21.51515\cdots$, where the dots indicate that the quotient can be extended indefinitely by successively annexing the digits 1 and 5 in that order at the right.

Solution We first notice that

$$\frac{28.4}{1.32} = \frac{284}{13.2} = \frac{2840}{132}$$

We shall now show three steps in the division process and then explain the conclusions we can draw from the results.

```
          2 1.5 · · ·
       _____
132│2 8 4 0.0
    2 6 4
    ─────
      2 0 0
      1 3 2
      ─────
        6 8 0
        6 6 0
        ─────
          2 0
```

We obtain a remainder of 20 after the first step. The next two digits in the quotient were 1 and 5, and we again obtained a remainder of 20. Hence, the digits 1 and 5 will be repeated indefinitely, and we write the quotient as $21.51515\cdots$, where the dots indicate that the digits 1 and 5 will be repeated indefinitely.

1.7
ZERO IN DIVISION

If we let $b = 0$ in $a \div b = x$ provided $bx = a$, we have $0(x) = a$. If a is different from 0, the statement becomes $0(x) = a$ and is clearly false since 0 times any number is 0; consequently, *division by zero is not a permissible operation*. For example, $4 \div 0$ is not defined.

If, in $bx = a$, $a = 0$ and $b \neq 0$, then $bx = 0$; hence $x = 0$. Therefore, *zero divided by any nonzero number is zero*.

DIVISION INTO ZERO If we think of division in terms of the subtraction process, we see that

$$\frac{0}{a} = 0 \qquad a \neq 0$$

since zero is the number of times that a must be subtracted from 0 to obtain 0.

Similarly, in $0(x) = a$, if $a = 0$, then by (1.8) any x could be used. For this reason, $0 \div 0$ is not defined.

EXERCISE 1.2 MULTIPLICATION AND DIVISION

0 Fill in the blanks in the diagram given below in such a way that each entry is the product of the number at the left end of the row and top of the column that contains the blank. Check your entries against those in the answers, and study any you miss.

×	7	12	5	10	3	8	0	6	11	4	9	2
2												
5												
9												
1												
7												
3												
11												
8												
6												
10												
4												
12												

16 | **Chapter 1: Review of Arithmetic**

Verify the statement and give the axiom or axioms that can be used to justify it.

1 $8 \times 5 = 5 \times 8$

2 $3 \times 7 = 7 \times 3$

3 $4 \times 9 = 9 \times 4$

4 $6 \times 8 = 8 \times 6$

5 $(2 \times 4) \times 7 = 2 \times (4 \times 7)$

6 $(3 \times 5) \times 2 = 3 \times (5 \times 2)$

7 $3 \times (5 \times 6) = (3 \times 5) \times 6$

8 $(4 \times 7) \times 8 = 4 \times (7 \times 8)$

9 $(2 \times 3) \times 5 = 2 \times (5 \times 3)$

10 $(5 \times 7) \times 9 = 5 \times (9 \times 7)$

11 $(3 \times 6) \times 4 = (4 \times 6) \times 3$

12 $(2 \times 8) \times 5 = (5 \times 8) \times 2$

13 $4 \times (2 + 3) = (4 \times 2) + (4 \times 3)$

14 $3 \times (1 + 3) = (3 \times 1) + (3 \times 3)$

15 $5 \times (4 + 1 + 7) = (5 \times 4) + (5 \times 1) + (5 \times 7)$

16 $(9 + 2 + 4) \times 6 = (6 \times 4) + (2 \times 6) + (6 \times 9)$

Perform the multiplications indicated.

17 83
    $\times 2$

18 47
    $\times 3$

19 56
    $\times 5$

20 29
    $\times 7$

21 358
    $\times 38$

22 429
    $\times 57$

23 673
    $\times 46$

24 940
    $\times 89$

25 1369
    $\times 257$

26 2578
    $\times 483$

27 4032
    $\times 509$

28 7304
    $\times 603$

Perform the divisions indicated.

29 3)783   30 7)784

31 6)954   32 7)875

33 8)2592   34 9)3807

35 4)2172   36 5)9145

Find the quotient and remainder.

37 6)3403   38 7)5967

39 4)2379   40 8)7988

41 29)32,831   42 37)58,436

43 53)43,597   44 76)80,977

Perform the operations indicated and give the product or quotient and remainder. Add zeros to the dividend if needed.

45   7.62   46   89.3
    × 1.3   × 2.7

47   7.39   48   0.803
    × 0.45   × 8.9

49   94.2   50   0.394
    × 1.73   × 8.65

51   5.04   52   78.7
    × 3.94   × 21.9

53 1.23)473.42   54 7.85)603.87

18 | **Chapter 1: Review of Arithmetic**

55 $0.352\overline{)7.3493}$ 56 $504\overline{)123.76}$

57 $4.62\overline{)8621.4}$ 58 $2.37\overline{)5734.1}$

59 $0.386\overline{)792.47}$ 60 $0.495\overline{)3472.4}$

Add zeros to the dividend and find each quotient to two decimal places.

61 $7.4\overline{)86}$ 62 $87\overline{)234}$

63 $9.3\overline{)437}$ 64 $0.59\overline{)8.713}$

By adding zeros at the right of the dividend, carry the division forward in each of the following problems until a remainder of 0 is obtained or until a portion of the quotient starts repeating.

65 $33\overline{)237}$ 66 $11\overline{)321}$

67 $136\overline{)561}$ 68 $238\overline{)630.7}$

Find the products and quotients indicated.

69 $3 \times 0 =$ 70 $0 \times 0 =$

71 $0 \times 17 =$ 72 $29 \times 0 =$

73 $0 \div 3 =$ 74 $4 \div 0 =$

75 $9 \div 0 =$ 76 $0 \div 13 =$

Factor the number into prime factors.

77 $77 =$ 78 $78 =$

79 $79 =$ 80 $80 =$

Express the number as the sum of two primes.

81 $18 =$ 82 $24 =$

83 $30 =$ 84 $36 =$

1.8 FRACTIONS

We have pointed out that an extension of the principle of place value led to decimal fractions, but we did not discuss the general nature of fractions or how the concept of a fraction evolved. Chronologically the natural numbers 1, 2, 3, 4, ... came into existence first, probably as scratches on a cliff or as some other method of tallying by means of which people could keep track of their possessions or their accomplishments. As the natural numbers developed, there grew up rudimentary methods of addition and multiplication. These two processes always produced natural numbers because the sum of two natural numbers is a natural number and the product of two natural numbers is a natural number. When, however, in the course of time the early thinkers considered the inverse of these operations, subtraction and division, they ran into trouble because it was not always possible to subtract one natural number from another or to divide one natural number by another and get a natural number as a result. The problem of subtraction made necessary the invention of negative numbers which will be treated further in a later chapter. In the operation of division it eventually became customary to write the results of such operations as $3 \div 7$ or $4 \div 5$ as $\frac{3}{7}$ or $\frac{4}{5}$ and to call them fractions. These indicated quotients were, in time, recognized as numbers, and methods for dealing with them were developed. We shall discuss numbers of this type in this section.

A **rational number** is the quotient of two natural numbers.

For example, $\frac{2}{3}$, $\frac{7}{8}$, and $\frac{13}{5}$ are rational numbers.

The number above the line in a fraction is called the **numerator** and the number below the line is the **denominator.**

If we divide a line segment into three equal parts, two of these parts, or *two one-thirds*, is $\frac{2}{3}$ of the line. Since *two one-thirds* is $2 \times \frac{1}{3}$, we have $2 \times \frac{1}{3} = \frac{2}{3}$. Similarly, we can show that $5 \times \frac{1}{8} = \frac{5}{8}$ and $3 \times \frac{1}{7} = \frac{3}{7}$. Therefore, if A and B stand for natural numbers, it is reasonable to define the product $A \times 1/B$ as A/B. Furthermore, since by the commutative law, $A \times 1/B = 1/B \times A$, we have

$$A \times \frac{1}{B} = \frac{1}{B} \times A = \frac{A}{B} \qquad (1.9)$$

Now we return to the line segment and divide each of the three equal parts into two equal parts. Then each of the new subdivisions is $\frac{1}{6}$ of the whole line segment and is also equal to $\frac{1}{2}$ of each of the first three subdivisions. Hence, $\frac{1}{2}$ of $\frac{1}{3}$ is $\frac{1}{6}$. We now interpret $\frac{1}{2}$ of $\frac{1}{3}$ to mean $\frac{1}{2} \times \frac{1}{3}$ and thus have $\frac{1}{2} \times \frac{1}{3} = \frac{1}{6}$. Hence, if C and D are natural numbers, it is reasonable to define the product of $1/C$ and $1/D$ as below.

$$\frac{1}{C} \times \frac{1}{D} = \frac{1}{CD}$$

We are now in a position to prove that

$$\frac{A}{B} \times \frac{C}{D} = \frac{AC}{BD} \qquad (1.10)$$

The proof follows:

$$\frac{A}{B} \times \frac{C}{D} = A \times \frac{1}{B} \times C \times \frac{1}{D} \quad \text{by (1.9)}$$

$$= A \times C \times \frac{1}{B} \times \frac{1}{D} \quad \text{by commutative law}$$

$$= AC \times \frac{1}{BD}$$

$$= \frac{AC}{BD} \quad \text{by (1.9)}$$

Hence, we have the following rule for obtaining the **product of two fractions:**

The numerator of the product of two fractions is the product of the two numerators, and the denominator is the product of the two denominators.

As examples, we have

$$\frac{2}{3} \frac{4}{5} = \frac{8}{15} \quad \text{and} \quad \frac{3}{4} \frac{5}{8} = \frac{15}{32}$$

where the times sign is understood between fractions printed side by side.

Frequently in a computation involving fractions, we use the **fundamental principle of fractions** stated below.

If the numerator and denominator of a fraction are multiplied or divided by the same nonzero number, the fraction thus obtained is equal to the original fraction.

For example, by the definition of the product of two fractions, we have $\frac{2}{3} \times \frac{5}{5} = \frac{10}{15}$. But since $\frac{5}{5} = 1$, it follows that $\frac{2}{3} \times \frac{5}{5} = \frac{2}{3} \times 1 = \frac{2}{3}$, by (1.6). Consequently, $\frac{2}{3} = \frac{10}{15}$.

CONVERSION OF FRACTIONS A fraction is said to be expressed in lowest terms if the numerator and denominator have no exact common divisor greater than 1. We express a fraction in lowest terms by dividing the numerator and denominator by the greatest divisor that is common to the two.

EXAMPLE 1 Reduce each of the fractions $\frac{8}{24}$ and $\frac{27}{72}$ to lowest terms.

Solution

$\frac{8}{24} = \frac{1}{3}$ dividing each member of fraction by 8

$\frac{27}{72} = \frac{3}{8}$ dividing each member of fraction by 9

If the result of an arithmetical operation is a fraction, it should, in general, be expressed in lowest terms.

Frequently in an operation, it is necessary to convert a given fraction to one in which the denominator is some multiple of the given denominator:

$$\frac{a}{b} = \frac{ka}{kb}$$

EXAMPLE 2 Convert $\frac{2}{3}$ to an equal fraction whose denominator is 48, and $\frac{7}{8}$ to an equal fraction whose denominator is 48.

Solution

$\frac{2}{3} = \frac{32}{48}$ multiplying each member of fraction by 16

$\frac{7}{8} = \frac{42}{48}$ multiplying each member of fraction by 6

If two or more fractions have equal denominators, we say that the fractions have a **common denominator.** The process of adding two or more fractions with a common denominator depends upon the distributive law, as is illustrated in the following example:

$$\frac{2}{9} + \frac{4}{9} + \frac{5}{9} = \left(2 \times \frac{1}{9}\right) + \left(4 \times \frac{1}{9}\right) + \left(5 \times \frac{1}{9}\right)$$
by (1.9)

$$= (2 + 4 + 5) \times \frac{1}{9} \quad \text{by distributive law}$$

$$= 11 \times \frac{1}{9}$$

$$= \frac{11}{9} \quad \text{by (1.9)}$$

In order to get the sum of two or more fractions with different denominators, we must first convert each fraction to an equal fraction so that the resulting fractions have equal denominators. Then, as in the above example, we can use the distributive law to get the sum. This process will yield, as the sum, a fraction whose numerator is the sum of the numerators of the converted fractions, and whose denominator is the common denominator. In the interest of efficiency, the common denominator should be as small as possible. We call such a denominator the **least common denominator** and abbreviate the name to lcd. Accordingly, the lcd of two or more fractions is defined thus:

The **lcd** of two or more fractions is the least number that is exactly divisible by each of the given denominators.

In order to satisfy this definition, the lcd must be a multiple of each of the given denominators; furthermore, any other number that is a multiple of the given denominators must be a multiple of the lcd.

In the method for finding the lcd, we use the terms defined below.

A **factor** of a given number is another number that is an exact divisor of the given number.

A **prime number** is a number greater than 1 that has no factors other than itself and 1.

If a number is factored, it is expressed as the indicated product of its prime factors. For example, $4 = 2 \times 2$, $36 = 2 \times 2 \times 3 \times 3$, and $30 = 2 \times 3 \times 5$. Note that in some cases a number in the indicated product appears more than once.

We shall now illustrate the method for obtaining the lcd of $\frac{3}{4}, \frac{5}{12}, \frac{7}{27}$, and $\frac{3}{10}$ and shall express the lcd as the product of its prime factors. The lcd of these fractions is the least number that is exactly divisible by 4, 12, 27, and 10. Now since $4 = 2 \times 2$, $12 = 2 \times 2 \times 3$, $27 = 3 \times 3 \times 3$, and $10 = 2 \times 5$, the lcd is $2 \times 2 \times 3 \times 3 \times 3 \times 5 = 540$ since the lcd must contain the greatest number of factors of 2, 3, and 5 respectively which occur in any of the given numbers. It should be observed that the lcd contains all required combinations of factors, 2×2, $2 \times 2 \times 3$, $3 \times 3 \times 3$, and 2×5. Furthermore, no one of the factors in $2 \times 2 \times 3 \times 3 \times 3 \times 5$ can be omitted. For example, if 5 were omitted, the number obtained by multiplying the remaining factors together would not be divisible by 10. Also, no number, other than those indicated, can appear in the factored form of the lcd, for otherwise the product would be a multiple of 540.

As a second example, we shall obtain without explanation the lcd of $\frac{1}{8}, \frac{1}{24}, \frac{1}{36}$, and $\frac{1}{45}$.

$$
\begin{aligned}
8 &= 2 \times 2 \times 2 \\
24 &= 2 \times 2 \times 2 \times 3 \\
36 &= 2 \times 2 \qquad \times 3 \times 3 \\
45 &= \qquad\qquad\quad 3 \times 3 \times 5 \\
\text{lcd} &= 2 \times 2 \times 2 \times 3 \times 3 \times 5 = 360
\end{aligned}
$$

EXAMPLE 3 Convert each of the fractions $\frac{3}{4}, \frac{7}{9}, \frac{3}{10}$, and $\frac{5}{12}$ to an equal fraction having the lcd.

Solution We first find that the lcd is 180. Consequently, the given denominators must be multiplied by 45, 20, 18, and 15 individually to obtain 180. Therefore, since the numerators must also be multiplied individually by these numbers, we have

$$\frac{3}{4} = \frac{135}{180} \qquad \frac{7}{9} = \frac{140}{180} \qquad \frac{3}{10} = \frac{54}{180} \qquad \frac{5}{12} = \frac{75}{180}$$

MIXED NUMBER A mixed number is a number composed of a whole number and a fraction; it is a short form for the sum of the whole number and the fraction. Examples are $3\frac{1}{4} = 3 + \frac{1}{4}$, $25\frac{3}{4} = 25 + \frac{3}{4}$. According to the definition

$$25\frac{3}{4} = 25 + \frac{3}{4} = \frac{4 \times 25}{4} + \frac{3}{4} = \frac{100}{4} + \frac{3}{4} = \frac{103}{4}$$

which illustrates the following procedure for changing a mixed number to an improper fraction.

To change a mixed number to an improper fraction,* we multiply the whole number by the denominator of the fraction, add the numerator, and write the result as the numerator of the improper fraction. The denominator of the improper fraction is the denominator in the given fraction.

EXAMPLE 4

$$18\frac{2}{3} = \frac{(3 \times 18) + 2}{3} = \frac{56}{3}$$

REDUCTION OF DECIMAL FRACTIONS TO COMMON FRACTIONS AND VICE VERSA The definition of a decimal fraction leads directly to the following rules:

To express a decimal fraction between 0 and 1 as a fraction (quotient of integers), we write as the nu-

---
* An improper fraction is a fraction in which the numerator is greater than the denominator.

merator the number obtained by dropping the decimal point in the given decimal fraction, and as the denominator the number obtained by writing 1 followed by as many zeros as there are digits in the given decimal. The fraction thus obtained should be reduced to lowest terms.

EXAMPLE 5

$$0.264 = \frac{264}{1000} = \frac{33}{125}$$

To convert a fraction to a decimal fraction, we place a decimal point after the numerator, add zeros, and divide by the denominator until the quotient either comes out even or starts repeating. There must be as many decimal places in the quotient as there are zeros added.

EXAMPLE 6

$$\frac{5}{8} = \frac{5.000}{8} = 0.625$$

If we apply the method of Example 6 to the fraction $\frac{4}{11}$, the division does not come out even, and for the quotient we get $0.363636\cdots$, where the dots mean that if the division process is continued, the sequence of digits 36 will recur indefinitely in the quotient.

A decimal fraction of the type $0.363636\cdots$ is called a **repeating decimal**, the sequence of digits that recur is called the **repetend**, and $\frac{4}{11}$ is called the **generating fraction.**

We shall now show that by extending 0.363636 to the right by annexing the repetend a sufficient number of times, we obtain a decimal fraction that differs from $\frac{4}{11}$ by an amount that is as small as we please. For this purpose we consider the following differences:

$$\frac{4}{11} - 0.36 = \frac{4}{11} - \frac{36}{100} = \frac{400 - 396}{1100} = \frac{4}{1100}$$

$$\frac{4}{11} - 0.3636 = \frac{4}{11} - \frac{3636}{10{,}000} = \frac{4}{110{,}000} = \frac{1}{100}\frac{4}{1100}$$

$$\frac{4}{11} - 0.363636 = \frac{4}{11{,}000{,}000} = \frac{1}{100}\left(\frac{1}{100}\frac{4}{1100}\right)$$

Each of these differences is $\frac{1}{100}$ times the difference in the preceding line. Hence, as we proceed, this difference becomes smaller and smaller, and by continuing the process, we can make it as small as we please. We describe this situation in mathematical language by stating that $\frac{4}{11} =$ the limit of $0.363636\cdots$. Usually the phrase "the limit of" is omitted, and we write

$$\frac{4}{11} = 0.363636\cdots$$

EXAMPLE 7

$$\frac{13}{37} = \frac{13.000000000\cdots}{37} = 0.351351351\cdots$$

EXAMPLE 8

$$\frac{5}{6} = \frac{5.0000}{6} = 0.8333\cdots$$

Our next task is to find the generating fraction of a repeating decimal. We illustrate the method by finding the generating fraction of $0.121212\cdots$. We start by letting

$$F = 0.121212\cdots \quad (1)$$

Then

$$100F = 12.1212\cdots \quad (2)$$

Now we subtract each member of (1) from the corresponding member of (2), remembering that the repetend 12 is repeated indefinitely, and proceed as indicated.

$$\begin{aligned} 100F &= 12.121212\cdots \\ F &= 0.121212\cdots \\ \hline 100F - F &= 12.000000\cdots \\ (100-1)F &= 12 \quad \text{by distributive law} \\ 99F &= 12 \\ F &= \frac{12}{99} = \frac{4}{33} \end{aligned}$$

By a similar argument, it can be verified that for any repeating decimal fraction in which all digits recur, the following statement is true:

The numerator of the generating fraction of a repeating decimal in which all digits recur is equal to the repetend. The denominator is the number consisting of as many 9s as there are digits in the repetend.

Instead of remembering this rule, we can apply the method used above the rule to each such problem.

EXAMPLE 9 Find the generating fraction of $0.351351351\cdots$.

Solution Here, the repetend is 351, and it contains three digits. Hence, the generating fraction is

$$\frac{351}{999} = \frac{13}{37} \quad \text{by rule}$$

or

$$\begin{aligned} F &= 0.351351351\cdots \\ 1000F &= 351.351351\cdots \\ 999F &= 351 \end{aligned}$$

$$F = \frac{351}{999} = \frac{13}{37}$$

EXAMPLE 10

$$0.212121\cdots = \frac{21}{99} = \frac{7}{33} \quad \text{by rule}$$

or

$$\begin{aligned} F &= 0.212121\cdots \\ 100F &= 21.2121\cdots \\ 99F &= 21 \quad F = \frac{21}{99} = \frac{7}{33} \end{aligned}$$

We shall next explain the method for obtaining the generating fraction of a repeating decimal in which certain digits at the right of the decimal point do not recur.

EXAMPLE 11 Find the generating fraction of $0.5666\cdots$.

Solution We first express $0.5666\cdots$ as $0.5 + 0.0666\cdots$. Note that the first addend in this sum is 0.5, which does not recur, and the second is the decimal fraction obtained by replacing the nonrecurring digit by zero. Now we proceed as follows:

$$\begin{aligned} 0.5666\cdots &= 0.5 + 0.0666\cdots \\ &= \frac{5}{10} + \frac{1}{10}(0.666\cdots) \\ &= \frac{5}{10} + \frac{1}{10}\cdot\frac{6}{9} \\ &= \frac{5}{10} + \frac{6}{90} = \frac{51}{90} = \frac{17}{30} \end{aligned}$$

Other more advanced methods, such as infinite series, can also be used to convert repeating decimals to fractions.

Usually the first step in any arithmetical process involving mixed numbers is to reduce them to improper fractions or mixed decimals.

EXAMPLE 12 Convert the mixed numbers $2\frac{7}{8}$, $12\frac{3}{4}$, and $3\frac{1}{16}$ to mixed decimals and find their sum.

Solution

$$\begin{aligned} 2\tfrac{7}{8} + 12\tfrac{3}{4} + 3\tfrac{1}{16} &= 2.875 + 12.75 + 3.0625 \\ &= 18.6875 \end{aligned}$$

EXAMPLE 13 Convert $15\frac{6}{25}$ and $24\frac{3}{8}$ to mixed decimals, find their product, and convert it to a mixed number.

Solution

$$\begin{aligned} 15\tfrac{6}{25} \times 24\tfrac{3}{8} &= 15.24 \times 24.375 = 371.47500 \\ &= 371\tfrac{475}{1000} = 371\tfrac{19}{40} \end{aligned}$$

EXERCISE 1.3

Fill in the blanks with the appropriate word or number.

1. The fraction $\frac{2}{7}$ is a(n) _____ fraction; 2 is the _____, and 7 is the _____.

2. The fraction 9.5 is a(n) _____ fraction; 9 is the _____, and .5 is the _____.

3. The number $2\frac{3}{5}$ is a(n) _____ number; 2 is the _____, and $\frac{3}{5}$ is the _____.

4. The fraction $\frac{8}{5}$ is a(n) _____ fraction; 8 is the _____, and 5 is the _____.

Fill in the missing number so that the equality is justified.

5. $\frac{3}{7} = \frac{6}{-}$

6. $\frac{2}{3} = \frac{6}{-}$

7. $\frac{3}{4} = \frac{12}{-}$

8. $\frac{1}{5} = \frac{3}{-}$

9. $\frac{2}{5} = \frac{-}{10}$

10. $\frac{3}{8} = \frac{-}{24}$

11. $\frac{5}{9} = \frac{-}{36}$

12. $\frac{7}{2} = \frac{-}{8}$

Convert the fractions into equal fractions with a lcd.

13. $\frac{1}{2}, \frac{2}{3}, \frac{4}{5}$

14. $\frac{1}{3}, \frac{3}{4}, \frac{2}{5}$

15. $\frac{3}{4}, \frac{2}{5}, \frac{1}{7}$

16 $\frac{1}{2}, \frac{3}{5}, \frac{2}{7}$

17 $\frac{1}{2}, \frac{3}{4}, \frac{2}{5}$

18 $\frac{1}{2}, \frac{5}{6}, \frac{1}{5}$

19 $\frac{3}{4}, \frac{5}{6}, \frac{4}{7}$

20 $\frac{2}{3}, \frac{1}{6}, \frac{3}{7}$

Reduce each of the following fractions to lowest terms.

21 $\frac{2}{6} =$ **22** $\frac{6}{9} =$

23 $\frac{5}{20} =$ **24** $\frac{7}{21} =$

25 $\frac{68}{84} =$ **26** $\frac{95}{115} =$

27 $\frac{91}{119} =$ **28** $\frac{88}{152} =$

29 $\frac{124}{148} =$ **30** $\frac{207}{261} =$

31 $\frac{497}{644} =$ **32** $\frac{517}{825} =$

Convert the improper fractions into mixed numbers.

33 $\dfrac{22}{7} =$ 	 	34 $\dfrac{8}{3} =$

35 $\dfrac{7}{5} =$ 	 	36 $\dfrac{17}{6} =$

Convert the mixed numbers into improper fractions.

37 $2\tfrac{3}{5} =$ 	 	38 $1\tfrac{1}{4} =$

39 $2\tfrac{7}{9} =$ 	 	40 $1\tfrac{4}{7} =$

Change the fractions into three-digit decimal fractions.

41 $\dfrac{21}{32} =$ 	 	42 $\dfrac{13}{31} =$

43 $\dfrac{49}{73} =$ 	 	44 $\dfrac{67}{77} =$

45 $\dfrac{7}{11} =$ 	 	46 $\dfrac{9}{25} =$

47 $\dfrac{7}{8} =$ 	 	48 $\dfrac{5}{6} =$

Chapter 1: Review of Arithmetic

Change the decimal fractions to common fractions and reduce to lowest terms.

49 $0.3 =$

50 $0.8 =$

51 $0.7 =$

52 $0.6 =$

53 $0.32 =$

54 $0.165 =$

55 $0.236 =$

56 $0.81 =$

Find the generating fraction of each of the following repeating decimals.

57 $0.555\cdots =$

58 $0.666\cdots =$

59 $0.999\cdots =$

60 $0.6363\cdots =$

61 $0.7575\cdots =$

62 $0.2727\cdots =$

63 $0.258258\cdots =$

64 $0.627627\cdots =$

65 $0.38181\cdots =$

66 $0.27373\cdots =$

67 $0.3873636\cdots =$

68 $0.316464\cdots =$

1.9 ADDITION AND SUBTRACTION OF FRACTIONS

The denominator of a fraction can be thought of as the number of equal parts into which the unit is divided. If the denominator is thought of in this way, it is readily seen that *to add fractions with the same denominator we add the numerators and put the sum over the common denominator.* Thus

$$\frac{1}{9} + \frac{3}{9} + \frac{4}{9} = \frac{1+3+4}{9} = \frac{8}{9}$$

This is just the distributive law since

$$\frac{1}{9} + \frac{3}{9} + \frac{4}{9} = \frac{1}{9} \times 1 + \frac{1}{9} \times 3 + \frac{1}{9} \times 4$$
$$= \frac{1}{9} \times (1 + 3 + 4) = \frac{8}{9}$$

MORE ON ADDITION Although we cannot add two fractions with different denominators by this method, this does not present an insurmountable difficulty. In fact, we can change two fractions into equal fractions with the same denominator by the fundamental principle of fractions, as done in Sec. 1.8, and then add the fractions in the new form.

EXAMPLE 1 Find the sum of $\frac{1}{3}$, $\frac{2}{5}$, and $\frac{3}{4}$.

Solution Since these fractions do not have the same denominator, we must reduce them to a common denominator as a first step. We readily see that $3 \times 5 \times 4 = 60$ is a common denominator. In fact, it is the least common denominator. Converting each given fraction to an equal fraction with the common denominator as denominator, we have

$$\frac{1}{3} + \frac{2}{5} + \frac{3}{4} = \frac{1}{3}\frac{20}{20} + \frac{2}{5}\frac{12}{12} + \frac{3}{4}\frac{15}{15}$$
$$= \frac{20 + 24 + 45}{60} = \frac{89}{60}$$

EXAMPLE 2 Find the sum of $\frac{3}{8}$, $3\frac{5}{6}$, and $2\frac{1}{12}$.

Solution We first reduce the mixed numbers to improper fractions and then complete the addition.

$$3\frac{5}{6} = \frac{23}{6} \qquad 2\frac{1}{12} = \frac{25}{12}$$

Hence,

$$\frac{3}{8} + 3\frac{5}{6} + 2\frac{1}{12} = \frac{3}{8} + \frac{23}{6} + \frac{25}{12}$$
$$= \frac{9 + 92 + 50}{24} = \frac{151}{24} = 6\frac{7}{24}$$

EXAMPLE 3 Find the sum of 2, $\frac{5}{9}$, $3\frac{1}{2}$, and $2\frac{1}{3}$.

Solution Reducing the mixed numbers to improper fractions, we have

$$3\frac{1}{2} = \frac{7}{2} \qquad 2\frac{1}{3} = \frac{7}{3}$$

and so

$$2 + \frac{5}{9} + 3\frac{1}{2} + 2\frac{1}{3} = 2 + \frac{5}{9} + \frac{7}{2} + \frac{7}{3}$$
$$= \frac{36 + 10 + 63 + 42}{18} = \frac{151}{18}$$
$$= 8\frac{7}{18}$$

SUBTRACTION OF FRACTIONS As illustrated below, essentially the same procedure is used in subtracting one fraction from another as in adding two fractions. In fact, the two procedures differ only in that in subtracting we get the difference of the numerators instead of their sum.

EXAMPLE 4 Subtract $\frac{2}{5}$ from $\frac{7}{8}$.

Solution Since the common denominator is 40, we have

$$\frac{7}{8} - \frac{2}{5} = \frac{35 - 16}{40} = \frac{19}{40}$$

STUDENT'S NOTES

EXERCISE 1.4 ADDITION AND SUBTRACTION OF FRACTIONS

Perform the indicated additions and subtractions and reduce the results to lowest terms.

1. $\dfrac{1}{6} + \dfrac{5}{6} + \dfrac{2}{6} =$

2. $\dfrac{1}{12} + \dfrac{7}{12} - \dfrac{5}{12} =$

3. $\dfrac{4}{7} + \dfrac{6}{7} - \dfrac{3}{7} =$

4. $\dfrac{4}{15} - \dfrac{3}{15} + \dfrac{8}{15} =$

5. $\dfrac{1}{2} + \dfrac{1}{3} - \dfrac{1}{4} = \dfrac{}{12}$

6. $\dfrac{1}{4} + \dfrac{2}{3} - \dfrac{5}{6} = \dfrac{}{12}$

7. $\dfrac{3}{8} - \dfrac{5}{6} + \dfrac{3}{4} = \dfrac{}{24}$

8. $\dfrac{2}{3} + \dfrac{3}{5} - \dfrac{7}{15} = \dfrac{}{15}$

9. $\dfrac{3}{4} - \dfrac{5}{12} =$

10. $\dfrac{2}{3} - \dfrac{5}{9} =$

11. $\dfrac{7}{8} - \dfrac{3}{4} =$

12. $\dfrac{3}{5} - \dfrac{7}{15} =$

13. $\dfrac{2}{5} + \dfrac{11}{15} - \dfrac{1}{3} =$

14. $\dfrac{17}{18} - \dfrac{5}{6} - \dfrac{1}{9} =$

15. $\dfrac{1}{3} - \dfrac{7}{24} + \dfrac{3}{8} =$

16. $\dfrac{2}{5} - \dfrac{6}{35} + \dfrac{4}{7} =$

17. $\dfrac{1}{6} + \dfrac{5}{7} - \dfrac{8}{21} - \dfrac{1}{14} =$

18. $\dfrac{4}{5} - \dfrac{2}{9} - \dfrac{7}{15} + \dfrac{1}{3} =$

19. $\dfrac{1}{2} + \dfrac{2}{3} - \dfrac{7}{8} - \dfrac{1}{6} =$

20. $\dfrac{29}{45} - \dfrac{7}{30} - \dfrac{5}{18} - \dfrac{1}{15} =$

Perform the indicated additions and subtractions on the mixed numbers by first changing each to an improper fraction and then using a common denominator. Reduce to lowest terms if numerator and denominator of the result have a common factor.

21 $2\frac{3}{4} + 1\frac{1}{2} =$

22 $3\frac{7}{10} - 2\frac{1}{5} =$

23 $4\frac{1}{2} - 2\frac{1}{3} =$

24 $5\frac{7}{8} - 3\frac{2}{5} =$

25 $\frac{2}{5} + 1\frac{5}{7} - 1\frac{3}{35} =$

26 $1\frac{1}{3} + 2\frac{3}{5} - 1\frac{7}{10} =$

27 $1\frac{5}{9} - 1\frac{1}{6} + 2\frac{1}{2} =$

28 $4\frac{2}{3} - 1\frac{4}{7} - 2\frac{3}{5} =$

Change each fraction to three-decimal-places or an exact decimal form, perform the indicated additions and subtractions, and give the answer to three decimal places.

29 $5\frac{2}{3} + 1\frac{4}{7} - 2\frac{3}{5} =$

30 $3\frac{2}{5} - 1\frac{7}{11} + 2\frac{3}{8} =$

31 $4\frac{1}{4} + 2\frac{1}{3} - 5\frac{1}{2} =$

32 $1\frac{7}{9} + 2\frac{3}{13} - 1\frac{1}{7} =$

33 $6\frac{3}{16} - 2\frac{7}{11} - 3\frac{1}{14} =$

34 $5\frac{7}{12} - 1\frac{4}{11} - 2\frac{3}{19} =$

35 $7\frac{4}{17} - 2\frac{13}{21} - 4\frac{1}{23} =$

36 $4\frac{1}{27} + 3\frac{3}{35} - 6\frac{7}{31} =$

1.10 MULTIPLICATION AND DIVISION OF FRACTIONS

PRODUCT OF TWO FRACTIONS

$$\frac{A}{B} \times \frac{C}{D} = \frac{AC}{BD}$$

As stated in Sec. 1.8, Eq. (1.10), the numerator of the product of two or more fractions is the product of the numerators of the given fractions, and the denominator is the product of the denominators. We can cancel a number which is a *factor* of *both* the numerator and denominator.

EXAMPLE 1 Find the product of $\frac{2}{3}$ and $\frac{4}{5}$.

Solution

$$\frac{2}{3} \times \frac{4}{5} = \frac{2 \times 4}{3 \times 5} = \frac{8}{15}$$

If the numerator and denominator of the product of two or more fractions are divisible by the same number, we reduce the product to lowest terms.

EXAMPLE 2 Find the product of $\frac{3}{8}$, $\frac{2}{9}$, and $\frac{6}{7}$.

Solution

$$\frac{3}{8} \times \frac{2}{9} \times \frac{6}{7} = \frac{3 \times 2 \times 6}{8 \times 9 \times 7}$$
$$= \frac{\cancel{3} \times \cancel{2} \times \cancel{3} \times \cancel{2}}{\cancel{2} \times \cancel{2} \times 2 \times \cancel{3} \times \cancel{3} \times 7}$$
$$= \frac{1}{2 \times 7} \quad \text{canceling 2, 2, 3, and 3}$$
$$= \frac{1}{14}$$

EXAMPLE 3 Find the product of $\frac{4}{5}$, $3\frac{1}{3}$, and $4\frac{1}{8}$.

Solution To get the product of $\frac{4}{5}$, $3\frac{1}{3}$, and $4\frac{1}{8}$, we first reduce the mixed numbers to fractions and obtain $3\frac{1}{3} = \frac{10}{3}$, $4\frac{1}{8} = \frac{33}{8}$; then we perform the required multiplication and get

$$\frac{4}{5} \times 3\frac{1}{3} \times 4\frac{1}{8} = \frac{4}{5} \times \frac{10}{3} \times \frac{33}{8}$$
$$= \frac{\cancel{2} \times \cancel{2} \times \cancel{5} \times \cancel{2} \times \cancel{3} \times 11}{\cancel{5} \times \cancel{3} \times 2 \times \cancel{2} \times \cancel{2}}$$
$$= 11 \quad \text{canceling 2, 2, 2, 3, and 5}$$

EXAMPLE 4 Find the product of $2\frac{2}{3}$, $2\frac{1}{4}$, and $5\frac{5}{6}$.

Solution

$$2\frac{2}{3} = \frac{8}{3} \quad 2\frac{1}{4} = \frac{9}{4} \quad 5\frac{5}{6} = \frac{35}{6}$$

Hence,

$$2\frac{2}{3} \times 2\frac{1}{4} \times 5\frac{5}{6} = \frac{8}{3} \times \frac{9}{4} \times \frac{35}{6}$$
$$= \frac{\cancel{2} \times \cancel{2} \times \cancel{2} \times \cancel{3} \times \cancel{3} \times 5 \times 7}{\cancel{3} \times \cancel{2} \times \cancel{2} \times \cancel{2} \times \cancel{3}}$$
$$= 35$$

In Sec. 1.6 we defined the quotient obtained by dividing the dividend by the divisor as the number that satisfied the relation

$$\text{Divisor} \times \text{quotient} = \text{dividend} \quad (1)$$

The quotient of two fractions is defined in the same way, but the method for computing the quotient requires some explanation. We shall first show how the method is derived by using relation (1) to obtain the quotient yielded by $\frac{2}{3} \div \frac{5}{7}$. If we replace "divisor" and "dividend" in (1) by $\frac{5}{7}$ and $\frac{2}{3}$, respectively, we get

$$\frac{5}{7} \times \text{quotient} = \frac{2}{3} \quad (2)$$

and we must find the fraction to put in the place of the word "quotient" so that the product on the left will be equal to $\frac{2}{3}$. Hence, if we multiply each member of (2) by $\frac{7}{5}$, we get

$$\frac{7}{5} \times \frac{5}{7} \times \text{quotient} = \frac{2}{3} \times \frac{7}{5}$$

Consequently, since $\frac{7}{5} \times \frac{5}{7} = 1$ and $1 \times \text{quotient} = \text{quotient}$, we have

$$\text{Quotient} = \frac{2}{3} \times \frac{7}{5} = \frac{14}{15}$$

Therefore,

$$\frac{2}{3} \div \frac{5}{7} = \frac{2}{3} \times \frac{7}{5} = \frac{14}{15}$$

We could also have written

$$\frac{\frac{2}{3}}{\frac{5}{7}} = \frac{\frac{2}{3}\frac{7}{5}}{\frac{5}{7}\frac{7}{5}} = \frac{\frac{2}{3}\frac{7}{5}}{1} = \frac{2}{3} \times \frac{7}{5}$$

The above example illustrates the following rule for obtaining the **quotient of two fractions:**

32 | Chapter 1: Review of Arithmetic

To get the quotient of two fractions, invert the terms of the divisor and multiply by the dividend.

In other words, multiply the numerator by the reciprocal of the denominator.

$$\frac{A/B}{C/D} = \frac{A}{B} \times \frac{D}{C}$$

EXAMPLE 5

$$\frac{7}{8} \div \frac{3}{4} = \frac{7}{8} \times \frac{4}{3} = \frac{28}{24} = \frac{7}{6}$$

EXAMPLE 6

$$3\tfrac{1}{3} \div 2\tfrac{1}{4} = \frac{10}{3} \div \frac{9}{4} = \frac{10}{3} \times \frac{4}{9} = \frac{40}{27} = 1\tfrac{13}{27}$$

EXAMPLE 7

$$\frac{2}{5} \times \frac{5}{6} \div \frac{3}{4} = \frac{2}{5} \times \frac{5}{6} \times \frac{4}{3} = \frac{4}{9}$$

EXAMPLE 8

$$2\tfrac{2}{5} \times 3\tfrac{1}{3} \div 4\tfrac{1}{6} = \frac{12}{5} \times \frac{10}{3} \times \frac{6}{25} = \frac{48}{25} = 1\tfrac{23}{25}$$

EXERCISE 1.5 MULTIPLICATION AND DIVISION OF FRACTIONS

Perform the indicated multiplications and divisions.

1. $\dfrac{1}{3} \times \dfrac{3}{5} =$

2. $\dfrac{1}{4} \times \dfrac{4}{21} =$

3. $\dfrac{2}{3} \times \dfrac{3}{7} =$

4. $\dfrac{5}{7} \times \dfrac{7}{11} =$

5. $\dfrac{3}{8} \div \dfrac{3}{7} =$

6. $\dfrac{4}{9} \div \dfrac{4}{11} =$

7. $\dfrac{2}{7} \div \dfrac{5}{7} =$

8. $\dfrac{5}{12} \div \dfrac{5}{11} =$

9. $\dfrac{2}{7} \times \dfrac{21}{5} =$

10. $\dfrac{13}{8} \times \dfrac{16}{11} =$

11. $\dfrac{6}{13} \times \dfrac{7}{8} =$

12. $\dfrac{17}{9} \times \dfrac{12}{7} =$

13. $\dfrac{3}{14} \div \dfrac{6}{7} =$

14. $\dfrac{13}{15} \div \dfrac{26}{5} =$

15. $\dfrac{21}{10} \div \dfrac{14}{15} =$

16. $\dfrac{35}{26} \div \dfrac{14}{39} =$

17. $\dfrac{1}{2} \times \dfrac{6}{5} \times \dfrac{16}{3} =$

18. $\dfrac{2}{3} \times \dfrac{15}{12} \times \dfrac{18}{5} =$

19. $\dfrac{6}{35} \times \dfrac{21}{15} \times \dfrac{50}{24} =$

20. $\dfrac{10}{21} \times \dfrac{14}{25} \times \dfrac{45}{6} =$

21. $\dfrac{8}{5} \times \dfrac{2}{3} \div \dfrac{4}{15} =$

22. $\dfrac{7}{5} \times \dfrac{25}{14} \div \dfrac{10}{3} =$

23. $\dfrac{56}{63} \times \dfrac{42}{45} \div \dfrac{32}{105} =$

24. $\dfrac{15}{14} \times \dfrac{35}{54} \div \dfrac{10}{9} =$

25. $\dfrac{48}{27} \times \dfrac{42}{55} \div \dfrac{56}{45} =$

26. $\dfrac{49}{54} \times \dfrac{27}{40} \div \dfrac{35}{48} =$

27. $\dfrac{51}{32} \times \dfrac{24}{34} \div \dfrac{27}{64} =$

28. $\dfrac{6}{35} \times \dfrac{55}{42} \div \dfrac{33}{98} =$

34 | Chapter 1: Review of Arithmetic

Change the mixed numbers to improper fractions and perform the indicated operations.

29 $(3\frac{1}{7})(1\frac{3}{11}) =$

30 $(4\frac{1}{11})(2\frac{4}{9}) =$

31 $(3\frac{1}{9})(3\frac{3}{4}) =$

32 $(4\frac{5}{7})(3\frac{2}{11}) =$

33 $9\frac{1}{2} \div 14\frac{1}{4} =$

34 $5\frac{1}{3} \div 3\frac{5}{9} =$

35 $1\frac{7}{8} \div 2\frac{3}{16} =$

36 $6\frac{4}{5} \div 5\frac{1}{10} =$

Convert the mixed numbers to decimal fractions and then perform the indicated operations exactly or to three decimal places.

37 $(3\frac{1}{2})(5\frac{1}{5}) =$

38 $(6\frac{3}{8})(3\frac{7}{10}) =$

39 $(3\frac{2}{25})(1\frac{2}{5}) =$

40 $(5\frac{3}{4})(2\frac{1}{8}) =$

41 $10\frac{1}{2} \div \frac{3}{8} =$

42 $6\frac{6}{25} \div \frac{12}{25} =$

43 $2\frac{3}{4} \div 4\frac{1}{8} =$

44 $12\frac{3}{5} \div 2\frac{1}{4} =$

1.11 PERCENTAGE

We studied decimal fractions earlier and now continue that study but with a change in name. The term **percent** is used to mean the same thing as hundredths. The Latin words *per centum* mean "by the hundred." For example, we use 0.07 and 7 percent, also written 7%, interchangeably. Consequently, we *move the decimal point two places to the right and write % to change a decimal fraction to percent* since the decimal point is moved two places to the right to find the number of hundredths in a decimal fraction. A comparable procedure is used in changing from percent to a decimal fraction.

EXAMPLE 1 $0.435 = 43.5\%$, moving the decimal point two places to the right and adding %.

EXAMPLE 2 $8.65\% = 0.0865$, moving the decimal point two places to the left and dropping %.

EXAMPLE 3 $0.0031 = \frac{0.31}{100} = 0.31\%$

If $r\%$ of a number B is a number P, we write

$$\frac{r}{100} \times B = P$$

and call B the **base**, P the **percentage**, and $r/100$ the **rate**, which we designate by R. Note that, if r is 25, then R is $\frac{25}{100}$, or 0.25. Then we have

Rate $\times$ base = percentage

or more briefly

$$R \times B = P \tag{1.11}$$

EXAMPLE 4 To get 16% of $2000, we first notice that $R = 0.16$ and $B = \$2000$. Therefore, by (1.11), $P = 0.16 \times \$2000 = \$320$.

It follows from (1.11) that $B = P \div R$ and $R = P \div B$; hence, if we know the percentage and the rate, we can get the base B by means of the formula

$$B = \frac{P}{R} \tag{1.12}$$

Furthermore, if we know the percentage and the base, we can get the rate by using

$$R = \frac{P}{B} \tag{1.13}$$

EXAMPLE 5 To illustrate the use of formula (1.12), we shall solve this problem: 36 is 18% of what number? Here we are given the percentage $P = 36$, the rate $R = 0.18$, and we want B. By using (1.12), we get

$$B = \frac{36}{0.18} = 200$$

EXAMPLE 6 Suppose we want the answer to this question: 36 is what percent of 132? Here we have the percentage $P = 36$, the base $B = 132$, and we want r. We shall therefore use (1.13) and get

$$R = \frac{36}{132} = \frac{3}{11} = 0.27\frac{3}{11}$$

Hence,

$$r = 27\frac{3}{11}$$

In situations involving profit and loss, the cost of the property involved is the base and the amount gained is the percentage.

EXAMPLE 7 A man bought a house for $5200 and 5 years later sold it for $7800. Find the percent gained.

Solution The amount gained is $\$7800 - \$5200 = \$2600$; hence $P = \$2600$ and $B = \$5200$. Therefore,

$$R = \frac{2600}{5200} = 0.5 = 50\%$$

EXAMPLE 8 A man sold a lot for $6000 and gained 20% on the transaction. Find the cost of the lot.

Solution Since the man gained 20% on the transaction, he sold the lot for 120% of the cost; hence $6000 is 120% of the cost. Thus, we have $P = \$6000$ and $R = 1.2$, and we must find B. By formula (1.12)

$$B = \frac{\$6000}{1.20} = \$5000 \quad \text{cost}$$

STUDENT'S NOTES

EXERCISE 1.6 PERCENT AND DECIMALS

Convert the percent to a decimal fraction.

1 31% =

2 14% =

3 78% =

4 45% =

5 125% =

6 37.5% =

7 0.3% =

8 2.7% =

Change the percent to a common fraction.

9 75% =

10 20% =

11 80% =

12 13% =

13 $33\frac{1}{3}$% =

14 18.75% =

15 $28\frac{4}{7}$% =

16 $11\frac{1}{9}$% =

Convert the number to a percent.

17 0.23 =

18 0.41 =

19 0.374 =

20 0.803 =

21 0.061 =

22 0.0047 =

23 7.69 =

24 81.3 =

25 $\frac{1}{5}$ =

26 $\frac{3}{4}$ =

27 $\frac{3}{8}$ =

28 $\frac{1}{2}$ =

Chapter 1: Review of Arithmetic

29 $\frac{2}{3} =$

30 $\frac{5}{6} =$

31 $\frac{4}{7} =$

32 $\frac{7}{9} =$

33 21% of 38 is _____

34 36% of 143 is _____

35 What is 45% of 309?

36 What is 72% of 463?

37 28 is what percent of 71?

38 53 is what percent of 84?

39 147 is what percent of 58?

40 279 is what percent of 89?

41 170 is 80% of what number?

42 288 is 72% of what number?

43 23 is 5% of what number?

44 96 is 32% of what number?

45 139 is 23% of what number?

46 231 is 37% of what number?

47 26 is 13.7% of what number?

48 409 is 84.6% of what number?

49 A boys' club purchased nuts that cost 80 cents a can. At what price did they sell the nuts if they made a profit of 25%?

50 A merchant sold a radio for $33. If she made a 32% profit on the radio, what had it cost her?

51 Collins' salary is $1400 a month, but each month the payroll office deducts 15% for income tax, 7% for retirement, and $17 for hospitalization. What is Collins' monthly take-home pay?

52 A woman bought a house for $16,000 and later sold it for 25% more than she paid for it. If the realtor who handled the sale charged a commission of 6% on the selling price, what was his fee?

53 The retail price of a certain color television set is advertised to be $525. During the spring sale, a store advertises that they will lower the retail price by 8%. What is the sale price of the television set?

54 Seward bought two blocks of stock, one for $6500 and one for $7200. He sold the first block at a profit of 9%, but was forced to sell the second at a loss of 8%. Did he gain or lose by his transaction and by how much?

55 Green owned 1245 shares of stock in a certain company. Through a stock dividend he received 249 more shares of stock. What percent stock dividend was declared?

56 If Tom bought a jalopy for $250 and a month later sold it to Sue for $325, what percent did he gain on the transaction?

57 At the end of a term an instructor who taught two classes of 32 and 36 students found that 8 students in the first class and 6 students in the second class had earned the grade of A. Find the percent of A students in each class and in both classes.

58 If 40 oz of a solution that is 6% glacial acetic acid is mixed with 50 oz of solution that contains 9 oz of glacial acetic acid, what percent of the mixture is glacial acetic acid?

59 The tank of a garden sprayer contains $1\tfrac{1}{2}$ gal of insecticide and $8\tfrac{1}{2}$ gal of water. What percent of the solution is insecticide?

60 If $\tfrac{3}{4}$ lb of salt is dissolved in 3 gal of water that weighs 24.9 lb, what is the percent of salt in the solution?

61 What should a woman pay for a share of stock that yields an annual dividend of $3.25 if she is to earn 5.2% on her investment?

62 A man earned 4.5% interest on an investment of $2400. How much additional money must he invest at 6% to bring his rate of return from his investment to 5%?

63 In the 1970 census it was found that the population of Cartersville had increased by 8500 people. If this represented an increase of 40% in 10 years, what was the population of Cartersville in 1960?

64 Dorothy borrowed a sum of money for 1 year at 7% interest. If at the end of the year she paid $4494 to retire the note, how much did she borrow?

65 The retail price of a chair was $156. If the merchant is allowed a 30% profit and the manufacturer earns a 20% profit, how much does the chair cost the manufacturer?

66 Henry received $126 in dividends from the shares he owned of a certain stock. If this represented a return of $4\tfrac{1}{2}\%$ on the investment, what had he paid for the stock?

67 If an automobile was worth $2380 at the end of a year and during the year it had depreciated by 32%, what was it worth at the beginning of the year?

68 Smith bought a house for $28,000 and later sold it to Thompson and gained 15% on the transaction. When Thompson's company transferred him to another city, he was forced to sell the house at a loss of 5%. How much did Thompson receive for the house?

1.12 SCIENTIFIC NOTATION AND APPROXIMATIONS

Since most measurements and many other numbers are only approximations, it is worthwhile for us to know how to indicate the accuracy of such numbers as well as to know how many digits or decimal places to retain in the result of a calculation that involves approximations.

In this section and the next we shall use exponential notation for powers of 10. Section 2.9 contains a complete development of positive integer exponents, but for our purposes all we need to know is that

10^1 means 10
10^2 means $10 \times 10 = 100$
10^3 means $10 \times 10 \times 10 = 1000$

etc.,

10^{-1} means $\frac{1}{10} = 0.1$

10^{-2} means $\frac{1}{10 \times 10} = 0.01$

10^{-3} means $\frac{1}{10 \times 10 \times 10} = 0.001$

etc.

This notation gives us a simple way to move the decimal point to the right or left. For example

$$843 = 8.43 \times 100 = 8.43 \times 10^2$$
$$6400 = 6.4 \times 1000 = 6.4 \times 10^3$$
$$0.2 = 2 \times 0.1 = 2 \times 10^{-1}$$
$$0.0034 = 3.4 \times 0.001 = 3.4 \times 10^{-3}$$

We also have

$$4083 = 408.3 \times 10 = 4.083 \times 10^1$$
$$= 40.83 \times 100 = 40.83 \times 10^2$$
$$= 4.083 \times 1000 = 4.083 \times 10^3$$
$$= 0.483 \times 10{,}000 = 0.4083 \times 10^4$$

A number is said to be in **scientific notation** if there is a decimal point just to the right of the first nonzero digit (this is often called the **reference position** of the decimal point) and the resulting number is multiplied by the integral power of 10 that is required to make a number of the proper size. Thus 3.1416 is in scientific notation, as is every other number equal to or greater than 1 and less than 10.

EXAMPLE 1 Write 1776 and 0.1066 in scientific notation.

Solution By placing a decimal just after the 1 in 1776, we have 1.776, and that must be multiplied by 10^3 to obtain the given number 1776; hence, 1.776×10^3 is the scientific notation for 1776. Since $1.066 \times 10^{-1} = 0.1066$ and there is a decimal point just to the right of the first nonzero digit, it follows that 1.066×10^{-1} is the scientific notation for 0.1066.

We shall consider a method for reducing the number of digits in a number that is given to more digits than we are to use. This is called **rounding off**. The usual procedure for rounding off a number of more than n digits to n digits is to put the number in scientific notation, omit all digits beyond the nth, and increase the nth by 1 if the $(n + 1)$th is 5, 6, 7, 8, or 9.

EXAMPLE 2 Round off 8124 and 0.042187 to three digits.

Solution The scientific notation of 8124 is 8.124×10^3; furthermore, the first digit dropped to keep only three digits is 4 and that is less than 5; hence, 8124 rounded to three digits is 8.12×10^3. In scientific notation, 0.042187 is 4.2187×10^{-2}, and since $8 > 5$, it follows that 4.22×10^{-2} is the three-digit form of 0.042187.

PRODUCTS AND QUOTIENTS OF APPROXIMATE NUMBERS

The following rules for multiplying and dividing approximate numbers can be justified in general, although we shall not do so here. In Sec. 1.14, after Example 3, we give a detailed justification for one typical case.

> If each of two numbers is correct to n digits, we carry their product or quotient to n digits. If one number is correct to n digits and another to more than n, we round the second one off to $n + 1$ digits, if given to more than that, then find the product or quotient and round it off to n digits.

EXAMPLE 3 Find the product P and the quotient Q of $F = 27.1$ and $S = 3.13$.

Solution By the usual procedures we find that $P = 27.1 \times 3.13 = 84.823$; hence, to three digits $P = 84.8$. Similarly, $Q = F/S = 27.1/3.13 = 8.6581$, and to three digits this is 8.66.

ADDITION AND SUBTRACTION OF APPROXIMATE NUMBERS In adding approximate numbers, we concentrate on their decimal portions since we are interested in the precision of the calculations.

If the decimal portions of the addends contain the same number of decimal places, the numbers are added and all the decimal portion is kept. If one or more addends have n decimal places and the others more than n, we round off the latter to $n + 1$ places, if given to more than that, perform the addition and round off the sum to n decimal places.

EXAMPLE 4 Find the sum of 2.76, 5.814, and 7.3.

Solution Since one addend contains three decimal places and another only one, we round off the one with three to two decimal places, then add and get 15.87. We now round off to one decimal place since there is only one in some addend. Thus, we get 15.9.

```
 2.7 6
 5.8 1
 7.3
1 5.8 7
```

A similar procedure is followed in subtraction.

EXAMPLE 5 Subtract 42.233 from 58.59.

Solution $58.59 - 42.233 = 16.357$. Now rounding off to two decimal places gives 16.36 as the difference.

Problems 1 to 40 in Exercise 1.7 may be done now.

1.13 THE METRIC SYSTEM

The metric system is a system of units based on powers of 10 (see Sec. 2.9 for a fuller discussion of powers). This makes it possible to change units by simply moving the decimal point to the left or right. Most of the world now uses the metric system. SI (for Système International d'Unités) is the name given the subdivision of the metric system used for most scientific work.

There are certain **basic units** in the SI, e.g., the **meter** (m) for length, the **kilogram** (kg) for mass, the **second** (s) for time, and the **ampere** (A) for electric current. Certain other **derived units** are then defined in terms of the basic units, e.g., square meters (m^2) for area, cubic meters (m^3) for volume, meters per second (m/s) for velocity, and kilogram-meters per second squared ($kg \cdot m/s^2$) for force. Some of these derived units have special names such as newton (N) for kilogram-meter per second squared and **liter** (L) for 1000 cubic centimeters (cm^3).

This brings us to the heart of the metric system, the names of multiples of units. These are formed by prefixes.

Prefix	Symbol	Multiplying factor
mega	M	$1,000,000 = 10^6$
kilo	k	$1000 = 10^3$
centi	c	$0.01 = 10^{-2}$
milli	m	$0.001 = 10^{-3}$
micro	μ	$0.000001 = 10^{-6}$

We thus have 1000 millimeters = 1000 mm = 1 meter = 1 m, 10 mm = 10 millimeters = 1 centimeter = 1 cm, 100 cm = 1 m, 1000 m = 1 kilometer = 1 km. Similarly 1,000,000 grams = 1000 kilograms = 1 megagram = 1 metric ton. See Table 1.2.

Table 1.2 Metric Conversion

Length	Volume
10 mm = 1 cm	$1000 \text{ mm}^3 = 1 \text{ cm}^3$
100 cm = 1 m	$10^6 \text{ cm}^3 = 1 \text{ m}^3$
1000 m = 1 km	1000 ml = 1 L

Area	Mass
$100 \text{ mm}^2 = 1 \text{ cm}^2$	10 mg = 1 cg
$10000 \text{ cm}^2 = 1 \text{ m}^2$	100 cg = 1 g
	1000 g = 1 kg

The ease in changing units is further illustrated by writing

45.71 cm = 457.1 mm
0.00298 kg = 2.98 g

Originally a meter was defined in terms of the earth's circumference (10 million meters was one-fourth the circumference), but because greater accuracy and reproducibility were necessary, since 1960 it has been defined in terms of the krypton atom. For similar reasons the *second* (s) is defined by means of the cesium atom. In 1901, a kilogram was defined as the mass of 1 liter (L) of pure water at its maximum density and at standard atmospheric pressure. The *kilogram* (kg) is now defined as the mass of a platinum-iridium bar kept in Paris.

Temperature is defined in terms of a certain property of water and is measured in *kelvins* (K). We have 273.16 K equal to 0°C (Celsius). Note that K = 273.16 + C. If C is degrees Celsius and F degrees *Fahrenheit,* then

$$F = \frac{9C}{5} + 32 \quad \text{and} \quad C = \frac{5}{9}(F - 32)$$

At F = −40, C = −40, at F = 32, C = 0, at F = 50, C = 10, at F = 86, C = 30, and at F = 212, C = 100. A change of 9°F equals a change of 5°C or 5 K.

Problems 41 to 56 in Exercise 1.7 may be done now.

1.14
DENOMINATE NUMBERS

A **denominate number** is a number obtained by measurement and is expressed in terms of some unit of measure. Examples are 4 yd, 11 cm, 100 lb, and 5.2 g. Any denominate number is at best an approximation.

Measuring instruments are calibrated in terms of some fractional unit. For example, a length obtained by use of a yardstick can be expressed to the nearest sixteenth of an inch, and if a meterstick is used, the length can be expressed to the nearest millimeter.

If a length is reported as 3.2 cm, it is understood that the exact length is between 3.15 and 3.25 cm. Now, since 3.2 is halfway between 3.15 and 3.25, the error in 3.2 cm cannot be greater than $\frac{1}{2}$(3.25 − 3.15) cm = $\frac{1}{2}$(0.10) cm = 0.05 cm. Similarly, a reported length of $2\frac{3}{16}$ in means that the exact length in between $2\frac{2.5}{16}$ in and $2\frac{3.5}{16}$ in, so the maximum error is

$\frac{1}{2}(2\frac{3.5}{16} - 2\frac{2.5}{16})$ in = $\frac{1}{2}(\frac{1}{16})$ in = $\frac{1}{32}$ in

Consequently, we define the **maximum error** in any measurement as one-half of the smallest fractional unit in which the measurement is expressed.

EXAMPLE 1

Measurement	Maximum error
4.26 cm	$\frac{1}{2}$(0.01 cm) = 0.005 cm
0.215 g	$\frac{1}{2}$(0.001 g) = 0.0005 g
3 ft 6 in	$\frac{1}{2}$(1 in) = 0.5 in
$5\frac{2}{3}$ yd	$\frac{1}{2}(\frac{1}{3}$ yd) = $\frac{1}{6}$ yd

There are two concepts associated with a denominate number, the precision of the number and the accuracy. The **precision** is related to the smallest fractional unit in terms of which the number is expressed. For example, we say that 4.26 cm is precise to hundredths of a centimeter, $2\frac{3}{16}$ in is precise to $\frac{1}{16}$ of 1 in, and 5 lb 6 oz is precise to 1 oz.

Frequently, when a high degree of precision is not necessary, a denominate number is expressed in multiples of 10 units, 100 units, or 1000 units, etc. For example, the diameter of the earth may be stated as

7917 mi, precise to the nearest mile
7900 mi, precise to the nearest 100 mi
8000 mi, precise to the nearest 1000 mi

The digit that indicates the degree of precision in 7917 is the last 7, in 7900 it is 9, and in 8000 it is 8.

The **accuracy** of a denominate number is related to the number of **significant digits** in it. We define the significant digits as follows.

The **significant digits** in a denominate number (or an approximate number) are the first nonzero digit in it and all other digits to the right of the first up to and including the digit that indicates the precision of the number.

EXAMPLE 2

Number	Significant digits
3.206 g	3, 2, 0, 6
0.00312 cm	3, 1, 2
47.09 in	4, 7, 0, 9
5200 precise to 100 mi	5, 2
3600 precise to 1 mi	3, 6, 0, 0

We say that the first number in the above example is accurate to four digits, the second is accurate to three digits, the third to four digits, the fourth to two digits, and the fifth to four digits. As the last two numbers illustrate, we cannot determine the significance of the final zeros in a number unless we know the precision. In this book, we shall assume that *the final zeros are significant unless otherwise stated.*

Since denominate numbers are approximations, the rules stated in Sec. 1.12 apply when rounding, adding, and multiplying them.

EXAMPLE 3 126.53 rounded off to four digits is 126.5; 126.53 rounded off to three digits is 127; and 126.53 rounded off to two digits is 130.

The following discussion explains the reason behind the rule for multiplying approximate numbers, as given in Sec. 1.12.

If the length of a rectangle is given as 43.27 cm and the width as 1.4 cm, and *if* both numbers are exact measurements, the exact area of the rectangle is 43.27 × 1.4 = 60.578 cm². However, we are not justified in *assuming* that the final digit in either of these measurements is exact, and this doubtfulness affects all digits in the product except possibly the first. All we know about the dimensions is that the length is between 43.265 and 43.275 cm and the width is between 1.35 and 1.45 cm. Hence, the area is between 43.265 × 1.35 = 58.40775 cm² and 43.275 × 1.45 = 62.74875 cm², and it could be any area between these limits. Hence, we surely are not justified in assuming an accuracy of more than two digits in the area, and it is reasonable to assume that it would be near the average of the two limits, which rounded off to two digits is 61 cm². Now if we round 43.27 off to 43.3 and multiply 43.3 by 1.4, we get 60.62; if we round 60.62 off to two digits, we get 61, as given by the rule.

EXAMPLE 4 To get the product of the denominate numbers 42.651 and 0.437, we first notice that the less accurate of these numbers, 0.437, contains *three* significant digits. Hence, we round 42.651 off to *four* significant digits to obtain 42.65 and then complete the problem:

42.65(0.437) = 18.63805
= 18.6   to three significant digits

EXAMPLE 5 To add 6.75, 43.8, 321.876, and 0.639, we round off the addends to two decimal places since the least precise addend is 43.8, and add the resulting numbers. We get

6.75 + 43.8 + 321.88 + 0.64 = 373.07 = 373.1
to one decimal place

CONVERSION OF DENOMINATE NUMBERS

One of the important problems in connection with denominate numbers is that of converting a given denominate number to another denominate number that expresses the same measurement in terms of a different unit. For example, it is frequently necessary to convert yards and feet to inches or miles per hour to feet per second. A partial set of tables of weights and measures, as well as some important conversion factors, is printed at the end of this section. The following examples and the practice problems are based on these tables.

EXAMPLE 6 Convert 8 yd 2 ft 9 in to inches.

Solution   1 yd = 3 ft, and 1 ft = 12 in; therefore,

8 yd = 8(3)(12) = 288 in
2 ft =   2(12) = 24 in
9 in =   9 in
Sum =   321 in

EXAMPLE 7 Convert 15 mi/h to feet per second.

Solution

1 mi = 5280 ft   1 h = 3600 s

Hence,

$$15 \text{ mi/h} = \frac{15(5280)}{3600} \text{ ft/s} = \frac{15}{3600} 5280 \text{ ft/s}$$

$$= \frac{1}{240} 5280 \text{ ft/s} = \frac{5280}{240} \text{ ft/s} = 22 \text{ ft/s}$$

EXAMPLE 8 Convert 8 in to metric units.

Solution   1 in = 2.54 cm so

8 in = 8(2.54) cm = 20.32 cm

EXAMPLE 9 Under proper conditions, 1 cm² of water weighs 1 g. Find the weight in the metric system of 8 in³ of water.

Solution   1 in³ = 16.39 cm³; therefore, 8 in³ = 8(16.39) cm³ = 131.12 cm³. Hence, 8 in³ of water weighs 131.12 g.

Table 1.3   Tables of Weights and Measures

Length	Liquid measure
12 in = 1 ft	2 cups = 1 pt
3 ft = 1 yd	2 pt = 1 qt
1 mi = 1760 yd	4 qt = 1 gal
= 5280 ft	1 gal = 231 in³
Area	**Avoirdupois weight**
144 in² = 1 ft²	16 oz = 1 lb
9 ft² = 1 yd²	2000 lb = 1 ton
1 acre = 4840 yd²	
= 43,560 ft²	
Volume	
1728 in³ = 1 ft³	
27 ft³ = 1 yd³	

Table 1.4   Conversion Factors

1 in = 2.54 cm	1 L = 1.0567 liquid qt
1 cm = 0.0328 ft	= 61.0250 in³
1 mi = 1.609 km	1 lb = 0.45359 kg
1 m = 39.37 in	1 gal = 3.7853 L
1 km = 0.621 mi	1 in³ = 16.3872 cm³
1 kg = 2.205 lb	

1.14 Denominate Numbers | 45

EXERCISE 1.7 APPROXIMATE NUMBERS AND THE METRIC SYSTEM

Put the number in scientific notation.

1 387 =

2 25.3 =

3 40.35 =

4 815.9 =

5 0.137 =

6 0.0202 =

7 0.9009 =

8 0.001863 =

Round off the number to two digits, using scientific notation as needed.

9 814 =

10 0.816 =

11 81.5 =

12 7.77 =

Round off the number to three digits, using scientific notation as needed.

13 0.02596 =

14 0.003052 =

15 60073 =

16 60037 =

Perform the indicated operations. Carry each result to the justified number of digits or decimal places under the assumption that all numbers involved are approximations. Use scientific notation as needed.

17 2.2(3.4) = 7.48
             7.5

18 46(0.73) = 33.58
               34

19 47.1(.302) = 14.2242
                 14.2

20 60.9(71.3) = 4342.17
                 4340

46 | Chapter 1: Review of Arithmetic

21 2.7(48.3) = 130.41
                130

22 723(5.476) = 3959.148
                 396

23 3.429(76) = 260.604
                260

24 12.3(543.21) = 6681.483
                   668

25 $\dfrac{37}{49}$ = .75

26 $\dfrac{7.3}{9.4}$ = .776595?
                        .78

27 $\dfrac{53.7}{8.1}$ = 6.6296296
                         (6.6)

28 $\dfrac{725}{0.84}$ =

29 $\dfrac{386.7}{3.1}$ = 124.7494
                          120

30 $\dfrac{598.4}{0.261}$ =

31 $\dfrac{1.8005}{0.0026}$ =

32 $\dfrac{809.6}{0.34}$ =

33   3.821
    + 1.76

     5.581
     5.58

34   5.9716
    + 2.34

     8.3116
     8.31

35   98.33
    + 31.78

     130.11
     130.1

36   852.384
    + 76.21

     928.594
     928.6

37   97.68
    − 23.45

     121.13

38   4.367
    − 2.184

39   4.5977
    − 2.389

40   12.703
    − 5.96

41 Express 512 cm in millimeters; in meters.

42 Express 2135 cg in grams; in kilograms.

43 14 cg = _____ g = _____ mg.

44 75 m = _____ km = _____ cm

45 0.314 L = _____ mL

46 0.01555 kg = _____ mg

47 How many kilometers are there in 876 cm?

48 How many centigrams are there in 0.00531 mg?

Problems 49 to 52 are True or False.

49 4 cm³ = 40 mm³

50 0.3 cm² = 30 mm²

51 There are as many cubic centimeters in a cubic meter as there are milliliters in a kiloliter.

52 There are as many square centimeters in a square meter as there are kilometers in 10 cm.

53 41°F = _____ °C

54 131°F = _____ °C

55 25°C = _____ °F

56 75°C = _____ °F

57 Convert 4 mi 6 yd 4 ft to inches.

58 Convert 6 gal 2 qt 3 pt to cups.

59 Convert 4 yd 2 ft 1 in to feet.

60 Convert ¼ gal 12 qt 6 cups to pints.

48 | Chapter 1: Review of Arithmetic

61 Add 2 yd 6 ft 1 in, 3 yd 7 ft 1 in, and 3 yd 5 ft 1 in, and express the sum in terms of the largest possible units without the use of fractions.

62 Add 528 acres 123 yd^2 29 ft^2, 213 acres 117 yd^2 21 ft^2, and 112 acres 91 yd^2 15 ft^2, and express the sum in terms of the largest possible units without the use of fractions.

63 A 1 N solution of potassium hydroxide contains 53 g of potassium hydroxide per liter of solution. How many pounds of potassium hydroxide would be found in 2.0 gal of 1 N solution?

64 If a car averages 18 mi/gal of gasoline, how many gallons of gasoline would be required for a trip of 60 km?

65 How many milliliters of acid could be stored in three 1.0-pt jars?

66 If 800 L of nitrogen under standard conditions weighs 1.0 kg, how many pounds would 50 gal of nitrogen weigh under standard conditions?

67 A standard recipe for making a large number of biscuits calls for 12 cups of flour per 2 cups of shortening. How many pounds of shortening must be used with a 5-lb sack of flour to make a batch of biscuit mix if 1 lb of shortening contains $2\frac{1}{2}$ cups and 1 lb of flour contains 4 cups.

68 If the formula for a color motion-picture developer calls for 5.0 g of developing agent per liter of developer, how many pounds of developing agent must be used to mix 130 gal of developer?

1.15 CALCULATORS

There are many hand-held or pocket calculators as well as computers that fill more than a room. Some calculators can be used for a large variety of operations, and essentially all can be used for addition, subtraction, multiplication, division, and taking square roots. The results of most operations are shown to more digits than can be justified and must be rounded off. Since not all calculators operate in the same way, you should read the directions before using one.

When using a calculator, remember that most of the time the number on the display is an approximation. Thus $\frac{1}{3}$ may be approximated by 0.33333333 and $\frac{2}{7}$ by 0.28571429. If we do a series of calculations, each involving approximations, the result may compound the individual errors, so that the last digit shown on the display may not be accurate.

The logic of the calculator must be understood clearly. On some, $3 + 4 \div 5$ will give $(3 + 4) \div 5 = 7 \div 5 = 1.4$, while on others it will give $3 + (4 \div 5) = 3 + 0.8 = 3.8$. In a series of calculations, you should check to make sure it is doing what you want it to do.

It is worthwhile to estimate the result (and even intermediate results) to avoid gross errors. Besides, your work in making the estimation will help keep your own calculating ability sharp or improve it if it is rusty.

You should not use a calculator if you don't need to. In Chap. 3, we shall show how to multiply 61×59 mentally and get the answer of 3599 quicker than several buttons can be pushed on a calculator. In addition to the satisfaction of such mental calculation, a calculator may not always be handy.

Even if your calculator has a memory, it may be necessary to think the calculation through thoroughly in order to have available the number needed for the next step. And having a calculator certainly does not eliminate the need to think mathematically; it is often necessary to solve an equation first to find the number to be calculated.

For example, to find $7/\sqrt{3}$ it may be necessary to find $\sqrt{3}$ first, put it in the memory, and then divide 7 by the memory. The result is 4.0414519. We shall find in Chap. 6 that

$$\frac{7}{\sqrt{3}} = \frac{7\sqrt{3}}{3} = \sqrt{3} \times 7 \div 3$$

which can be done without use of the memory.

Regardless of all eight or more digits sitting there in the display after each calculation, we cannot get more accuracy out of the calculation than we began with (see Example 2 below; also Sec. 1.12 for the rules concerning multiplying and adding approximate numbers).

EXAMPLE 1 Find the sum and the product of 24.6 and 32.1.

Solution To find the sum, we put 24.6 on the display, depress the + button, put 32.1 on the display, depress the = button, then read the sum 56.7 on the display.

To find the product we put 24.6 on the display, depress the × button, put 32.1 on the display, depress the = button, then read the product 789.66 on the display. This product is all the digits we would get if we multiplied the two numbers together in the usual manner. It is correct if the factors are exact numbers but must be rounded off if they are approximations. In this case, the product of the approximations 24.6 and 32.1 should be rounded off to $790 = 7.90 \times 10^2$.

EXAMPLE 2 Find the difference and quotient of 227.3 and 37.46 if they are approximations.

Solution The difference $227.3 - 37.46 = 189.84 = 189.8$ is obtained in a manner similar to addition. To find the quotient $227.3 \div 37.46$, we perform exactly the same operations as in getting the sum except for depressing the ÷ sign instead of the + sign. Thus, we get the quotient 6.0678057 on the display to eight digits. Some calculators give more digits, and others give less. The quotient given above is correct if the given numbers are exact but must be rounded off to 6.068 if they are approximations.

EXAMPLE 3 Find the square root of 266.3.

Solution In order to find $\sqrt{266.3}$ with a calculator, we put 266.3 on the display, depress the $\sqrt{x}$ button, then read the square root 16.318701 on the display. Rounding off, we use 16.32.

EXAMPLE 4 Evaluate $8.27 \sqrt{173}/4.263$ and round the result off to three digits.

Solution We begin by finding $\sqrt{173}$ to four digits, then multiply it by 8.27, divide by 4.263, and round off. We find $\sqrt{173}$ to be 13.152946, round it off to four digits, and get 13.15; this times 8.27 gives 108.8 to four digits. Now dividing by 4.263 gives 25.5 to three digits.

1.16 SETS

In this book we shall use the concept and terminology of sets where it is helpful. They are reviewed briefly here.

A **set** is a collection of well-defined objects called **elements.**

By "well-defined" we mean that there is a criterion that enables us to make the following decisions about an object or an element a:

1. a belongs to the set.
2. a does not belong to the set.

We next list three sets of well-defined objects. Each set is designated by the letter S, and the criterion that defines an element of S is stated.

1. S is the football squad of Mohawk College. The criterion that determines the membership of the squad is the list of names selected by the coach.
2. S is the herd of sheep in Mr. McGuire's pasture. Here the criterion is this: if an animal is a sheep and is in the specified pasture, it is an element of the set. An animal is not in the set if it is not a sheep or if it is not in the pasture.
3. S is the set of natural numbers less than 7 that are divisible by 2. Hence the numbers 2, 4, and 6 are elements of S since each is less than 7 and is divisible by 2. Numbers not divisible by 2 and those equal to or greater than 7 are not elements of S.

We frequently use a capital letter to stand for a set. A set is described in two ways. In one, we *tabulate the elements* of the set, e.g., letters, numerals, or the names of objects, and enclose the tabulation in braces { }. In the other, we enclose a *descriptive phrase* in braces and understand that the set contains those elements and only those elements which satisfy the description. For example, if W is the set of the names of the days of the week, then we designate W in tabular form as

W = {Sunday, Monday, Tuesday, Wednesday, Thursday, Friday, Saturday}

or in descriptive form as

W = {the names of the days of the week}

or

$W = \{x|x$ is the name of a day of the week$\}$

The last form is the method most often used in the descriptive method. The vertical line is read "such that." Furthermore, if T is the set of natural numbers less than 5, then

$T = \{1,2,3,4\}$

or

$T = \{x|x$ is a natural number and is less than 5$\}$

Two sets are **equal** if they have exactly the same elements.

The relation of equality does not require that the elements of the sets be arranged in the same order. For example, $\{s,t,a,r\} = \{r,a,t,s\} = \{t,a,r,s\}$.

In the sets $A = \{a,b,c,d,e\}$ and $B = \{a,c,e\}$, each element of B is an element of A. This situation illustrates the following definition.

If each element of a set B is an element of the set A, then B is a **subset** of A. Furthermore, if there are elements in A that are not elements of B, then B is a **proper subset** of A.

We use the notation $B \subseteq A$ to indicate that B is a subset of A, and $B \subset A$ to denote that B is a proper subset of A.

EXAMPLE 1 If $A = \{1,2,3,4,5\}$, $B = \{1,3,5\}$, and $C = \{3,5,1\}$, then $B \subset A$, and $C \subseteq B$.

EXAMPLE 2 If $T = \{x|x$ is a member of a football squad$\}$ and $S = \{x|x$ belongs to the squad and is a defensive back$\}$, then $S \subset T$.

If a is an element of the set S, we say that a belongs to S and express the statement by the notation $a \in S$. The notation $a \notin S$ means that a does not belong to S.

Now if $B \subseteq A$, and $A \subseteq B$, it follows that each element of B belongs to A and each element of A belongs to B. Hence $A = B$. Therefore the definition of equality of the sets A and B can be stated thus:

If $B \subseteq A$ and $A \subseteq B$, then $A = B$.

It may happen that a subset of A is also a subset of B. For example, if $A = \{1,2,3,4,5,6\}$ and $B =$

{2,4,6,8,10}, the set {2,4,6} is a subset of A and also of B and no other element appears in both A and B. This set is called the **intersection** of A and B and illustrates the following definition:

The intersection of the sets A and B is denoted by $A \cap B$ and consists of all elements of A that also belong to B.

EXAMPLE 3 If $A = \{c,a,g,e\}$ and $B = \{c,a,f,e\}$, then $A \cap B = \{a,c,e\}$.

EXAMPLE 4 If $A = \{x|x$ is an alderman of Wilton$\}$ and $B = \{x|x$ is a member of the Wilton Rotary Club$\}$, then $A \cap B = \{x|x$ is an alderman who is a Rotarian in Wilton$\}$.

If in Example 4 no alderman is a Rotarian, then $A \cap B$ contains no elements and is an example of the empty or null set, illustrating the following definition:

The **empty** or **null set** is designated by $\emptyset$, and is the set that contains no elements.

Other examples of the null set are:

1. $\{x|x$ is a woman who has been President of the United States$\}$
2. $\{x|x$ is a two-digit positive integer less than 10$\}$
3. $\{x|x$ is a former governor of California$\} \cap \{x|x$ is a former governor of Texas$\}$

If $S \cap T = \emptyset$, then the sets S and T are **disjoint** sets.

Another concept associated with the theory of sets is the **complement** of one set with respect to another. As an example, if $A = \{x|x$ is a student in a given college$\}$ and $B = \{x|x$ is on the football squad of that college$\}$, then the complement of B with respect to A is $C = \{x|x$ is a student of the college who is not on the football squad$\}$. This illustrates the following definition:

The complement of the set B with respect to A is designated by $A - B$, and

$A - B = \{x|x \in A$ and $x \notin B\}$

As a second example, we shall consider the sets $T = \{x|x$ is a girl in college $C\}$ and $S = \{x|x$ is a member of the senior class of $C\}$; then $T - S = \{x|x$ is a girl not classified as a senior$\}$. Note that in this case, S is not a subset of T.

The totality of elements involved in any specific discussion or situation is called the **universal set** and is designated by the capital letter U. For example, the states in the United States are frequently classified into sets, such as the New England states, the Midwestern states, the Southern states, and in several other ways. Each of these sets is a subset of the universal set, which in this example is composed of all the states in the United States. Each of the various clubs, athletic teams, academic classes, and other groups whose members are students of a given college is a subset of the universal set composed of the entire student body of the college.

A method for picturing sets and certain relations between them was devised by an Englishman, John Venn (1834–1923), who used a simple plane figure to represent a set. We shall illustrate the method by the use of circles and shall define the universal set U as all points within and on the circumference of a circle C. The various subsets of U will be represented by circles wholly within the circle C. In Fig. 1.4 we show the Venn diagrams for $A \cap B$, $A - B$, and the situation in which $A \cap B = \emptyset$.

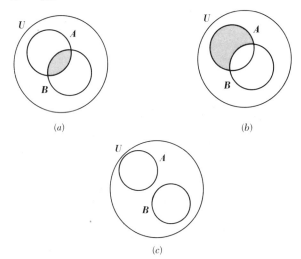

Fig. 1.4 (a) $A \cup B$; (b) $A - B$; (c) disjoint.

If $S = \{1,2,3,4,5,6\}$ and $T = \{2,4,6,8,10\}$, the elements 1, 3, and 5 belong to S but not to T; the elements 8 and 10 belong to T but not to S; and the elements 2, 4, and 6 belong to both S and T. Hence the elements of $V = \{1,2,3,4,5,6,8,10\}$ are in S or are in T or are in both S and T. The set V is called the **union** of the sets S and T, and illustrates the following definition:

The union of the sets S and T is the set whose elements belong to S or to T or to both S and T, and it is designated by $S \cup T$.

Figure 1.5 shows the Venn diagram for the union of two sets.

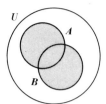

Fig. 1.5 $A \cup B$.

The following examples are illustrations of the union and the intersection of two sets, and of the complement of one set with respect to another.

EXAMPLE 5 If $A = \{m,r,t\}$ and $B = \{r,t,s\}$, then

$A \cup B = \{m,r,t,s\}$
$A \cap B = \{r,t\}$
$A - B = \{m\}$
$B - A = \{s\}$

EXAMPLE 6 If $A = \{1,3,5\}$, $B = \{2,4,6,7\}$, and $C = \{3,5,4\}$, then

$A \cup B = \{1,2,3,4,5,6,7\}$
$A \cap B = \emptyset$
$A - B = A$
$A \cup C = \{1,3,4,5\}$
$A \cap C = \{3,5\}$
$A - C = \{1\}$

Letters or other symbols that stand for numbers are called **variables** and are defined more precisely below.

A letter or a symbol that may be replaced by a number that is an element of a given set is a **variable**, and the given set is the **replacement set**.

If the replacement set for a letter or a symbol contains only one element, then that letter or symbol is a **constant**.

For example, the Greek letter pi (π) stands for the ratio of the circumference of a circle to its diameter and is equal to approximately 3.1416. Hence, π is a constant since there is only one number, $3.1416\cdots$, in the replacement set. Also, the symbol for each real number, such as 2, -3, $\frac{3}{4}$, or $\sqrt{2}$, is a constant. Furthermore, if n stands for the score made by the New York Yankees on May 5, 1979, n is a constant. However, if s stands for a score made by the Yankees in May, s is a variable.

EXAMPLE 7 Let $A = \{x | x$ is a counting number less than or equal to 11$\}$, $B = \{x | x \in A$ and x is not divisible by 2$\}$, $C = \{x | x \in A$ and x is divisible by 3$\}$, $D = \{x | x \in A$ and x is not divisible by 3$\}$, and $E = \{3,5,7,9\}$. Then it is true that $C \cup (D \cap E) = (C \cup D) \cap (C \cup E)$ since each side is equal to $\{3,5,6,7,9\}$.

EXAMPLE 8 If $U = \{x | x$ is a counting number less than or equal to 20$\}$, $S = \{x | x \in U$ and x is divisible by 2$\}$, and $T = \{x | x \in U$ and x is divisible by 3$\}$, the statements below are true.

$T \cup (T - S) = T$
$T \subset (T \cup S)$ and $(T \cap S) \subset T$
$T \cap (T \cup S) = T$

EXERCISE 1.8 CALCULATORS

Find the square root indicated exactly or to three digits by use of a calculator.

1 $\sqrt{169} =$

2 $\sqrt{1764} =$

3 $\sqrt{5041} =$

4 $\sqrt{8649} =$

5 $\sqrt{382} =$

6 $\sqrt{23.37} =$

7 $\sqrt{7.182} =$

8 $\sqrt{974.3} =$

Find each number indicated to three digits.

9 $58.1(2.06) =$

10 $2.84(18.7) =$

11 $59.7(60.3) =$

12 $7.32(28.9) =$

13 $0.276\sqrt{301} =$

14 $40.7\sqrt{53.8} =$

15 $39.72\sqrt{27.93} =$

16 $80.3\sqrt{987} =$

17 $\sqrt{409} \div 71.4 =$

18 $\sqrt{76.2} \div 76.2 =$

19 $\sqrt{928} \div 12.7 =$

20 $\sqrt{471} \div 8.43 =$

54 | Chapter 1: Review of Arithmetic

21 $125.6 \div \sqrt{813} =$

22 $71.43 \div \sqrt{614} =$

23 $54.36 \div \sqrt{27.41} =$

24 $875.6 \div \sqrt{447.3} =$

25 $\dfrac{23.8\sqrt{797}}{35.9} =$

26 $\dfrac{5.93\sqrt{0.871}}{1.76} =$

27 $\dfrac{2472}{76.4\sqrt{987.3}} =$

28 $\dfrac{79.43}{43.5\sqrt{3.712}} =$

29 $\dfrac{2.3 + 1.6}{1.7 + 3.1} =$

30 $\dfrac{3.6 - 1.1}{1.4 + 0.7} =$

31 $\dfrac{2.6 + 1.2}{5.7 - 2.3} =$

32 $\dfrac{4.1 - 2.4}{5.7 - 3.2} =$

33 $6.82 + \dfrac{3.15}{2.84} =$

34 $\dfrac{6.82 + 3.15}{2.84} =$

35 $\dfrac{8.593 + 6.422}{5.333 + 3.567} =$

36 $8.593 + \dfrac{6.422}{5.333} + 3.567 =$

37 $4.2 \div (3.6 \div 2.9) =$

38 $(4.2 \div 3.6) \div 2.9 =$

39 $48.1\sqrt{3.61} =$

40 $\sqrt{48.1}(3.61) =$

1.17 CHAPTER SUMMARY

We began the chapter with a graphical representation of positive integers and proceeded to a discussion of the Hindu-Arabic, or decimal, number system. We then presented the four fundamental operations as applied to it and to fractions. The next section was on base, rate, and percentage, followed by approximate numbers, the metric system, calculators, and sets. The axioms and postulates presented in the chapter are listed below.

$$A + B = B + A \qquad \text{commutative law for addition} \qquad (1.1)$$

$$(A + B) + C = A + (B + C) \qquad \text{associative law for addition} \qquad (1.2)$$

$$A + 0 = 0 + A = A \qquad \text{identity for addition} \qquad (1.3)$$

$$A \times B = B \times A \qquad \text{commutative law for multiplication} \qquad (1.4)$$

$$A \times (B \times C) = (A \times B) \times C \qquad \text{associative law for multiplication} \qquad (1.5)$$

$$A \times (B + C) = (A \times B) + (A \times C) \qquad \text{distributive law} \qquad (1.6)$$

$$A \times 1 = 1 \times A = A \qquad \text{identity for multiplication} \qquad (1.7)$$

$$A \times 0 = 0 \times A = 0 \qquad (1.8)$$

$$A \times \frac{1}{B} = \frac{1}{B} \times A = \frac{A}{B} \qquad (1.9)$$

$$\frac{A}{B} \times \frac{C}{D} = \frac{AC}{BD} \qquad \text{product of fractions} \qquad (1.10)$$

$$R \times B = P \qquad (1.11)$$

$$B = \frac{P}{R} \qquad (1.12)$$

$$R = \frac{P}{B} \qquad (1.13)$$

STUDENT'S NOTES

EXERCISE 1.9 REVIEW

Find the sum of the numbers.

1   738
   +265

2   431
   +649

Find the difference of the numbers.

3   875
   −243

4   726
   −147

5   Show that the following is a magic square:

$$\begin{bmatrix} 7 & 3 & 2 \\ -1 & 4 & 9 \\ 6 & 5 & 1 \end{bmatrix}$$

Perform the indicated operations.

6   347
   +259

7   738.2
   + 46.37

8   284.7
   − 12.6

9   58.43
   − 2.7

10   7⟌2891

11   0.8⟌70.23

12   7.4⟌3826

13   2.83⟌584.3

58 | Chapter 1: Review of Arithmetic

Find the products and quotients.

14 $3 \times 0 = $ 0

15 $0 \div 5 = $ 0

16 $7 \div 0 = $ 0

17 Find the prime factors of 210.

18 Express 26 as the sum of two primes.

19 Find the square root of 529.

20 Find the square root of 5.7132 to three decimals.

21 Evaluate

$$\frac{3.2 - 1.7}{1.3 + 2.4}$$

22 Convert $\frac{1}{2}, \frac{2}{5}, \frac{6}{7}$ into equal fractions with a common denominator.

23 Reduce $\frac{15}{24}$ and $\frac{21}{77}$ to lowest terms.

24 Change $\frac{13}{22}$ and $\frac{37}{23}$ to three-place decimal form.

25 Change 0.6 and 0.72 to common fractions and reduce each to lowest terms.

26 Find the generating fraction of $0.727272 \cdots$.

Perform the indicated operations.

27 $\dfrac{2}{3} + \dfrac{1}{4} - \dfrac{5}{6} =$

28 $\dfrac{3}{5} + 1\dfrac{2}{3} - 1\dfrac{2}{15} =$

29 $\dfrac{21}{15} \dfrac{5}{28} \dfrac{6}{7} =$

30 $\dfrac{3}{2} \dfrac{14}{5} \div \dfrac{126}{15} =$

Convert the numbers to three-digit decimal form and then perform the indicated operations.

31 $2\dfrac{1}{3} + 4\dfrac{6}{7} - 5\dfrac{5}{9} =$

32 $17\dfrac{4}{13} - 8\dfrac{5}{11} - 3\dfrac{5}{6} =$

33 $(7\dfrac{8}{15})(3\dfrac{4}{23}) =$

34 $13\dfrac{5}{19} \div 6\dfrac{11}{17} =$

35 Change 71.3% to decimal form.

36 Change 0.4715 to percent.

37 What is 26% of 317?

38 42.1 is 37% of what number?

39 27.9 is what percent of 863?

40 Find $123 + 45 - 67 + 8 - 9$.

41 Show that $1 = \frac{1}{2} + \frac{1}{3} + \frac{1}{8} + \frac{1}{24}$.

42 Show that $1 = \frac{1}{3} + \frac{1}{7} + \frac{1}{6} + \frac{1}{4} + \frac{1}{12} + \frac{1}{42}$.

43 Show that $\frac{7}{9} = \frac{1}{2} + \frac{1}{5} + \frac{1}{15} + \frac{1}{90}$.

44 3 mL is what percent of 4L?

45 15% of 12 cm is how many millimeters?

46 If one brick weighs 1.6 kg, how many grams are there in four bricks?

47 Find the sum of 4 yd 2 ft 6 in, 5 yd 1 ft 9 in, and 3 yd 2 ft 10 in. Express the sum in the largest possible units without the use of fractions.

48 How many liters of water must be heated to provide 1 cup of instant coffee for each of 18 workers in a laboratory?

49 Show that subtraction is not associative by verifying that
$15 - (9 - 2) \neq (15 - 9) - 2$

50 Show that division is not associative by verifying that
$48 \div (6 \div 2) \neq (48 \div 6) \div 2$

NAME _____ DATE _____ SCORE _____

EXERCISE 1.10 CHAPTER TEST

Treat the numbers in Probs. 1, 2, 6, 10, and 11 as approximations.

1 2.56(41.34) =

2 98.7 ÷ 3.16 =

3 Express 38 using exactly 4 fours.

4 Express 455 as the product of primes.

5 Change 0.86 to a common fraction and reduce to lowest terms.

6 Evaluate $\dfrac{2.70 + 6.13}{8.61 - 3.20} =$

7 $\dfrac{2}{5} \dfrac{20}{17} \div \dfrac{24}{51} =$

8 $\dfrac{2}{7} + \dfrac{3}{4} - \dfrac{5}{6} =$

9 Change $2\tfrac{7}{9}$ and $6\tfrac{4}{13}$ to three-digit decimal form and find their product and the quotient of the first divided by the second.

10 What is 31.6% of 503?

11 39.7 cm is what percent of 11.34 m?

CHAPTER 2

Polynomials

2.1 INTRODUCTION

In Chap. 1 we defined the set of real numbers and briefly discussed the four fundamental operations of addition, subtraction, multiplication, and division, but we left some questions unanswered or unasked. For example, is there a way other than associating numbers with points on a line for proving that $-5 + 2 = -3$? What is the meaning of multiplication if the multiplier is some number other than a positive integer; e.g., what does -3×-2 mean? In this chapter we shall develop the foundation for a logical structure that enables us to answer these questions.

Some material will be repeated in this chapter since it is important to realize that the properties of real numbers hold whether the numbers are 3 and -2 or $3b$ and $c + 4$, that is, numbers or expressions.

A good example of a logical structure in precollege mathematics is plane geometry. If you examine a book on plane geometry, you will find that the text starts with some definitions which are followed by statements called **axioms**. A definition ultimately depends upon a certain number of undefined terms. The axioms are statements, or agreements, that are accepted without proof, and they deal with both the defined and undefined terms and the relations that exist between them. No assumption is made about the defined or undefined terms that is not set forth in either a definition or an axiom. With the definitions and axioms as a basis, statements called **theorems** are proved by deductive reasoning. The first theorem, of course, depends solely on the axioms and definitions. After a theorem is proved, however, it becomes a part of the foundation for proving other theorems. We shall use the same procedure in establishing the logical foundation for the algebra of numbers, and in the next section we define some of the terms that are used.

2.2
DEFINITIONS

From this chapter on, when we use a letter such as a to stand for a number, we mean that a stands for a *real number*. It may be positive, negative, or zero.

We stated in Chap. 1 that the sum of a and b is written as $a + b$, the product is written as ab, and the quotient is written as

$$\frac{a}{b} \quad \text{or} \quad a/b \quad \text{or} \quad a \div b$$

The **fundamental operations** of algebra are **addition, subtraction, multiplication,** and **division.**

A number of the type a/b is called a **fraction.*** The number a is the **numerator** and b is the **denominator.** Later we shall refer to a numerical fraction, and by this we mean a fraction whose numerator and denominator are numerals.

SUBTRACTION The **difference** of the numbers a and b is expressed as $a - b$. Thus,

$a - b$ is equal to the number x such that $b + x = a$
(2.1)

The product $a \cdot a$ is expressed as a^2 and is read "a square." Likewise, $a \cdot a \cdot a = a^3$ and is read "a cube";† $a \cdot a \cdot a \cdot a \cdot a = a^5$ and is called the fifth power of a or more simply a to the fifth.

Any set of numbers and letters that are combined by one or more of the fundamental operations is called an **expression.** For example, $a - 4bc + 3ab/2d$ and $(3xy - 2z)/(4x^2 + 3y)$ are expressions.

In a product of two or more numbers, such as $2abc$, the numbers appearing in the product are called **factors.**

Expressions such as 3, $\frac{2}{3}$, a, $2a$, and $\frac{1}{2}a^2b$ are called **terms.** We define a term as a single number or as a product in which, excepting for numerical fractions, no factor is a fraction.‡

The portion of a term that involves letters only is called the **literal part** of the term. If a term is expressed as the product of two numbers, either numerical or literal, one of the numbers is called the **coefficient** of the other. For example, in $3a$, 3 is the coefficient of a; in $4bc$, 4 is the coefficient of bc, $4b$ is the coefficient of c, and $4c$ is the coefficient of b.

We shall use the phrase **similar terms** to refer to two or more terms in which the literal parts are the same. For example, in each of the following two sets the elements are similar terms: $\{a, 2a, 3a, \frac{1}{2}a\}$, $\{5x^2yz, \frac{1}{4}x^2yz, x^2yz\}$.

A **monomial** is an expression that contains exactly one term; a **binomial** contains exactly two terms; and a **trinomial,** exactly three terms. The word **polynomial** or **multinomial** means many terms, and hence a multinomial is an expression that contains more than one term. It usually means an expression with more than three terms.

Problems 1 to 8 of Exercise 2.1 may be worked now.

2.3
AXIOMS OF ADDITION

In Chap. 1, we interpreted the operation of addition graphically in the set of integers. The set of real numbers is closed under the operation of addition. This means that if a and b are any two real numbers, $a + b$ is also a real number. A set of integers that contains a finite number of elements, however, is not closed under addition. For example, $\{1,2,3,4,5,6\}$ is not closed under addition since $3 + 5 = 8$ and 8 is not an element of the set. Two other sets of integers that are not closed under addition are the set of odd numbers $S = \{3,5,7,9, \ldots\}$ and the set of prime integers $P = \{2,3,5,7,11, \ldots\}$. The set S is not closed under addition since $9 + 7 = 16$ is an even number, and hence is not an element of S. The set P is not closed under addition since $7 + 11 = 18$ is not a prime number.

---

* If a is an integral multiple of b, then a/b is equal to an integer. If, however, the numerical replacements for a and b are not known, it is customary to call a/b a fraction.

† The use of the terms "square" and "cube" to describe a^2 and a^3 originated because the area of a square whose side contains a linear units is a^2 square units, and the volume of a cube whose side contains a units is a^3 cubic units.

‡ According to this definition, the product $a(b + c)$ is a term, since neither factor is a fraction. We shall show later, however, that $a(b + c) = ab + ac$ and hence is a binomial.

CLOSURE UNDER ADDITION It seems intuitively obvious that the set of real numbers is closed under addition, and we shall assume the closure property in the following axiom.

If a and b are real numbers, there exists a real number c such that $a + b = c$.

Addition is an example of a binary operation defined below.

A **binary operation** in a set S is a rule that assigns to every pair of elements of S (taken in a prescribed order) another element of S.

We shall use parentheses ( ), brackets [ ], and sometimes braces { } to indicate the order in which addition is performed. For example, $[(3 + 2) + 5] + [1 + (2 + 4)]$ means that the numbers in parentheses are added first to get $[5 + 5] + [1 + 6]$, then the numbers in the brackets are added to get $10 + 7$, and finally 10 and 7 are added to get 17.

We now consider the sum of the term $2a$ and the binomial $2b + 4a$ and show in detail the steps necessary for obtaining the sum in the simplest form. First write the sum $2a + (3b + 4a)$ and proceed as follows:

1. $2a + (4a + 3b)$ interchanging $3b$ and $4a$
2. $(2a + 4a) + 3b$ grouping $4a$ with $2a$
3. $(2 + 4)a + 3b$
4. $6a + 3b$

We must point out that in this procedure we made the following assumptions. In step 1 we assumed that $4a + 3b = 3b + 4a$. In step 2 we assumed that we could detach $4a$ from $4a + 3b$ and combine it with $2a$. In step 3 we assumed that $(2a + 4a) = (2 + 4)a$. This example shows that the operation of addition in the set of algebraic expressions depends upon at least three assumptions. These will be stated as axioms.

$a + b = b + a$ commutative axiom   (2.2)

This axiom can be verified for any two numbers that we may try. Furthermore, it can be verified graphically, using the number line L. Hence it is reasonable to assume that it is true for all numbers.

The next axiom is the associative axiom, and it enables us to extend the binary operation of addition to an operation involving three or more numbers.

$(a + b) + c = a + (b + c)$ associative axiom   (2.3)

This axiom states that we can get the sum of three numbers either by adding the third number to the sum of the first two or by adding the first to the sum of the second and third. Furthermore, we can prove that $(a + b) + c = (a + c) + b$. To prove this statement, we start with (2.3) and proceed as follows:

$$\begin{aligned}(a + b) + c &= a + (b + c) & \text{by (2.3)} \\ &= a + (c + b) & \text{interchanging } b \\ & & \text{and } c \text{ by (2.2)} \\ &= (a + c) + b & \text{by (2.3)}\end{aligned}$$

Therefore, we can get the sum $a + b + c$ by combining any two of the numbers and then adding this result to the remaining number. For example,

$$\begin{aligned} 4 + 6 + 3 &= (4 + 6) + 3 = 10 + 3 = 13 \\ &= 4 + (6 + 3) = 4 + 9 = 13 \\ &= (4 + 3) + 6 = 7 + 6 = 13 \end{aligned}$$

The next axiom is the **distributive axiom.**

$(a + b)c = ac + bc$ distributive axiom   (2.4)

It can be extended to cover situations in which there are more than two terms in the parentheses. For example,

$$\begin{aligned}(a + b + d)c &= [(a + b) + d]c \\ &= (a + b)c + dc \\ &= ac + bc + dc\end{aligned}$$

We use (2.4) to combine similar terms in a polynomial. For example,

$$3a + 5a + 2b + 4b + 6b \\ = (3 + 5)a + (2 + 4 + 6)b = 8a + 12b$$

In Chap. 1 we defined 0 as the number such that $a + 0 = a$ and the negative of the number a as $-a$ such that $a + (-a) = 0$. By the commutative axiom, $a + 0 = 0 + a$, and $a + (-a) = (-a) + a$. Therefore,

$$a + 0 = 0 + a = a \qquad (2.5)$$
$$a + (-a) = (-a) + a = 0 \qquad (2.6)$$

We shall use these axioms and definitions repeatedly in the remainder of this book whenever real numbers are treated, either as numerals or as literal expressions.

2.4
LAW OF SIGNS FOR ADDITION

The extension of the number system to include negative numbers necessitates an extension of the arithmetical notion of addition. We must define the sum of two signed numbers and decide whether the sum is positive or negative. We shall use the number line L, the graphical interpretation of addition, and the definition of the absolute value of a number in the discussion of these two problems.

By the definition of **absolute value,**

$$|x| = \begin{cases} x & \text{if } x \geq 0 \\ -x & \text{if } x < 0 \end{cases}$$

Thus $|x| \geq 0$, and if n is a positive number, that is, if $n > 0$, then $|n|$ is the distance from (0) to (n) on L, and $|-n|$ is the distance from $(-n)$ to (0).

We first consider the two positive numbers a and b. In Fig. 2.1, we show the points (a), (b), $(a + b)$, $(-a)$, $(-b)$, and $(-a + (-b))$, as well as the distances represented by $|a|$, $|b|$, $|a + b|$, $|-a|$, $|-b|$, and $|-a + (-b)|$. From the figure, it is evident that $a + b$ is positive and $-a + (-b)$ is negative. Furthermore, by comparing the distances

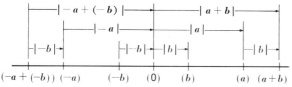

Fig. 2.1

shown in the figure, we see that $|a + b| = |a| + |b|$ and $|-a + (-b)| = |-a| + |-b|$. We therefore have the conclusion:

> The **absolute value of the sum** of two positive numbers or of two negative numbers is the sum of their absolute values. The sum is positive if the two addends are positive, and the sum is negative if the addends are negative.

Next we consider the sum $a + (-b)$ with $a > b > 0$, and show the points (a), (b), $(a + (-b))$, and the distances $|a|$, $|b|$, $|-b|$, and $|a + (-b)|$ in Fig. 2.2. Since the point $(a + (-b))$ is to the right of (0), $a + (-b)$ is positive. From the figure, we see that $|a| - |b| = |a + (-b)|$, since $|a + (-b)| + |b| = |a|$.

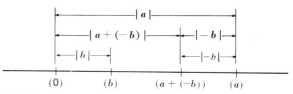

Fig. 2.2

Finally, we consider $c + (-d)$, where c and d are positive and $d > c$. The points associated with these numbers and the distances that represent their absolute values are shown in Fig. 2.3. The point $(c + (-d))$ is to the left of (0), and therefore $c + (-d)$ is negative. Furthermore, the distance

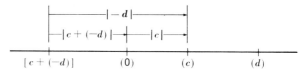

Fig. 2.3

from $(c + (-d))$ to (c) is $|-d|$, and also is $|c + (-d)| + |c|$. Hence, $|-d| = |c + (-d)| + |c|$, and it follows from the definition of the difference that

$$|c + (-d)| = |-d| - |c|$$

These conclusions are summarized in the following statement:

> The absolute value of the sum of a positive number and a negative number is equal to the difference of their absolute values. The sum is positive if the number with the greater absolute value is positive and is negative if the number with the greater absolute value is negative.

The sign of a signed numeral is the sign that precedes it. For example, the sign of 3 is positive, since the sign $+$ is understood to appear before it, and the sign of -3 is negative. In the case of signed numerals, we may state the law of signs as follows:

> The sum of two signed numbers with the same sign is the sum of their absolute values preceded by the common sign of the addends. The sum of two signed numbers with different signs is the difference of their absolute values preceded by the sign of the addend with the greater absolute value.

EXAMPLE 1 6 plus 3 = 9, since 6 and 3 are positive.

EXAMPLE 2 -2 plus $-4 = -6$, since -2 and -4 are negative.

EXAMPLE 3 -4 plus $9 = 5$, since -4 and 9 have different signs and 9 has the greater absolute value.

EXAMPLE 4 3 plus $-7 = -4$, since 3 and -7 have different signs and -7 has the greater absolute value.

There are three signs in any fraction, those of the numerator, the denominator, and the fraction itself. If any two are changed, the value of the fraction is not changed. Thus

$$-\frac{a}{b} = \frac{-a}{b} = \frac{a}{-b} = -\frac{-a}{-b}$$

and

$$\frac{a}{b} = \frac{-a}{-b} = -\frac{-a}{b} = -\frac{a}{-b}$$

Problems 9 to 28 may be done now.

2.5 ADDITION AND SUBTRACTION OF MONOMIALS AND POLYNOMIALS

In order to add similar monomials we use the distributive and commutative laws to write

$$a(b + c) = ab + ac \quad \text{by distributive law}$$

and

$$a(b + c) = (b + c)a \quad \text{by commutative law}$$

They can be extended to more than two terms, as in

$$a(b - c + d) = ab - ac + ad$$

EXAMPLE 1 Combine $4a + 6a - 3a$ into a single term.

Solution

$$4a + 6a - 3a = (4 + 6 - 3)a \quad \text{by distributive law}$$
$$= 7a \quad \text{since } 4 + 6 - 3 = 7$$

EXAMPLE 2 Find the sum of $2ab$, $5ab$, $-8ab$, and $-3ab$.

Solution The sum is

$$2ab + 5ab - 8ab - 3ab = (2 + 5 - 8 - 3)ab$$
$$= (7 - 11)ab = -4ab$$

If a polynomial contains two or more sets of similar terms, we use the commutative and distributive axioms to combine the terms in each set.

EXAMPLE 3

$$7x + 3y + 4xy + 5y - 6x - 8xy + 4x - 10y$$
$$= 7x - 6x + 4x + 3y + 5y - 10y + 4xy - 8xy$$
$$\quad \text{commutative law}$$
$$= (7 - 6 + 4)x + (3 + 5 - 10)y + (4 - 8)xy$$
$$\quad \text{distributive law}$$
$$= 5x - 2y - 4xy$$

To add two or more polynomials, we use the commutative axiom to rearrange the terms so that similar terms are together and use the distributive axiom to combine similar terms.

EXAMPLE 4 Find the sum of $3x^2 - 2xy + y^2$, $2xy - 3y^2 - 2x^2$, and $2y^2 - 5x^2 + 4xy$.

Solution We begin by indicating the sum of the three terms and then we rearrange and combine like terms.

$$3x^2 - 2xy + y^2 + 2xy - 3y^2 - 2x^2 + 2y^2 - 5x^2$$
$$+ 4xy = 3x^2 - 2x^2 - 5x^2 - 2xy + 2xy + 4xy$$
$$+ y^2 - 3y^2 + 2y^2 \quad \text{rearranging}$$
$$= (3 - 2 - 5)x^2 + (-2 + 2 + 4)xy$$
$$+ (1 - 3 + 2)y^2 \quad \text{combining}$$
$$= -4x^2 + 4xy$$

Quite often similar terms are arranged in a column as below.

$$\begin{array}{r} 3x^2 - 2xy + y^2 \\ -2x^2 + 2xy - 3y^2 \\ -5x^2 + 4xy + 2y^2 \\ \hline -4x^2 + 4xy \end{array}$$

EXAMPLE 5 Find the sum of $3a^3 - 2a^2 - 2a$, $2ab - 3b - 2a^3$, and $3a - 4a^2 + 4a^3 - 2b$.

Solution The procedure in this problem is essentially the same as in Example 4 but since there are terms in the second trinomial that are not similar to any in the first trinomial we put them to the right of the first set of terms. Thus, we have

$$\begin{array}{r} 3a^3 - 2a^2 - 2a \\ -2a^3 \qquad\qquad + 2ab - 3b \\ 4a^3 - 4a^2 + 3a \qquad\quad - 2b \\ \hline 5a^3 - 6a^2 + a \;\; + 2ab - 5b \end{array}$$

According to the definition of **subtraction**, $a - b = a + (-b)$. Consequently,

In order to subtract one number or polynomial from another, we change the sign or signs of the subtrahend and add.

EXAMPLE 6 Subtract $7a$ from $9a$.

Solution In keeping with the above

$9a - 7a = 9a + (-7a) = 2a$

EXAMPLE 7 $8x - (-3x) = 8x + (+3x) = 11x$

Often it is possible to change the signs mentally rather than physically.

EXAMPLE 8 Subtract $3x^3 - 5x^2 - 8x + 6$ from $9x^4 + 8x^3 - 6x + 3$.

Solution We put each term of the subtrahend below the corresponding similar term of the minuend, changing the signs of the subtrahend mentally, and add. Thus we have

$$\begin{array}{ll} 9x^4 + 8x^3 - 6x + 3 & \text{minuend} \\ \underline{3x^3 - 5x^2 - 8x + 6} & \text{subtrahend} \\ 9x^4 + 5x^3 + 5x^2 + 2x - 3 & \text{remainder} \end{array}$$

2.6 THE RELATION OF EQUALITY

The relation of equality states that two numbers are equal if they represent the same point on the line L. We shall not attempt a general definition of this relation but shall state some agreements or axioms that enable us to deal with it.

$a = a$ reflexive axiom   (2.7)

For example, if a is $x + y$, then $x + y = x + y$. Similarly, $2x^2 + y = 2x^2 + y$,

$\dfrac{3a}{4b} = \dfrac{3a}{4b}$ and $\dfrac{a - 2b}{c + 3d} = \dfrac{a - 2b}{c + 3d}$

If $a = b$, then $b = a$.   symmetric axiom   (2.8)

If $a = b$ and $b = c$, then $a = c$.
                               transitivity axiom   (2.9)

For example, in the rectangle in Fig. 2.4 $AB = CD$, and in the parallelogram $CD = EF$. Hence, $AB = EF$.

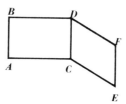

Fig. 2.4

If $a = b$, then $a + c = b + c$.   additivity axiom
                                       (2.10)

EXAMPLE If $x + 5 = 8$, prove that $x + 2 = 5$.

Proof

$$\begin{array}{ll} x + 5 = 8 & \text{given} \\ x + 5 + (-3) = 8 + (-3) & \text{by (2.10)} \\ x + 2 = 5 & \text{since } 5 + (-3) = \\ & 2 \text{ and} \\ & 8 + (-3) = 5 \end{array}$$

If $a = b$, then $ac = bc$.   multiplicativity axiom
                                (2.11)

If $a = b$, then a can be replaced by b in any statement involving algebraic expressions without affecting the truth or falsity of the statement.   (2.12)

We now prove two theorems that will be needed in later discussions.

If $a = b$ and $c = b$, then $a = c$.   transitivity
                                          axiom   (2.13)

PROOF If $c = b$, then $b = c$ by the symmetric axiom (2.8). Hence, we have $a = b$ and $b = c$, and it follows that $a = c$ by the transitivity axiom (2.9).

If $a = b$ and $c = d$, then $a + c = b + d$.   (2.14)

PROOF

$$\begin{array}{ll} a = b & \text{given} \\ a + c = b + c & \text{by additivity axiom (2.10)} \\ a + c = b + d & \text{replacing } c \text{ by } d \text{ in } b + c \end{array}$$

EXAMPLE 7 If $3x - y = 7$ and $y - 2x = 3$, prove that $x = 10$.

Proof

$$\begin{array}{ll} 3x - y + y - 2x = 7 + 3 & \text{by (2.14)} \\ 3x - 2x - y + y = 10 & \text{by commutative} \\ & \text{axiom (2.2)} \\ (3 - 2)x - y + y = 10 & \text{by distributive} \\ & \text{axiom (2.4)} \\ x + 0 = 10 & \text{since } -y + y = 0 \\ x = 10 & \text{since } x + 0 = x \end{array}$$

If $a + b = a + d$, then $b = d$.
                    cancellation theorem for addition   (2.15)

PROOF

$a + b = a + d$	given
$a + b + (-a) = a + d + (-a)$	by additivity axiom (2.10)
$[a + (-a)] + b = [a + (-a)] + d$	by commutative axiom (2.2) and associative axiom (2.3)
$0 + b = 0 + d$	since $a + (-a) = 0$
$b = d$	since $0 + b = b$ and $0 + d = d$

EXAMPLE 2 If in Fig. 2.5 area of $ACDB$ = area of $EFGB = k$ square units, then area of $ACHE$ = area of $HFGD$.

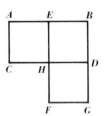

Fig. 2.5

Proof

$HDBE + ACHE = ACDB = k$ and
$HDBE + HFGD = EFGB = k$

Hence,

$HDBE + ACHE = HDBE + HFGD$

since each is equal to k. Therefore,

$ACHE = HFGD$ by (2.15)

In Sec. 1.4 we stated that $-(-a) = --a = a$. We now show that this conclusion necessarily follows from the above definitions, axioms, and theorems if a is any real number. If in (2.6) we replace a by $-d$ we have

$-d + (--d) = 0$

It also follows from (2.6) that

$-d + d = 0$

Hence, by (2.13),

$-d + (--d) = -d + d$

since each is equal to 0. Thus, by the cancellation theorem (2.15),

$$-(-d) = d \qquad (2.16)$$

NEGATIVE OF A SUM We can also prove the following theorem that will be used repeatedly in the remainder of this text:

$$-(d + c) = -d - c \qquad (2.17)$$

PROOF If we replace a by $d + c$ in (2.6), $a + (-a) = 0$, we have

$d + c + [-(d + c)] = 0$

Also, by the commutative axiom (2.2),

$d + c + (-d) + (-c)$
$\quad = d + (-d) + c + (-c) = d - d + c - c$
$\quad = 0 + 0 \qquad$ since $d - d = c - c = 0$
$\qquad\qquad\qquad$ by (2.6)
$\quad = 0$

Consequently, by (2.13),

$d + c + [-(d + c)] = d + c - d - c$

it follows from the cancellation theorem (2.15) that

$-(d + c) = -d - c$

If in (2.17) we replace c by $-e$, we have

$-[d + (-e)] = -d - (-e) = -d + e$

This theorem can be extended to cover situations in which there are more than two terms in parentheses. If we replace d by $e + f$ and c by $g + h$, we have

$-(e + f + g + h) = -(e + f) - (g + h)$
$\qquad\qquad\qquad = -e - f - g - h$

Consequently, if a polynomial is enclosed in parentheses preceded by a minus sign, the parentheses can be removed if the sign of every term in the polynomial is changed. For example,

$2x - (3y - 2z + 4w) = 2x - 3y + 2z - 4w$

and

$3a - (-2b + 5c - 4d) = 3a + 2b - 5c + 4d$

STUDENT'S NOTES

EXERCISE 2.1

Classify the expression as a monomial, binomial, trinomial, multinomial, or none of these.

1. $x^3 + 3x$

2. $3x$

3. $5x + 1/x$

4. $x^4 + 3x^2 - 5x$

5. $9x^2yz^3$

6. $xy + x/y$

7. $x + xy + 3yw$

8. $7xy - 5yw$

Use one or more of axioms (2.1) to (2.3) to prove the statement.

9. $(x + y) + w = (y + w) + x$

10. $3 + (x + 2) = x + 5$

11. $x + y + (-x) = y$

12. $x + (-y) + (-x) + y = 0$

13. $2a + 3c + 4a + 2c = 6a + 5c$

14. $3a + 4b - a + (-8b) + (-a) = a - 4b$

15. $5a - (-3a) - 4b + (-7a) + 5b = a + b$

16. $4a + (-5b) - (-2a) + 7b = 6a + 2b$

17. If $a = 2$ and $b = 3$, then $a + |b| =$

18. If $a = 4$ and $b = -3$, then $2a + |b| =$

19. If $a = -3$ and $b = -2$, then $|a| + |b| - |a + b| =$

20. If $a = -5$ and $b = -4$, then $a + |a| + b - |b| =$

Combine the numbers.

21. $7 + 6 - 9 =$

22. $8 - 5 - 7 =$

23. $14 - 5 + 2 - 11 =$

24. $-7 - 6 + 15 - 4 =$

25	-376	26	540
	$\underline{214}$		$\underline{-893}$

27	234	28	958
	456		-146
	$\underline{-678}$		$\underline{-623}$

Write similar terms together but do not add.

29 $5a + 7b + 6c - 2a - 3b + 2c - a =$

30 $9x - 3y - 4w - 8x + 5y + 6w - 4y =$

31 $8y - 9a + 13p + 6a - 11p - 10y + 4p =$

32 $15p - 8d - 6q - 14p + 8d - 3q - 5p + 4q =$

33 $3a^2b + 5ab^2 - (7a^2b - 2ab^2 + 5ab^2) =$

34 $7a^2b - 3ab - (8ab^2 - 5ab - 2ab^2) =$

35 $14x^3y - (17x^2y^2 + 13xy^3 - 21x^2y^2 - 9xy^3) =$

36 $19x^2y^2 + 7xy + 8xy^2 - (22x^2y^2 + 7xy + 9xy^2) =$

Combine similar terms.

37 $4a + 7b - 9c - 2a - 9b + 2b =$

38 $12t + 8u + 7m - 9t - 11u + 12m - 3m =$

39 $7w - 6h + 3l - 4w + 7h - 3l - 4h =$

40 $6r + 5a - 11w + 9w - 5a - 6r + 2w =$

41 $5x^3y + 7x^2y^2 - (3xy^3 + 7x^2y^2 - 4xy^3) =$

42 $9p^2r^2 - 8pr^3 + 7r^4 - (9p^2r^2 - 7pr^3 + 6r^4) =$

43 $11pq + 12pr - 6qr - (10p + 9pq - 7qr + 12pr) =$

44 $13ca - 10ab - 2bc + 3a - (9ac - 7ba - 2cb) =$

Subtract the second expression from the first.

45 $3x + 2y - 4w,\ 5x + y + 2w$

46 $5x - 3y + 7w,\ 2x - 4y + 6w$

47 $8x + 5y - 12w,\ 8x - 5y - 13w$

48 $7x + 7y - 5w,\ 6x - y - 6w$

49 $5a + 3b + 6c,\ 5a + 3b - c$

50 $9a + 6b - 8c,\ 8a + 5b - 9c$

51 $4p + 5d + 4q,\ 3p - 2d + 3q$

52 $7s - 3a - 5m,\ 6s - 4a - 7m$

Write similar terms in a column and then add.

53 $3a + 5b + 7c - 2a + 3b - c + a - 7b - 6c$

54 $7x + 3y - 4z - 5x - 2y + 3z - 2x - y + z$

55 $5p - 2i + 4e + 3i - 5e + 4p + 3e - 7p + i$

56 $7s - 3i + 6t + 8i - 4t - 5s - 2t + s - 3t$

57 If $a = 3$, $b = 2$, and $c = 1$, then $a + b + c =$

58 If $s = 7$, $a = -4$, and $m = 2$, then $4s - 8a + m =$

59 If $x = -2$, $y = -1$, and $w = 4$, then $3x + 2y - 2w =$

60 If $a = 3$, $t = 4$, and $s = -5$, then $2a - 3t + s =$

61 If $a = -1$, $b = 2$, and $c = 5$, then $ab + bc + ca =$

62 If $p = 7$, $a = -6$, and $t = 4$, then $pa + 2pt - 3at =$

63 If $p = 4$, $q = -3$, and $r = 7$, then $pqr + pq - qr =$

64 If $s = 6$, $t = 5$, and $u = -3$, then $2st + 3su - 4tu =$

65 It is true that $|a + b| \leq |a| + |b|$. Verify this for $a = 2$, $b = 5$ and for $a = 6$, $b = -2$.

2.7 AXIOMS AND THEOREMS OF MULTIPLICATION

In this section we shall state the essential axioms of multiplication and prove the theorems that govern the operation of multiplication when applied to signed numbers, to monomials, and to polynomials.

As stated in Sec. 2.2, the result of multiplying the number b by the number a is called the **product** of a and b. The operation is denoted by $a \times b$ (or $a \cdot b$) and the product is written ab. The numbers a and b are the *factors* of ab.

Parentheses and other signs of grouping are also used to indicate multiplication. For example, $a(-b)$ means $a \times (-b)$, $a(bc)$ means $a \times (bc) = a \times b \times c$.

CLOSURE UNDER MULTIPLICATION We assume that the set of real numbers is closed under the operation of multiplication; this means that the product of two real numbers is a unique real number. We state this assumption more precisely in the following axiom.

If a and b are real numbers, there exists a unique real number c such that $ab = c$.   closure axiom (2.18)

Hence, multiplication is a binary operation in the set of real numbers.

We can readily verify that $3 \times 5 = 5 \times 3 = 15$ and $4 \times 7 = 7 \times 4 = 28$. Furthermore, we get the product of three integers such as $8 \times 5 \times 3$ by obtaining the product of any two of them and then multiplying this product by the third. For example, it is easily verified that $8 \times 5 \times 3 = 8 \times (5 \times 3) = (8 \times 5) \times 3 = (8 \times 3) \times 5 = 120$. We assume that these properties of multiplication hold for all real numbers and state the assumptions in the axioms below. In these and in others to be stated presently, as well as in the theorems that will be proved, the letters used stand for real numbers.

$ab = ba$ commutative axiom for multiplication   (2.19)

$(ab)c = a(bc)$ associative axiom for multiplication   (2.20)

These axioms enable us to get the product $a \times b \times c$ by obtaining the product of two of the numbers and then multiplying this product by the third, and this product is unique, regardless of the choice of the first two numbers and the order in which the multiplication is performed. This means that

$$abc = a(bc) = (ab)c = a(cb)$$
$$= (ac)b = c(ba) = (cb)a = \cdots \quad (1)$$

where the list of equalities can be extended to include all possible orders in which a, b, and c can be arranged, and all choices of the two that are enclosed in parentheses. To prove that this statement is true, we must prove that each combination in Eq. (1) is equal to some one of them. To illustrate the method, we shall prove that $c(ba) = a(bc)$.

PROOF

$c(ba) = c(ab)$ by commutative axiom (2.19)
$\quad\quad = (ab)c$ by (2.19)
$\quad\quad = a(bc)$ by associative axiom (2.20)

In Sec. 2.3 we stated the distributive axiom. By the commutative axiom (2.19), we can rewrite this as

$c(a + b) = ca + cb$ distributive axiom   (2.21)

We next examine the role of 1 and 0 in multiplication. Since $1 \cdot a + 1 \cdot a + 1 \cdot a = (1 + 1 + 1)a = 3a$, and $1 \cdot a + 1 \cdot a = (1 + 1)a = 2a$, it is logical to define $1(a)$ as a, and since $1(a) = a(1)$, we have

$1(a) = a(1) = a$ multiplicative identity   (2.22)

For this reason, 1 is called the **identity element** for multiplication.

We next prove that $a(0) = 0(a) = 0$. We start with

$na + (-na) = [n + (-n)]a$
$\quad\quad\quad\quad\quad\quad\quad$ by distributive axiom (2.4)
$\quad\quad\quad = 0(a)$ since $n + (-n) = 0$
$\quad\quad\quad\quad\quad\quad\quad$ by (2.6)

However,

$na + (-na) = 0$ by (2.6)

Hence, $0(a) = 0$ by (2.8) since each is equal to $na + (-na)$. Therefore, by the commutative axiom, it follows that

$0(a) = a(0) = 0$ (2.23)

RECIPROCAL OR MULTIPLICATIVE INVERSE
If $a \neq 0$, the *reciprocal* of a is the number $1/a$ such that

$$a\frac{1}{a} = \frac{1}{a}a = 1 \qquad (2.24)$$

Now by use of (2.23) and (2.24) we can prove the following theorem, which we shall use frequently in solving equations.

If $a(b) = ab = 0$ and $a \neq 0$, then $b = 0$. (2.25)

PROOF

$ab = 0$	given
$\frac{1}{a}ab = \frac{1}{a}0$	by **multiplicativity** axiom (2.11)
$\left(\frac{1}{a}a\right)b = 0$	by associative axiom (2.20) and by (2.23)
$b = 0$	since $(1/a)(a) = 1$ by (2.24) and $1(b) = b$ by (2.22)

CANCELLATION LAW FOR MULTIPLICATION

We shall frequently have occasion to use the cancellation theorem:

If $ab = ac$ and $a \neq 0$, then $b = c$. (2.26)

PROOF

$ab = ac$	given
$\left(\frac{1}{a}a\right)b = \left(\frac{1}{a}a\right)c$	by multiplicativity axiom (2.11) and associative axiom (2.20)
$1(b) = 1(c)$	since $\frac{1}{a}a = 1$ by (2.24)
$b = c$	by (2.22)

2.8 LAW OF SIGNS FOR MULTIPLICATION

We have stated that the number a is positive if the point (a) is to the right of (0) on the number line L. We also stated that $a > b$ if the point (a) is to the right of (b) on L. We call the statement $a > b$ an **inequality**. Now $4 > 2$ and $4(3) > 2(3)$, since $12 > 6$. Furthermore, $4(\frac{1}{2}) > 2(\frac{1}{2})$ since $2 > 1$. We assume that this multiplicative property of an inequality holds in the set of all real numbers and state the assumption in the following axiom:

If $a > c$ and $b > 0$, then $ab > cb$. (2.27)

It follows from this axiom that the product of two positive real numbers is positive, since if $a > 0$ and $b > 0$, then

$a(b) > 0 \qquad$ by (2.27)

Hence,

$ab > 0 \qquad$ since $a(b) = ab$ and $0(b) = 0$

Therefore, *the product of two positive numbers is positive*.

The question of what happens if either or both factors are negative can be settled by the distributive axiom (2.4) and the definition of a negative number (2.6), which states that $a + (-a) = 0$. To show how this axiom and definition work, we shall prove that $-2(3) = -6$ and that $-2(-3) = 6$.

PROOF

$6 + [-2(3)] = [2 + (-2)](3)$	by distributive axiom (2.4)
$= 0(3)$	since $2 + (-2) = 0$
$= 0$	

Also

$6 + (-6) = 0 \qquad$ by (2.6)

Therefore,

$6 + (-2)(3) = 6 + (-6) \qquad$ since each is equal to 0

Hence, by the cancellation theorem for addition (2.15)

$(-2)(3) = -6$

It follows from the commutative axiom (2.19) that

$(-2)(3) = 3(-2) = -6$

Similarly, by the distributive theorem (2.21)

$-6 + (-2)(-3) = (-2)(3) + (-2)(-3)$
$= -2[3 + (-3)]$
$= (-2)(0)$
$= 0$

Also,

$-6 + 6 = 0$

Therefore,

$-6 + (-2)(-3) = -6 + 6 \qquad$ since each is equal to 0

Hence,

$(-2)(-3) = 6 \qquad$ by cancellation theorem (2.15)

If in each of the above proofs we replace 2 by $a > 0$ and 3 by $b > 0$ and use exactly the same argument, we get $(-a)(b) = b(-a) = -(ab)$ from the first and $(-a)(-b) = ab$ from the second.

Consequently, if a and b are positive numbers, $a(b) = ab$ and ab is positive, $(-a)(b) = b(-a) = -(ab)$, and $(-a)(-b) = ab$.

We summarize the above conclusions in the following statement:

Law of signs for multiplication: The product of two positive numbers or of two negative numbers is positive; the product of a positive number and a negative number is negative.

A number with no sign appearing before it is understood to be positive. A number preceded by a minus sign is negative. For example, 3, 6, and 8 are positive, whereas -3, -6, and -8 are negative.

The following examples illustrate the application of the law of signs to signed numbers.

EXAMPLE 1

$3(4) = 12 \quad 2(-5) = -10 \quad (-6)(-7) = 42$

EXAMPLE 2

$3(4)(-2) = [3(4)](-2) = 12(-2) = -24$
$-2 \cdot 3 \cdot -6 = (-2 \cdot 3) \cdot (-6)$
$\qquad = (-6)(-6) = 36$
$-3 \cdot -5 \cdot -7 = (-3 \cdot -5) \cdot (-7)$
$\qquad = 15(-7) = -105$

The sign of a variable is determined by the replacement set. Hence if the replacement set for x is $\{x | 0 < x < 20\}$, then x is positive. If the replacement set is $\{x | -15 < x < 0\}$, then x is negative. If the replacement set is the set of all real numbers, we do not know whether x is positive or negative. If no replacement set is specified for x, it is understood to be the set of all real numbers.

If the factors in a product involve one or more variables, we get the product by the following procedure: (1) apply the law of signs to the numerical coefficients; (2) write the product of the variables at the right of the product of the coefficients.

EXAMPLE 3

$3x(4y) = [3(4)](xy) = 12xy$

$-2ab(8cd) = (-2)(8)(ab)(cd) = -16abcd$
$-5c(-8df) = (-5)(-8)(c)(df) = 40cdf$

Problems 1 to 16 of Exercise 2.2 may be done now.

2.9 EXPONENTS IN MULTIPLICATION

We have seen that $a \cdot a = a^2$ and that $a \cdot a \cdot a = a^3$. If we use the same notation, we have

$$a \cdot a \cdot a \cdots a \text{ to } n \text{ factors is } a^n \qquad (2.28)$$

We called a^2 the second power of a and a^3 the third power of a. Similarly, a^n is called the nth power of a or merely a to the nth. The number a in a^n is called the **base** and n the **exponent** of the base. This concept can be used only if n is a positive integer, but we shall discuss other types of exponents in a later chapter.

PRODUCT OF POWERS If m and n are positive integers, $a^n a^m$ is the product of n a's and m a's and can be written as a^{m+n}. Hence, we have

$$a^n a^m = a^{m+n} \qquad (2.29)$$

for m and n positive integers.

EXAMPLE 1 $x^3 x^5 = x^{3+5} = x^8$

EXAMPLE 2 $x^2 x^4 x^7 = x^{2+4+7} = x^{13}$

We can use the associative and commutative axioms to rearrange the factors of a product in any desirable order.

EXAMPLE 3 $3a^2(4a^5) = (3)(4)(a^2)(a^5) = 12a^7$

EXAMPLE 4

$2x^3(3y^5)(2y^4)(4x^2) = 2(3)(2)(4)(x^3)(x^2)(y^5)(y^4)$
$\qquad = 48x^5 y^9$

EXAMPLE 5 $-4x^2(3x^5) = -4(3)(x^2)(x^5) = -12x^7$

Notice that $-4x^2$ means $(-4)(x^2)$, not $(-4x)^2$.

Problems 17 to 28 of Exercise 2.2 may be done now.

2.10 PRODUCTS INVOLVING POLYNOMIALS

We use the distributive axiom for finding the product of a monomial and a polynomial, as follows:

EXAMPLE 1

$4a^2b(2a^2 - 3ab + b^2)$
$\quad = (4a^2b)(2a^2) + 4a^2b(-3ab) + (4a^2b)(b^2)$
$\quad = 8a^4b - 12a^3b^2 + 4a^2b^3$

EXAMPLE 2

$(2x^3 - 4x^2y + 2xy^2 - 3y^2)(-3x^3y^4)$
$\quad = -6x^6y^4 + 12x^5y^5 - 6x^4y^6 + 9x^3y^6$

ZERO AS A FACTOR Since by (2.23), $0(a) = a(0) = 0$, the rule for a product involving zero as one of the factors is

If zero appears as one of the factors in any product, the product is equal to zero.

It follows from an extended application of the distributive axiom that the product of two polynomials is equal to the sum of the products obtained by multiplying each term in one of them by every term in the other. The plan to be used for this is illustrated in Examples 3 to 5.

EXAMPLE 3 In order to multiply $3x + 2y$ by $2x - y$, we perform the steps shown below:

- Multiply $3x + 2y$ by $2x$
- Multiply $3x + 2y$ by $-y$
- Add coefficients of similar terms

$3x + 2y$
$2x - y$
$6x^2 + 4xy$ 1st partial product
$\quad - 3xy - 2y^2$ 2d partial product
$6x^2 + xy - 2y^2$ product

EXAMPLE 4 To multiply $4x^2 - 3xy + 2y^2$ by $3x^2 + 4xy - 5y^2$, we proceed as follows:

- Multiply $4x^2 - 3xy + 2y^2$ by $3x^2$
- Multiply $4x^2 - 3xy + 2y^2$ by $4xy$
- Multiply $4x^2 - 3xy + 2y^2$ by $-5y^2$
- Add coefficients

$4x^2 - 3xy + 2y^2$
$3x^2 + 4xy - 5y^2$
$12x^4 - 9x^3y + 6x^2y^2$
$\quad\quad 16x^3y - 12x^2y^2 + 8xy^3$ partial
$\quad\quad\quad\quad - 20x^2y^2 + 15xy^3 - 10y^4$ products
$12x^4 + 7x^3y - 26x^2y^2 + 23xy^3 - 10y^4$ product

Note: In these examples the partial products are so written that similar terms form columns. Also, in any partial product after the first, any term that is not similar to one that has already appeared is written at the right. This practice is shown in Example 5.

EXAMPLE 5 If we use the principle of place value to express the two numbers 326 and 243 as trinomials in powers of 10, we have

$326 = 3 \times 100 + 2 \times 10 + 6$
$\quad\quad = 3 \times 10^2 + 2 \times 10 + 6$
$243 = 2 \times 100 + 4 \times 10 + 3$
$\quad\quad = 2 \times 10^2 + 4 \times 10 + 3$

We shall now multiply these two trinomials showing each step in this computation as well as the simplified form of the partial products:

$3 \times 10^2 + 2 \times 10 + 6$
$2 \times 10^2 + 4 \times 10 + 3$
$6 \times 10^4 + 4 \times 10^3 + 12 \times 10^2$
$\quad\quad\quad 12 \times 10^3 + 8 \times 10^2 + 24 \times 10$
$\quad\quad\quad\quad\quad\quad 9 \times 10^2 + 6 \times 10 + 18$
$6 \times 10^4 + 16 \times 10^3 + 29 \times 10^2 + 30 \times 10 + 18$
$= 60{,}000 + 16{,}000 + 2900 + 300 + 18 = 79{,}218$

Simplified form of partial products
$60{,}000 + 4000 + 1200 = 65{,}200$
$12{,}000 + 800 + 240 = 13{,}040$
$\quad\quad\quad 900 + 60 + 18 = \underline{\quad 978}$
$\quad\quad\quad\quad\quad\quad\quad\quad\quad\quad 79{,}218$

If we multiply the two numbers by the method of arithmetic, we have the calculation shown below. If we fill the blank spaces in the second and third partial-product columns with zeros, the partial products will be the same as those obtained in the first multiplication, although the order is reversed.

$\quad\quad 3\ 2\ 6$
$\quad\quad 2\ 4\ 3$
$\quad\quad \overline{9\ 7\ 8}$
$\quad 1\ 3\ 0\ 4$
$\quad 6\ 5\ 2$
$\overline{7\ 9{,}2\ 1\ 8}$

2.11
SYMBOLS OF GROUPING

Parentheses and brackets have been used in preceding sections to clarify the meaning of certain expressions and to indicate the order in which certain operations are to be performed. If parentheses and brackets are not sufficient for this purpose, braces { } are also used.

The rule given in Sec. 2.6 for the insertion or removal of parentheses applies to brackets and braces also. However, usually when brackets, braces, or both, appear together with parentheses, in an expression, there will be one or more sets enclosed in another set. For example, in

$$3x^2 - \{2x^2 - xy - [x(x - y) - y(2x - y)] + 4xy\} - 3y^2$$

two sets of parentheses are enclosed in the brackets and the braces enclose both the brackets and the parentheses. When the symbols of grouping are removed from an expression of this type, it is usually advisable to remove the innermost symbols first.

EXAMPLE We now show the process of removing symbols of grouping by applying it to the expression:

$$3x^2 - \{2x^2 - xy - [x(x - y) - y(2x - y)] + 4xy\} - 3y^2$$

Removing parentheses gives

$$3x^2 - \{2x^2 - xy - [x^2 - xy - 2xy + y^2] + 4xy\} - 3y^2$$

Combining similar terms in brackets leads to

$$3x^2 - \{2x^2 - xy - [x^2 - 3xy + y^2] + 4xy\} - 3y^2$$

Removing brackets, we have

$$3x^2 - \{2x^2 - xy - x^2 + 3xy - y^2 + 4xy\} - 3y^2$$

Combining similar terms in braces gives

$$3x^2 - \{x^2 + 6xy - y^2\} - 3y^2$$

Removing braces leads to

$$3x^2 - x^2 - 6xy + y^2 - 3y^2$$

and combining similar terms gives finally

$$2x^2 - 6xy - 2y^2$$

STUDENT'S NOTES

EXERCISE 2.2

Prove the statement.

1. If $a(b + c) = A + ac$, then $ab = A$.

2. If $a - b = -(a + b)$, then $a = 0$.

3. If $a \neq 0$ and $a(b - c) = 0$, then $b = c$.

4. If $a = -a$, then $a = 0$.

5. If $a(b - 1) = ab + c$, then $a = -c$.

6. If $ab - 2 = 2(b - 1)$, then $a = 2$.

7. If $(a + b)(c - 1) = (a + b)(d + 1)$ and $a \neq -b$, then $c = d + 2$.

8. If $(a - b)(c - d) = 2a(c - d)$ and $c \neq d$, then $a = -b$.

Fill in the blank.

9. $a \cdot a \cdot a \cdot a =$ _____

10. $-a \cdot a \cdot a \cdot a \cdot a =$ _____

11. $(-a)(-a)(a) =$ _____

12. $(-a)(-a)(-a)(a)(a)(a) =$ _____

13. If $-a$ is used 6 times as a factor, the product is _____.

14. If $-a$ is used 7 times as a factor, the product is _____.

15. If $-a$ is used 9 times as a factor, the product is _____.

16. If $-a$ is used 10 times as a factor, the product is _____.

Chapter 2: Polynomials

Indicate whether the statement is true or false. If it is false, give the correct product.

17. $2a^2(2a^2) = 4a^4$

18. $3a^3(3a^3) = 9a^9$

19. $3a^2(4a^3) = 12a^5$

20. $4a(5a^4) = 20a^4$

21. $2a^2(4a^3)(3a^4) = 24a^{24}$

22. $3a^3(4a^4)(2a^5) = 24a^{12}$

23. $2a^2(3a^3)(4a^4) = 24a^{24}$

24. $1a^3(2a^2)(3a) = 6a^6$

25. $2x^2y(4x)(5y^3) = 40x^3y^4$

26. $3xy^3(2x^2y)(4xy^2) = 24x^2y^5$

27. $3x^2y^2(2x^3y^3)(xy^2) = 60x^6y^7$

28. $4xy^2(3x^2y)(5x^3y^4) = 60x^6y^7$

Find the product indicated.

29. $a(b + c) =$

30. $a(b - 2c) =$

31. $2a(3b - c) =$

32. $-3a(2a - 3b) =$

33. $2rs^2(3r^2s + rs) =$

34. $(2a^2b - 3ab^2)2a =$

35. $3x^2y - 2xy^2)(-2xy) =$

36. $5a^2b^3(-2ab^2 + 3a^2b) =$

37 $2x - y$
   $\underline{x + 3y}$

38 $3x - 2y$
   $\underline{2x - 3y}$

39 $4x + 3y$
   $\underline{3x - 2y}$

40 $-5x + 2y$
   $\underline{2x - 5y}$

41 $a^2 + 2a + 3$
   $\underline{2a + 5}$

42 $3x^2 - 2x - 1$
   $\underline{5x - 3}$

43 $a^2 + 3ab + 4b^2$
   $\underline{2a - 3b}$

44 $5a^2 - 3ab - 2b^2$
   $\underline{3a - 4b}$

45 $2a + b - 3c$
   $\underline{3a - 2b + c}$

46 $5a + 6b - 2c$
   $\underline{3a - b - 2c}$

47 $3x^2 - 2xy - y^2$
   $\underline{x^2 - xy + y^2}$

48 $4x^3 - 3x^2y + xy^2$
   $\underline{3x^2 - 2xy + y^2}$

Remove the symbols of grouping and collect like terms.

49 $[x + 3(y - w)] + 2[x - 3(y - w)] =$

50 $3[x - 2(y + 2w) - 4] + 5[y - 3(x - w) + 5] =$

51 $3[x - 2(y + w)] + 5[x - 3(y - w)] =$

52 $[2(x - 2y) + 3(x - y)] - 2[x - 5(x + y)] =$

53 $3x^2 - \{3x^2 - xy - [5x^2 - 5xy - (3x^2 - 3y^2) + 4xy]\} - 3y^2 =$

54 $3x - \{2y - w + [6x - (4y - w)] - 2x\} + y =$

55 $2w - \{3y - [4w - (2y - 4x) + 2w] - 5x\} =$

56 $4a - \{3a - 2[3a - 2(3a - 2) + 3a] + 2\} =$

2.12 DIVISION

In Sec. 1.6 we stated that if $b \neq 0$, the quotient of the numbers a and b is indicated by $a \div b$, $\frac{a}{b}$, or a/b, and we defined the quotient as the unique number x that satisfies the following:

If $b \neq 0$, then $a \div b = \frac{a}{b} = x$ if and only if $bx = a$. (2.30)

The number a is the **dividend,** and b is the **divisor.**

If we apply the law of signs for multiplication to the product bx in $bx = a$, we have the possibilities given in the following tabulation:

a dividend	b divisor	x quotient
Positive	Positive	Positive
Negative	Negative	Positive
Negative	Positive	Negative
Positive	Negative	Negative

Hence, if a and b are both positive or both negative, x is positive. If a is positive and b is negative, or if a is negative and b is positive, x is negative.

SIGN OF A QUOTIENT Consequently, we have the following law of signs for division:

The quotient of two positive numbers or of two negative numbers is positive. The quotient of a negative and a positive number, or of a positive and a negative number, is negative.

EXAMPLE 1 $12 \div 3 = 4$, since 12 and 3 are both positive.

EXAMPLE 2 $-18 \div (-2) = 9$, since -18 and -2 are both negative.

EXAMPLE 3 $12a \div -6a = -2$, since 12 and -2 are of opposite signs.

EXAMPLE 4 $-21x \div 3x = -7$ since $-21x$ and $3x$ are of opposite signs.

DIVIDEND ZERO As in Sec. 1.6, if $a/b = x$, then, by (2.30) we have $bx = a$. Therefore, if $a = 0$ and $b \neq 0$, we have $bx = 0$ and it follows by (2.25) that $x = 0$. Therefore:

If the dividend is zero and the divisor is not zero, the quotient is zero.

If, however, $b = 0$ and $a \neq 0$, we have $0(x) = a$. Since, by (2.23) $0(x) = 0$ if x is replaced by any number, there is no replacement for x such that $0(x) = a$. Hence, we conclude:

If the dividend is not zero and the divisor is zero, the quotient does not exist.

Finally, if $a = 0$ and $b = 0$, we have $0(x) = 0$, and this statement is true if x is replaced by any number. Therefore $\frac{0}{0}$ can represent any number whatsoever, and hence does not exist as a unique number. Therefore, division by zero is never defined.

2.13 MONOMIAL DIVISORS

The following theorems are used when the divisor is a monomial:

$$\frac{a}{a} = 1 \quad (2.31)$$

$$\frac{a}{1} = a \quad (2.32)$$

PROOF Since $1(a) = a(1) = a$, it follows by (2.30) that $a/1 = a$ and $a/a = 1$.

$$\frac{a}{b} = a\frac{1}{b} = \frac{1}{b}a \quad (2.33)$$

PROOF If we let

$$\frac{a}{b} = x$$

then

$a = bx$ by (2.30)

Furthermore,

$\frac{1}{b}a = \frac{1}{b}bx$ by multiplicativity axiom for equalities (2.11)

$\phantom{\frac{1}{b}a} = \left(\frac{1}{b}b\right)x$ by associative axiom for multiplication (2.19)

$\phantom{\frac{1}{b}a} = x$ since $(1/b)(b) = 1$ by (2.24), and $1(x) = x$ by (2.22)

Hence,

$$\frac{1}{b}a = \frac{a}{b} \quad \text{since each is equal to } x$$

Furthermore, by the commutative axiom $(1/b)(a) = a(1/b)$

$$\frac{ab}{cd} = \frac{a}{c}\frac{b}{d} \qquad (2.34)$$

PROOF If we let

$$\frac{a}{c} = x \quad \text{and} \quad \frac{b}{d} = y$$

then

$$a = cx \quad \text{and} \quad b = dy \quad \text{by (2.30)}$$

Hence,

$$\frac{ab}{cd} = \frac{cx(dy)}{cd}$$

$$= \frac{cd(xy)}{cd} \qquad \text{by commutative and associative axioms (2.18) and (2.19)}$$

$$= \left(\frac{1}{cd} cd\right) xy \qquad \text{by associative axiom}$$

$$= xy \qquad \text{since } (1/cd)(cd) = 1 \text{ and } 1(xy) = xy$$

$$= \frac{a}{c}\frac{b}{d} \qquad \text{replacing } x \text{ by } a/c \text{ and } y \text{ by } b/d$$

We can extend (2.34) to cover cases in which the dividend and divisor are the products of more than two numbers by an argument similar to the following:

$$\frac{acx}{bdy} = \frac{ac(x)}{bd(y)} = \frac{ac}{bd}\frac{x}{y} = \frac{a}{b}\frac{c}{d}\frac{x}{y}$$

We use statement (2.34) or its extension, together with (2.22) and (2.31), to get the quotient of two monomials, as illustrated in the following examples.

EXAMPLE 1

$$\frac{15abc}{5ab} = \frac{15c}{5}\frac{a}{a}\frac{b}{b} \qquad \text{by commutative axiom (2.19) and by (2.34)}$$
$$= 3c(1)(1) \qquad \text{by (2.34)}$$
$$= 3c \qquad \text{by (2.22)}$$

EXAMPLE 2

$$\frac{18a}{6} = \frac{18(a)}{6(1)} \qquad \text{since } 6 = 6 \cdot 1 \text{ by (2.22)}$$
$$= \frac{18}{6}\frac{a}{1} \qquad \text{by (2.34)}$$
$$= 3a \qquad \text{by (2.32)}$$

EXAMPLE 3

$$\frac{4}{6a} = \frac{4(1)}{6(a)} = \frac{4}{6}\frac{1}{a} = \frac{2}{3}\frac{1}{a} = \frac{2}{3a}$$

Ordinarily, the intermediate steps are omitted and we write $4/6a = 2/3a$.

EXAMPLE 4

$$\frac{6(2x + 3y)}{2(2x + 3y)} = \frac{6}{2}\frac{2x + 3y}{2x + 3y} \qquad \text{by (2.34)}$$
$$= 3(1) \qquad \text{since } \tfrac{6}{2} = 3 \text{ and } (2x + 3y)/(2x + 3y) = 1$$
$$= 3 \qquad \text{by (2.31) with } a = 2x + 3y$$

EXPONENTS IN DIVISION We shall first illustrate and then give the general rule for obtaining the quotient of two powers of the same number. We shall consider a^6/a^4, with $a \neq 0$. Since $a^6 = a^4(a^2)$, we have

$$\frac{a^6}{a^4} = \frac{a^4(a^2)}{a^4(1)}$$
$$= \frac{a^4}{a^4}\frac{a^2}{1} \qquad \text{by (2.34)}$$
$$= a^2 \qquad \text{since } a^4/a^4 = 1, a^2/1 = a^2, \text{ and } 1(a^2) = a^2$$

Similarly, if m and n are positive integers and $m > n$, we have $m - n + n = m$. Hence, by the law of exponents for multiplication (2.29)

$$a^m = a^{m-n+n} = a^{m-n}a^n$$

Consequently, if $a \neq 0$, we have

$$\frac{a^m}{a^n} = \frac{a^n a^{m-n}}{a^n(1)}$$
$$= \frac{a^n}{a^n} a^{m-n} \qquad \text{by (2.34)}$$
$$= a^{m-n} \qquad \text{since } a^n/a^n = 1$$

Therefore, we have the following **law of exponents for division:**

$$\frac{a^m}{a^n} = a^{m-n} \quad \text{if } a \neq 0 \qquad (2.35)$$

provided that m and n are integers, $m > n$, and $a \neq 0$.

If in (2.35), $n = m$ and $a \neq 0$, we have $a^m/a^m = a^{m-m} = a^0$, and this brings us to a result a^0 that has no meaning, according to our previous definition of exponents. However, since by (2.31) $a^m/a^m = 1$, we define a^0 to be 1. Thus,

$$a^0 = 1 \quad \text{provided that } a \neq 0 \quad (2.36)$$

The following examples further illustrate the method of obtaining the quotient of two monomials.

EXAMPLE 5

$$\frac{12a^4b^6}{3a^2b^3} = \frac{12}{3} \frac{a^4}{a^2} \frac{b^6}{b^3}$$
$$= 4a^2b^3$$

EXAMPLE 6

$$\frac{24x^4y^7z^3w}{8x^3y^2z^3} = \frac{24}{8} \frac{x^4}{x^3} \frac{y^7}{y^2} \frac{z^3}{z^3} \frac{w}{1}$$
$$= 3xy^5z^0w = 3xy^5w \quad \text{since } z^0 = 1$$

THE QUOTIENT OF A POLYNOMIAL AND A MONOMIAL   We use (2.31) and the distributive axiom to obtain the quotient of the binomial $a + b$ and the monomial n, as shown below:

$$\frac{a+b}{n} = \frac{1}{n}(a+b) \quad \text{by (2.33)}$$
$$= \frac{a}{n} + \frac{b}{n} \quad \text{by distributive axiom and (2.33)}$$

By extending the above argument, we can show that

$$\frac{a+b+c+\cdots+m}{n}$$
$$= \frac{a}{n} + \frac{b}{n} + \frac{c}{n} + \cdots + \frac{m}{n} \quad (2.37)$$

EXAMPLE 7

$$\frac{4x^3y^7 + 9x^5y^4 + 12x^8y^3}{3x^2y^3} = \frac{4x^3y^7}{3x^2y^3} + \frac{9x^5y^4}{3x^2y^3} + \frac{12x^8y^3}{3x^2y^3}$$
$$= \tfrac{4}{3}xy^4 + 3x^3y + 4x^6$$

Problems 1 to 28 of Exercise 2.3 may be done now.

2.14 QUOTIENT OF TWO POLYNOMIALS

We shall discuss the procedure for obtaining the quotient of two polynomials in this section, restricting our discussion to polynomials with integral coefficients.

We begin by considering the quotient $\frac{234}{5}$. If we divide 234 by 5, we obtain the quotient 46 and the remainder 4 and thus $234 = 5(46) + 4$. Similarly,

$$\frac{6x^2 + 12x + 5}{3x} = \frac{6x^2}{3x} + \frac{12x}{3x} + \frac{5}{3x} = 2x + 4 + \frac{5}{3x}$$

or

$$6x^2 + 12x + 5 = 3x(2x + 4) + 5$$

in which the quotient is $2x + 4$ and the remainder is 5.

We can readily verify the fact that the relation

Dividend = (divisor)(quotient) + remainder   **(1)**

is satisfied by each of the above problems.

DIVISION OF POLYNOMIALS   In order to divide one polynomial by another, we begin by arranging the terms in each polynomial in descending powers of some variable that occurs in each. We then seek the quotient, a polynomial that satisfies Eq. (1) where the degree of the remainder (in the variable chosen as the basis of the arrangement of terms) is less than the degree of the divisor (in that variable).

EXAMPLE 1   Find the quotient obtained by dividing $6x^2 + 5x - 1$ by $2x - 1$.

Solution   The dividend is $6x^2 + 5x - 1$, the divisor is $2x - 1$, and we are looking for the quotient that satisfies

$$6x^2 + 5x - 1 = (2x - 1)(\text{quotient}) + \text{remainder} \quad (2)$$

The degree of the quotient must be 1 and that of the remainder must be 0 since the dividend is of degree 2 and the divisor of degree 1. Hence (2) takes the form

$$6x^2 + 5x - 1 = (2x - 1)(ax + b) + c \quad (3)$$

If we now divide $6x^2 + 5x - 1$ by $2x - 1$ by the usual procedure, we find that the quotient $ax + b$ is $3x + 4$ and the remainder c is 3, as seen from

$$
\begin{array}{r}
3x + 4 \text{quotient}\\
2x - 1 \overline{)6x^2 + 5x - 1}\\
\underline{6x^2 - 3x} (2x-1)(3x)\\
8x - 1 \text{subtracting}\\
\underline{8x - 4} (2x-1)(4)\\
3 \text{remainder}
\end{array}
$$

EXAMPLE 2 Divide $8x^2 + 26x + 15$ by $2x + 5$.

Solution

$$
\begin{array}{r}
4x + 3 \text{quotient}\\
2x + 5 \overline{)8x^2 + 26x + 15}\\
\underline{8x^2 + 20x} (2x+5)(4x)\\
6x + 15 \\
\underline{6x + 15} (2x+5)(3)\\
0 \text{remainder}
\end{array}
$$

We further illustrate the long division process in the following example.

EXAMPLE 3

$$
\begin{array}{r}
x^3 + 2x^2 - 3x + 4 \text{quotient}\\
\text{divisor} \text{dividend}\\
3x^2 - 4x - 3 \overline{)3x^5 + 2x^4 - 20x^3 + 18x^2 - 7x - 12}\\
\underline{3x^5 - 4x^4 - 3x^3}\\
6x^4 - 17x^3 + 18x^2 \\
\underline{6x^4 - 8x^3 - 6x^2}\\
- 9x^3 + 24x^2 - 7x \\
\underline{- 9x^3 + 12x^2 + 9x}\\
12x^2 - 16x - 12\\
\underline{12x^2 - 16x - 12}
\end{array}
$$

Note: In the above example the terms in the dividend and divisor are so arranged that the exponents of x are in descending numerical order. If in any division problem the terms of the dividend and divisor are not already in descending order of exponents, it is advisable to so arrange them.

In each of the above examples the final difference in the long division process is 0, and we say that the division "comes out even"; i.e., the dividend is a multiple of the divisor. Usually, however, this is not the case, and we ultimately obtain a difference in which the greatest exponent of the variable chosen as the basis for arranging the terms is less than the greatest exponent of that variable in the divisor. When this happens, the process terminates and the final difference is called the **remainder.**

We illustrate such a situation in the following example.

EXAMPLE 4 Find the quotient and remainder obtained by dividing $6x^4 + 8x^3y - 5x^2y^2 + 2xy^3 - y^4$ by $3x^2 + xy - 2y^2$.

Solution Since the exponents of x are in descending order in the dividend and divisor, we proceed at once with the division shown below:

$$
\begin{array}{r}
2x^2 + 2xy - y^2 \text{quotient}\\
3x^2 + xy - 2y^2 \overline{)6x^4 + 8x^3y - 5x^2y^2 + 2xy^3 - y^4}\\
\underline{6x^4 + 2x^3y - 4x^2y^2}\\
6x^3y - x^2y^2 + 2xy^3 \\
\underline{6x^3y + 2x^2y^2 - 4xy^3}\\
- 3x^2y^2 + 6xy^3 - y^4\\
\underline{- 3x^2y^2 - xy^3 + 2y^4}\\
\text{remainder} 7xy^3 - 3y^4
\end{array}
$$

Since the greatest exponent of x in $7xy^3 - 3y^4$ is less than the greatest exponent of x in the divisor, the division process terminates and $7xy^3 - 3y^4$ is the remainder.

EXERCISE 2.3

Perform the indicated divisions.

1. $a^5 \div a^3 =$

2. $a^9 \div a^4 =$

3. $b^7 \div b =$

4. $b^5 \div b^4 =$

5. $x^3 y^4 \div x^2 y^2 =$

6. $x^8 y^5 \div x^6 y^3 =$

7. $a^3 b^2 \div a^3 b =$

8. $a^6 b^7 \div a^5 b^4 =$

Indicate whether the statement is true or false. If false, give the correct answer.

9. $\dfrac{6x^4 y^2}{x^2 y} = 6x^2 y$

10. $\dfrac{15 a^7 b^4}{5 a^5 b^2} = 3 a^2 b^2$

11. $\dfrac{4 x^6 y^2}{2 x^3 y^2} = 2x^2$

12. $\dfrac{6 x^5 y^4}{3 x^2 y} = 3 x^3 y^3$

13. $\dfrac{12 c^6 d^5}{6 c^3 d} = 2 c^2 d^5$

14. $\dfrac{12 a^5 b^3}{4 a^2 b} = 3 a^3 b^2$

15. $\dfrac{10 x^7 y^6}{5 x^6 y^6} = 2x$

16. $\dfrac{18 x^5 y^8}{6 x^2 y^6} = 12 x^3 y^2$

17. $\dfrac{0}{17 a^2 b} = 0$

18. $\dfrac{13 a b^3}{0} = 13 a b^3$

19. $\dfrac{4 x^5 y^4}{2 x^0 y^4} = 2 x^5$

20. $\dfrac{6 a^0 b^5 c}{3 b^2 c^0} = 2 b^3 c$

Find the indicated quotient.

21. $\dfrac{x^4y^5 - x^6y^4}{x^2y^3} =$

22. $\dfrac{x^7y^3w^2 + x^6y^5w^4}{x^3y^3w} =$

23. $\dfrac{r^4s^5t^3 - 2r^5s^2t^2}{r^4s^2t^2} =$

24. $\dfrac{3a^4b^5c^6 + 4a^7b^6c^5}{a^3b^5c^4} =$

25. $\dfrac{20a^5b^7c^4 - 15a^3b^4c^3}{5a^3bc^3} =$

26. $\dfrac{18x^4y^7t^5 + 12x^5y^4t^3}{6x^4y^4t^2} =$

27. $\dfrac{15a^7b^6c^4 - 9a^4b^5c^6}{3a^3bc^4} =$

28. $\dfrac{15x^7y^5w^3 - 30x^6y^6w^4}{15x^5y^5w^2} =$

29. $x + 2 \overline{\smash{)}x^2 + 5x + 6}$

30. $x - 3 \overline{\smash{)}x^2 + 2x - 15}$

31. $2x - 3 \overline{\smash{)}6x^2 - 7x - 3}$

32. $4x + 3 \overline{\smash{)}12x^2 - 7x - 12}$

33. $2x - 1 \overline{\smash{)}2x^3 - 7x^2 + 11x - 4}$

34. $x - 5 \overline{\smash{)}3x^3 - 14x^2 - 6x + 5}$

35. $2x + 3 \overline{\smash{)}6x^3 + 17x^2 + 8x - 6}$

36. $3x + 5 \overline{\smash{)}6x^3 + 7x^2 - 8x - 5}$

37 $3x^2 - x + 2 \overline{\smash{\big)}3x^3 + 11x^2 - 2x + 8}$

38 $2x^2 + x + 3 \overline{\smash{\big)}4x^3 + 4x^2 + 7x + 3}$

39 $5x^2 - 7x + 4 \overline{\smash{\big)}5x^3 + 3x^2 - 10x + 8}$

40 $3x^2 + 2x - 2 \overline{\smash{\big)}6x^3 + 13x^2 + 2x - 6}$

41 $2x - 1 \overline{\smash{\big)}2x^4 - 5x^3 + 4x^2 + 5x - 3}$

42 $x - 4 \overline{\smash{\big)}2x^4 - 8x^3 + 3x^2 - 13x + 4}$

43 $3x + 2 \overline{\smash{\big)}9x^4 + 18x^3 + 8x^2 + 15x + 10}$

44 $5x + 1 \overline{\smash{\big)}5x^4 + 16x^3 - 17x^2 - 14x - 2}$

Find the quotient and write the remainder.

45 $x^2 + 3x - 1 \overline{\smash{\big)}2x^4 + 5x^3 - 2x^2 + 10x - 1}$

46 $2x^2 - x + 2 \overline{\smash{\big)}2x^4 + 3x^3 - 2x^2 + 5x - 5}$

47 $2x^2 + 3x - 1 \overline{)6x^4 + 11x^3 + 4x^2 + 5x - 7}$

48 $3x^2 + 2x + 1 \overline{)6x^4 - 5x^3 - 16x^2 - 11x - 6}$

49 $p + 2q \overline{)2p^3 + 3p^2q - 3pq^2 - 4q^3}$

50 $p - 3q \overline{)2p^3 - 4p^2q - 7pq^2 + 7q^3}$

51 $2p + 3q \overline{)6p^3 + 11p^2q + pq^2 - 8q^3}$

52 $3p - 2q \overline{)6p^3 + 5p^2q - 9q^3}$

53 $2r - 3s \overline{)4r^4 - 11r^2s^2 + 5rs^3 - 5s^4}$

54 $3r + 2s \overline{)6r^4 + r^3s - 2r^2s^2 + 3rs^3 + 9s^4}$

55 $5r + s \overline{)5r^4 + r^3s - 10r^2s^2 + 13rs^3 + 9s^4}$

56 $r + 4s \overline{)4r^4 + 13r^3s - 11r^2s^2 + rs^3 - 4s^4}$

2.15
NUMBERS TO VARIOUS BASES

If a number is expressed as the sum of multiples of integral powers of a number b, we say that the number is expressed to the base b. For example, if $b > 7$, then

$$7 + 3b + 5b^2 + 0b^3 + 4b^4$$

is expressed to the base b. If $b = 10$, then

$$7 + 3 \times 10 + 5 \times 10^2 + 0 \times 10^3 + 4 \times 10^4$$

is in base 10, and if $b = 8$, then

$$7 + 3(8) + 5(8^2) + 0(8^3) + 4(8^4)$$

is in base 8.

We shall write b as a subscript to indicate that the number N is to the base b other than 10; hence we write N_b. Accordingly

$$3102_4 = 3(4^3) + 1(4^2) + 0(4) + 2$$
$$= 3(64) + 1(16) + 0 + 2 = 210$$

We shall now consider changing a number from base 10 to another base. We begin by finding the largest integral power of the base that is less than or equal to the number, the number of times this power goes into the number, and the remainder when this multiple of the base to a power is subtracted from the number; we then treat this remainder just as the number was treated and continue the process until we obtain a remainder smaller than the base.

EXAMPLE 1 Write 210 in terms of base 4.

Solution The largest power of 4 that is smaller or equal to 210 is $4^3 = 64$, and this power goes into 210 three times since $3(64) = 192$. Now we have $210 = 192 + 18 = 3(4^3) + 18$. The largest power of 4 that is equal to or less than 18 is 4^2, and it goes into 18 only once with a remainder of 2, which is less than the base. We now know that 210 is $3(4^3) + 1(4^2) + 0(4) + 2$; hence $210 = 3102_4$ as we saw above.

DIVISION PROCESS FOR CHANGING BASE
We shall now see how to get this result by another method that may be simpler for some to follow. It consists of dividing the number N by the base b, dividing the quotient by b, dividing the next quotient by b, and so on until a *quotient* of 0 is obtained, being certain to keep track of the re-

mainder at each stage. We shall now find 210 in terms of the base 4 again.

	Remainder	Use	Number
4⌐210			
4⌐ 52	2	2	
4⌐ 13	0	$0(4^1)$	
4⌐ 3	1	$1(4^2)$	
0	3	$3(4^3)$	3102_4

Consequently $210_{10} = 3102_4$. Note that $10_4 = 1(4) + 0 = 4 + 0 = 4 = 4_{10}$.

EXAMPLE 2 Use the division process to change 1978_{10} to base 9.

Solution

	Remainder	Use	Number
9⌐1978			
9⌐ 219	7	7	
9⌐ 24	3	$3(9^1)$	
9⌐ 2	6	$6(9^2)$	
0	2	$2(9^3)$	$2637_9 = 1978_{10}$

It is a rather simple matter to change a number from any other base to base 10.

EXAMPLE 3 $1325_6 = 1(6^3) + 3(6^2) + 2(6) + 5$
$= 216 + 108 + 12 + 5 = 341$

If we have a number to some base other than 10 and want it to another non-10 base, we need only change to base 10 and then to the desired base.

EXAMPLE 4 Express 356_7 in terms of base 2.

Solution

$356_7 = 3(7^2) + 5(7) + 6 = 147 + 35 + 6$
$= 188$ base 10

We now use the division process to change to base 2. Thus

	Remainder	Use
2⌐188		
2⌐ 94	0	0
2⌐ 47	0	0
2⌐ 23	1	$1(2^2)$
2⌐ 11	1	$1(2^3)$
2⌐ 5	1	$1(2^4)$
2⌐ 2	1	$1(2^5)$
2⌐ 1	0	0
0	1	$1(2^7)$

Therefore, $356_7 = 188 = 10111100_2$.

Since the base of the binary system is 2, only the

digits 0 and 1 are used. For clarity we shall show binary numbers in italics.

If in any system the sum of 2 one-digit numerals is equal to the base, then this sum is denoted by 10. In the following tabulation, we show all pairs of one-digit numerals whose sum is 10 to the base indicated at the head of each column.

10 (base 10)	*10 (base 5)*	*10 (base 3)*	*Binary 10 (base 2)*
9 + 1	4 + 1	2 + 1	1 + 1
8 + 2	3 + 2		
7 + 3			
6 + 4			
5 + 5			

Since the one-digit numerals in any system are less than the base, the one-digit numerals in the binary system are 0 and 1, and a binary numeral is expressed with one or more of these digits. The following tabulation shows the binary equivalent for each of the first 20 decimal numerals and for each power of 2 through 2^6.

Decimal	*Binary*	*Decimal*	*Binary*
1	*1*	12	*1100*
2	*10*	13	*1101*
3	*11*	14	*1110*
4 = 2^2	*100*	15	*1111*
5	*101*	16 = 2^4	*10000*
6	*110*	17	*10001*
7	*111*	18	*10010*
8 = 2^3	*1000*	19	*10011*
9	*1001*	20	*10100*
10	*1010*	32 = 2^5	*100000*
11	*1011*	64 = 2^6	*1000000*

As we can see from the above tabulation, most binary numerals are quite long, but since only two digits are required, a binary numeral can be expressed by a line of properly spaced holes punched in a card. For example, the binary number *1100101* can be represented by Fig. 2.6,

Fig. 2.6

where the holes represent 1 and the blank intervals represent 0. A binary numeral can also be "flashed" by a properly spaced string of electric pulses. For this reason, high-speed electronic computers are designed to use the binary system, and an understanding of the binary system is necessary for computer programming.

We next illustrate the addition and multiplication of binary numerals. The following addition table shows the sums of all pairs of binary numerals that contain three or fewer digits.

$0 + 0 = 0$
$0 + 1 = 1$ $\qquad$ $10 + 10 = 100$
$1 + 1 = 10$ $\qquad$ $10 + 11 = 101$
$1 + 10 = 11$ $\qquad$ $11 + 11 = 110$
$1 + 11 = 100$

The reader should verify that the above results are correct by using the method shown below to get *11 + 11*. We write the two numerals as indicated below and proceed according to the directions under the problem.

$$\begin{array}{r} 11 \\ + \ 11 \\ \hline 110 \end{array}$$

We add the digits in the second column and get $1 + 1 = 10$. Record *0* and carry *1*. Add the digits in the first column and get *10*; then add the digit *1* that was carried and get *11*. Record *11* as the first two digits in the sum.

EXAMPLE 5 Find the following binary sums:

(a) $\;\; 1 + 1 + 1 \quad$ (b) $\;\; 1 + 1 + 1 + 1$

Solution

(a) $\quad 1 + 1 + 1 = (1 + 1) + 1 = 10 + 1 = 11$
(b) $\; 1 + 1 + 1 + 1 = (1 + 1) + (1 + 1)$
$\qquad\qquad\qquad = 10 + 10 = 100$

EXAMPLE 6 Find the sum of the binary numerals *10, 11, 110, 101,* and *111*.

Solution We write the numerals in a column as indicated below and proceed as directed in steps 1 to 3.

1. Add digits in first column on the right and get $1 + 1 + 0 + 1 + 0 = (1 + 1) + 1 = 10 + 1 = 11$. Record 1 below the column; carry 1.

2. Add digits in second column to the carried 1 and get $1 + 0 + 1 + 1 + 1 + 1 = (1 + 1) + (1 + 1) + 1 = 10 + 10 + 1 = 101$. Record 1; carry 10.

3. Add digits in third column to the carried 10 and get $1 + 1 + 1 + 10 = (1 + 1) + (1 + 10) = 10 + 11 = 101$. Record 101 as first three digits in sum.

```
                    numbers
                    carried
                10 1
                   10
                   11
                   110
                   101
                   111
                 10111
```

Check

Binary	Decimal
111 =	7
101 =	5
110 =	6
11 =	3
10 =	2
	23 = binary 10111

EXAMPLE 7 Find the product of the binary numerals 101 and 111.

Solution We write the multiplicand, multiplier, and partial products as indicated below, and then add the partial products. Note that each partial product is $1(101) = 101$.

```
    101       multiplicand
    111       multiplier
    101 ⎫
    101 ⎬    partial products
    101 ⎭
  100011     product
```

Check

Binary	Decimal
101	5
111	7
	35 = binary 100011

Numerals to the base 3 are called *ternary* numerals and are expressed with one or more of the digits $0, 1,$ and 2. Since the sum of the ternary numerals 1 and 2 is equal to the base, $1 + 2 = 10$.

The following addition table shows the sums of all pairs of ternary numbers that contain fewer than four digits.

$1 + 1 = 2$	$10 + 10 = 20$	$12 + 1 = 20$
$1 + 2 = 10$	$11 + 1 = 12$	$12 + 2 = 21$
$2 + 2 = 11$	$11 + 2 = 20$	$12 + 10 = 22$
$10 + 1 = 11$	$11 + 10 = 21$	$12 + 11 = 100$
$10 + 2 = 12$	$11 + 11 = 22$	$12 + 12 = 101$

Note: In obtaining the sum $12 + 12$, we add the last two digits to get $2 + 2 = 11$. We record 1 and carry 1. Then add the first two digits and the 1 that was carried and get $1 + 1 + 1 = (1 + 1) + 1 = 2 + 1 = 10$, so the sum is 101. The reader should verify that the other results are correct.

EXAMPLE 8 Find each of the following sums, if the numerals are in the ternary system:

(a) $1 + 1 + 1$ (b) $1 + 1 + 1 + 1$
(c) $1 + 1 + 2 + 2$ (d) $2 + 2 + 2$

Solution

(a) $1 + 1 + 1 = (1 + 1) + 1 = 2 + 1 = 10$
(b) $1 + 1 + 1 + 1 = (1 + 1) + (1 + 1)$
$\qquad = 2 + 2 = 11$
(c) $1 + 1 + 2 + 2 = (1 + 2) + (1 + 2)$
$\qquad = 10 + 10 = 20$
(d) $2 + 2 + 2 = (2 + 2) + 2 = 11 + 2 = 20$

EXAMPLE 9 Add the following ternary numerals: $102, 121, 212,$ and 112.

Solution

1. Add numerals in right-hand column to get $2 + 2 + 1 + 2 = (2 + 2) + (1 + 2) = 11 + 10 = 21$. Record 1; carry 2.

2. Add the middle column to the carried 2 to get $1 + 1 + 2 + 0 + 2 = (1 + 1) + (2 + 2) = 2 + 11 = 20$. Record 0; carry 2.

3. Add left-hand column, include carried 2 to get $1 + 2 + 1 + 1 + 2 = (1 + 2) + (1 + 1) + 2 = 10 + 2 + 2 = 12 + 2 = 21$. Record 21 as first two digits in sum.

```
    22
   102
   121
   212
   112
  2101
```

Check

Ternary	Decimal
$102 =$	11
$121 =$	16
$212 =$	23
$112 =$	14
sum $64 = 2101$	ternary

In the ternary system,

$2(1) = 2$
$2(2) = 2 + 2 = 11$
$2(10) = 20$
$2(11) = 22$
$2(12) = 101$

To verify the last product, we express *12* as *10 + 2*, and then have $2(10 + 2) = 20 + 11 = 101$.

EXAMPLE 10 Find the product of the ternary numerals *1212* and *2*.

Solution

1. $2(2) = 11$. Record *1* and carry *1*.

2. $2(1) = 2$, $2 + 1$ carried $= 10$. Record *0* and carry *1*.
$$\begin{array}{r} 1212 \\ 2 \\ \hline 10201 \end{array}$$

3. $2(2) = 11$, $11 + 1$ carried $= 12$. Record *2* and carry *1*.

4. $2(1) = 2$, $2 + 1$ carried $= 10$. Record *10* as first two digits in product.

Check

Ternary $1212 = 3^3 + 2(3^2) + 1(3) + 2$
$= 27 + 18 + 3 + 2$ decimal
$= 50$ decimal

Thus, $2(50) = 100$, and decimal $100 =$ ternary *10201*.

EXAMPLE 11 Obtain the product of the ternary numerals *212* and *122*.

Solution

1. Partial products are $2(212) = 1201$ and $1(212) = 212$.

2. Write partial products as indicated and find sum by method of Example 5.

$$\begin{array}{r} 212 \\ 122 \\ \hline 1201 \\ 1201 \\ 212 \\ \hline 112111 \end{array}$$

Check

Ternary $212 =$ decimal 23
ternary $122 =$ decimal 17
$$\begin{array}{r} 161 \\ 23 \\ \hline 391 \end{array} = \text{ternary } 112111$$

EXERCISE 2.4

Find N.

1. $384 = N_5$
2. $384 = N_8$
3. $384 = N_2$
4. $384 = N_3$
5. $1728 = N_{12}$
6. $2521 = N_9$
7. $1023 = N_4$
8. $1117 = N_{10}$
9. $4592 = N_8$
10. $2345 = N_7$
11. $256 = N_2$
12. $2189 = N_3$
13. $357_8 = N_9$
14. $2761_9 = N_6$
15. $10101_2 = N_8$
16. $10045_8 = N_3$

98 | Chapter 2: Polynomials

17 $14203_5 = N_9$

18 $10111011_2 = N_4$

19 $111100101_2 = N_3$

20 $1001021_3 = N_9$

Find the sum of the binary numbers.

21   *11*
      + 10

22   *11*
      + 11

23   *10*
      11
      11
     + 11

24   *11*
      10
      10
     + 11

25   *101*
      110
      111
     + 100

26   *111*
      101
      10
     + 100

27   *1111*
      1110
      1101
      1011
     + 110

28   *11011*
      10111
      11101
      11110
     + 1111

Find the product of the pair of binary numbers.

29   101
       × 11

30   110
       ×101

31   1011
       × 111

32   1001
       × 101

33   10101
       ×   111

34   10011
       ×   101

35   1111
       ×1111

36   10011
       × 1001

100 | Chapter 2: Polynomials

Find the sums of the ternary numbers.

37	102		**38**	211
	210			122
	+ 21			+111

39	2210		**40**	2221
	1102			1212
	1222			1010
	+2101			+2112

Find the product of the pair of ternary numbers.

41	201		**42**	102
	× 12			× 22

43	21012		**44**	12212
	× 201			× 2112

2.16 CHAPTER SUMMARY

The chapter began with several definitions and then launched into a discussion of addition and subtraction, including the necessary axioms. This was followed by a comparable discussion of multiplication and division. Symbols of grouping were also discussed. The final section dealt with numbers to bases different from 10, with particular attention to base 2. The essential axioms are listed below.

$a - b$ is equal to the number x such that $b + x = a$ (2.1)

$a + b = b + a$ commutative (2.2)

$(a + b) + c = a + (b + c)$ associative (2.3)

$(a + b)c = ac + bc$ distributive (2.4)

$a + 0 = 0 + a = a$ additive identity (2.5)

$a + (-a) = (-a) + a = 0$ additive inverse (2.6)

$a = a$ reflexive (2.7)

If $a = b$, then $b = a$ (2.8)

If $a = b$ and $b = c$, then $a = c$ (2.9)

If $a = b$, then $a + c = b + c$ (2.10)

If $a = b$, then $ac = bc$ (2.11)

If $a = b$, then a can be replaced by b in any algebraic expression without affecting the truth or falsity of the statement (2.12)

If $a = b$ and $c = b$, then $a = c$ (2.13)

If $a = b$ and $c = d$, then $a + c = b + d$ (2.14)

If $a + b = a + d$, then $b = d$ (2.15)

$-(-d) = d$ (2.16)

$-(d + c) = -d - c$ (2.17)

If a and b are real numbers, there exists a unique real number c such that

$ab = c$ closure (2.18)

$ab = ba$ commutative (2.19)

$(ab)c = a(bc)$ associative (2.20)

$c(a + b) = ca + cb$ distributive (2.21)

$1(a) = a(1) = a$ multiplicative identity (2.22)

$0(a) = a(0) = 0$ zero in multiplication (2.23)

$a \dfrac{1}{a} = \dfrac{1}{a} a = 1$ multiplicative inverse (2.24)

If $a(b) = ab = 0$ and $a \neq 0$, then $b = 0$ (2.25)

If $ab = ac$ and $a \neq 0$, then $b = c$ (2.26)

If $a > c$ and $b > 0$, then $ab > cb$ (2.27)

$a \cdot a \cdot a \cdots a$ to n factors is a^n (2.28)

$a^n a^m = a^{n+m}$ (2.29)

$\dfrac{a}{b} = x$ if and only if $bx = a$ for $b \neq 0$ (2.30)

$\dfrac{a}{a} = 1$ if $a \neq 0$ (2.31)

$\dfrac{a}{1} = a$ (2.32)

$\dfrac{a}{b} = a \dfrac{1}{b} = \dfrac{1}{b} a$ (2.33)

$\dfrac{ab}{cd} = \dfrac{a}{c} \dfrac{b}{d}$ (2.34)

$$\frac{a^m}{a^n} = a^{m-n} \qquad a \neq 0 \tag{2.35}$$

$$a^0 = 1 \qquad \text{provided } a \neq 0, \tag{2.36}$$

$$\frac{a + b + c + \cdots + m}{n} = \frac{a}{n} + \frac{b}{n} + \frac{c}{n} + \cdots + \frac{m}{n} \tag{2.37}$$

EXERCISE 2.5 REVIEW

Use one or more of axioms (2.1) to (2.3) to prove the statements in Probs. 1 to 3.

1. $(x - y) + w = w + (x - y)$

2. $a - b - a - (-b) = 0$

3. $3a - (-2a) - 4b - (-3b) = 5a - b$

4. If $a = 2$, and $b = -1$, then $a - |a| + b - |b| =$ _____

5. $3a + 2a - 7b - a + 6b =$

6. $4a + 3b - [7c - (2a - 7b + 6c) - a] =$

7. Subtract $2a - 3b - 4c$ from $5a + b - 2c$.

8. Prove that if $a(b - c) = aA$ and $a \neq 0$, then $b - A = c$.

9. Prove that if $(a + 2b)(c - 2) = (a + 2b)(d + 3)$ and $a \neq -2b$, then $c = d + 5$.

10. If $-x$ is used 4 times as a factor, the product is _____

11. If $-x$ is used 7 times as a factor, the product is _____

12. Is $2a^2(2a^2) = 4a^4$? _____

13. Is $3a^3(3a^3) = 9a^9$? _____

14. $2x^2y^3(3xy^0)(5x^3y^4) =$

15. $3x^2y(2xy^2 - 5x^2y^4) =$

16. Find the product of $a^2 + 2a + 3$ and $2a - 5$.

17. Remove the signs of grouping and collect similar terms in $2\{x^2 - 2x[x + 3(2x - 1) - 2x] + x^2\}$.

18. $x^5y^3 \div x^2y =$

19. $\dfrac{25a^4b^3c^5 - 15a^7b^4c^2}{5a^3b^2c} =$

20 Find the quotient and remainder if $2x^3 - 5x^2 - 3x + 4$ is divided by $x - 2$.

21 Find the quotient and remainder if $3x^3 + 5x^2y + 3xy^2 + 3y^3$ is divided by $x + 2y$.

22 Find the quotient and remainder if $6x^4 + 5x^3 - 10x^2 + 7x - 2$ is divided by $2x^2 + 3x - 2$.

23 Find N if $275 = N_2$.

24 Find N if $314_5 = N_2$.

25 Find the product of the binary numbers *101* and *110*.

26 Find the product of the ternary numbers *201* and *12*.

27 Find the sum of the binary numbers *111*, *101*, and *110*.

28 Find the sum of the ternary numbers *111*, *101*, and *110*.

NAME _____ DATE _____ SCORE _____

EXERCISE 2.6 CHAPTER TEST

1 Prove that $(a + b) + c = a + (b + c)$ by use of **(2.1)** to **(2.3)**.

2 If $a = 3$ and $b = -2$, then $|a + b| - a + (-b) = $ _____

3 $3a^2b^3(2a^3b)(5a^0b^4) =$

4 $2x^2y^4(3xy^0 - 2x^2y^3) =$

5 Find the product of $2a^2 - 3a + 1$ and $3a + 2$.

6 Remove signs and then collect like terms $x\{x + 2[x + x(x + 2) + x] + 2\}$.

7 Find the quotient and remainder if $3x^3 + 5x^2 - 8x - 6$ is divided by $x + 3$.

8 $\dfrac{21x^4y^3w^5 - 14x^5y^5w^3}{7x^3y^3w^2} =$

9 Find N if $256_7 = N_2$

10 Find the product of the binary numbers *111* and *101*.

11 If *2120*, *1202*, and *1022* are ternary numbers, find their sum.

CHAPTER 3

Special Products and Factoring

In this chapter we develop and illustrate methods of procedure that will enable us to find many products without actually having to multiply the factors together. We shall also see how to find the factors of a variety of expressions.

3.1
THE PRODUCT OF TWO SIMILAR BINOMIALS

The binomials $ax + by$ and $cx + dy$ are similar since each contains the sum of a constant times x and a constant times y. We shall find the product $(ax + by)(cx + dy)$, which we can use as a formula for finding the product of other similar binomials. If we multiply ax by $cx + dy$, we get $acx^2 + adxy$; furthermore, the product of by and $cx + dy$ is $bcxy + bdy^2$. Consequently, we see that

$$(ax + by)(cx + dy) = acx^2 + (ad + bc)(xy) + bdy^2 \quad (3.1)$$

Instead of remembering (3.1) as a formula, we can obtain the product by noting that

1. The product of the two first terms is the first term acx^2 of the product.
2. The sum of the two products obtained by multiplying the first term of each factor by the second term of the other is the second term $(ad + bc)(xy)$ of the product, called the **cross products.**
3. The product of the two second terms is the third term bdy^2 of the product.

EXAMPLE 1 Find the product of $3x + 5y$ and $4x - 7y$.

Solution We shall follow the procedure outlined below (3.1).

107

1. The product of the first terms is $3x(4x) = 12x^2$.
2. The sum of the cross products is $3x(-7y) + 4x(5y) = -xy$.
3. The product of the second terms is $5y(-7y) = -35y^2$; hence

$$(3x + 5y)(4x - 7y) = 12x^2 - xy - 35y^2$$

EXAMPLE 2 Find the product of $2x + 3y$ and $5x - 6y$.

Solution

$$\begin{aligned}(2x + 3y)(5x - 6y) &= 2x(5x) \\ &\quad + 2x(-6y) + 5x(3y) \\ &\quad + 3y(-6y) \\ &= 10x^2 + 3xy - 18y^2\end{aligned}$$

You should prove to yourself that you understand the procedure by verifying each of the following examples.

EXAMPLE 3

$(5x - 2y)(4x - 3y) = 20x^2 - 23xy + 6y^2$

EXAMPLE 4

$(7a + 3b)(3a + 5b) = 21a^2 + 44ab + 15b^2$

EXAMPLE 5

$(4b - 3c)(3b + 4c) = 12b^2 + 7bc - 12c^2$

Problems 1 to 28 of Exercise 3.1 may be worked now.

3.2 THE SQUARE OF A BINOMIAL

The **square of a binomial** is a special case of the product of two binomials in which the two binomials are equal. By actual multiplication or use of the procedure of Sec. 3.1 we find that

$$(a + b)^2 = (a + b)(a + b) = a^2 + 2ab + b^2 \quad (3.2)$$

and that

$$(a - b)^2 = (a - b)(a - b) = a^2 - 2ab + b^2 \quad (3.3)$$

Thus, we know that

The **square of a binomial** is the sum of the squares of the two terms plus or minus twice the product of the terms according as we are finding the square of a sum or a difference.

EXAMPLE 1

$$\begin{aligned}(2x + 3)^2 &= (2x)^2 + 2(2x)(3) + 3^2 \\ &= 4x^2 + 12x + 9\end{aligned}$$

EXAMPLE 2

$$\begin{aligned}(3x - 4y)^2 &= (3x)^2 + 2(3x)(-4y) + (-4y)^2 \\ &= 9x^2 - 24xy + 16y^2\end{aligned}$$

EXAMPLE 3 $(5x + 2y)^2 = 25x^2 + 20xy + 4y^2$

EXERCISE 3.1

Find the product.

1. $(x + 2)(x + 3) =$

2. $(x + 5)(x + 1) =$

3. $(x + 3)(x - 2) =$

4. $(x - 4)(x + 1) =$

5. $(x - 7)(x + 3) =$

6. $(x + 5)(x + 2) =$

7. $(x - 4)(x - 3) =$

8. $(x - 5)(x - 6) =$

9. $(2x + y)(3x + y) =$

10. $(5x + y)(4x + y) =$

11. $(x + 3y)(x + 4y) =$

12. $(x + 2y)(x + 7y) =$

13. $(x - 3y)(x + 2y) =$

14. $(x - 2y)(x + 4y) =$

15. $(2x - y)(5x - y) =$

16. $(3x - y)(4x - y) =$

17. $(2x + 3y)(3x + 4y) =$

18. $(2x + 5y)(2x + 3y) =$

19. $(3x + 4y)(4x + 3y) =$

20. $(7x + 2y)(5x + 4y) =$

21. $(4x - 3y)(3x - 4y) =$

22. $(5x - 2y)(4x - 3y) =$

23. $(7x - 5y)(3x - 8y) =$

24. $(9x - 4y)(4x - 7y) =$

25 $(2x + 3y)(4x - 5y) =$

26 $(5x + 2y)(3x - 4y) =$

27 $(6x - 5y)(7x + 4y) =$

28 $(3x - 7y)(5x + 2y) =$

Find each of the following squares.

29 $(a + 3)^2 =$

30 $(a + 4)^2 =$

31 $(2a + 1)^2 =$

32 $(5a + 1)^2 =$

33 $(b - 2)^2 =$

34 $(b - 5)^2 =$

35 $(2b - 1)^2 =$

36 $(3b - 1)^2 =$

37 $(2d + 3)^2 =$

38 $(3d + 4)^2 =$

39 $(4d - 5)^2 =$

40 $(6d - 7)^2 =$

41 $(2x + 3y)^2 =$

42 $(5x + 4y)^2 =$

43 $(4x + 7y)^2 =$

44 $(6x + 5y)^2 =$

45 $(3x - 2y)^2 =$

46 $(4x - 5y)^2 =$

47 $(7x - 3y)^2 =$

48 $(6x - 11y)^2 =$

3.3 THE PRODUCT OF THE SUM AND DIFFERENCE OF THE SAME TWO NUMBERS

We can represent the product of the sum and difference of the same two numbers by $(a + b)(a - b)$; furthermore by use of (3.1) that product is

$$(a + b)(a - b) = a^2 + (-ab + ab) - b^2$$

Hence,

$$(a + b)(a - b) = a^2 - b^2 \qquad (3.4)$$

This can be put in words as

> The **product of the sum and difference** of the same two numbers is the square of the first minus the square of the second.

EXAMPLE 1 $(x + 2)(x - 2) = x^2 - 4$

EXAMPLE 2 $(3x + 4y)(3x - 4y) = (3x)^2 - (4y)^2$
$$= 9x^2 - 16y^2$$

Problems 1 to 20 of Exercise 3.2 may be worked now.

3.4 PRODUCT OF SELECTED POLYNOMIALS

At times we can think of the product of two polynomials as the square of a binomial or as the product of the sum and the difference of the same two numbers by properly grouping the terms of the polynomials. If that is done, we can then obtain the product by using (3.2), (3.3), or (3.4).

EXAMPLE 1

$$(2x - y + 3w)^2 = [(2x - y) + 3w]^2$$
$$= (2x - y)^2 + 2(2x - y)(3w) + (3w)^2$$
$$= 4x^2 - 4xy + y^2 + 12xw - 6yw + 9w^2$$

EXAMPLE 2

$$(x + 2y - 3z + 4w)^2 = [(x + 2y) - (3z - 4w)]^2$$
$$= (x + 2y)^2 - 2(x + 2y)(3z - 4w) + (3z - 4w)^2$$
$$= x^2 + 4xy + 4y^2 - 6xz + 8xw - 12yz + 16yw + 9z^2 - 24zw + 16w^2$$
$$= x^2 + 4y^2 + 9z^2 + 16w^2 + 4xy - 6xz + 8xw - 12yz + 16yw - 24zw$$

If we examine the last form of the expansion in Example 2, we find that

> The **square of a polynomial** is the sum of the squares of the separate terms plus the sum of the terms obtained by multiplying twice each term by those that follow it.

EXAMPLE 3

$$(p - 2q + 3w + x - 5y)^2$$
$$= p^2 + (2q)^2 + (3w)^2 + x^2 + (5y)^2$$
$$+ 2p(-2q + 3w + x - 5y)$$
$$+ 2(-2q)(3w + x - 5y)$$
$$+ 2(3w)(x - 5y)$$
$$+ 2x(-5y)$$
$$= p^2 + 4q^2 + 9w^2 + x^2$$
$$+ 25y^2 - 4pq + 6pw$$
$$+ 2px - 10py - 12qw$$
$$- 4qx + 20qy + 6wx$$
$$- 30wy - 10xy$$

We shall now look at two products that can be considered as the product of the sum and differences of the same two numbers.

EXAMPLE 4 Find the product of $2x + y - 3z$ and $2x - y + 3z$.

Solution By proper grouping we have

$$(2x + y - 3z)(2x - y + 3z)$$
$$= [2x + (y - 3z)][2x - (y - 3z)]$$
$$= (2x)^2 - (y - 3z)^2 = 4x^2 - y^2 + 6yz - 9z^2$$

EXAMPLE 5 Find the product of $2a - 3b + 5c - d$ and $2a - 3b - 5c + d$.

Solution If properly grouped, this becomes

$$[(2a - 3b) + (5c - d)][(2a - 3b) - (5c - d)]$$
$$= (2a - 3b)^2 - (5c - d)^2$$
$$= 4a^2 - 12ab + 9b^2 - 25c^2 + 10cd - d^2$$

STUDENT'S NOTES

EXERCISE 3.2

Find the product.

1. $(x + 3)(x - 3) =$ $x^2 - 9$

2. $(x + 4)(x - 4) =$ $x^2 - 16$

3. $(x + 7)(x - 7) =$ $x^2 - 49$

4. $(x + 5)(x - 5) =$ $x^2 - 25$

5. $(2x + 1)(2x - 1) =$ $4x^2 - 1$

6. $(3y + 1)(3y - 1) =$ $9y^2 - 1$

7. $(5b + 1)(5b - 1) =$ $25b^2 - 1$

8. $(7b - 1)(7b + 1) =$ $49b^2 - 1$

9. $(2x + 3y)(2x - 3y) =$ $4x^2 - 3y^2$

10. $(5x + 7y)(5x - 7y) =$ $25x^2 - 49y^2$

11. $(6x - 5y)(6x + 5y) =$ $36x^2 - 25y^2$

12. $(8x - 11y)(8x + 11y) =$ $64x^2 - 121y^2$

13. $7(13) = (10 - 3)(10 + 3) =$ $100 - 9 = 91$

14. $22(18) =$

15. $26(34) =$

16. $31(29) =$

17. $27(53) =$

18. $61(39) =$

19. $53(67) =$

20. $74(46) =$

Find the product in each of the following problems after grouping the terms of the factors if necessary.

21 $(x + 2y + 3z)^2 =$

22 $(2x - y - 5z)^2 =$

23 $(3x - 2y - 4z)^2 =$

24 $(2x + 3y - z)^2 =$

25 $(a + b + c - d)^2 =$

26 $(2a - b - 3c + 4d)^2 =$

27 $(2a + 3b + 5c - d)^2 =$

28 $(3a - 2b - 2c - 3d)^2 =$

29 $(4p + 3q - 2r)(4p + 3q + 2r) =$

30 $(2p - q + 3r)(2p - q - 3r) =$

31 $(3p - 5q - 2r)(3p + 5q + 2r) =$

32 $(5p - 2q + 3r)(5p + 2q - 3r) =$

33 $(a + b + c + d)(a + b - c - d) =$

34 $(p - q + r - s)(p + q + r + s) =$

35 $(2p - 3q - 4r + 5s)(2p - 3q + 4r - 5s) =$

36 $(3b + 2c + 5d - f)(3b + 2c - 5d + f) =$

37 $(a + b + c - d - f)(a + b - c + d + f) =$

38 $(2p + 3q - r + 3s - 2t)(2p + 3q - r - 3s + 2t) =$

39 $(3b + 2a - 2d - c + 3f)(3b + 2a - 2d + c - 3f) =$

40 $(4p - 3d + q - u + 2t)(4p - 3d + q + u - 2t) =$

41 Show that the sum of two squares times the sum of two squares is again a sum of two squares by verifying that

$(a^2 + b^2)(c^2 + d^2) = (ac + bd)^2 + (ad - bc)^2$

3.5 COMMON FACTORS

If one expression is expressed as the product of two or more others (not including 1), we say it is **factored**. If an expression cannot be factored, it is said to be **irreducible**. If each factor is irreducible, the expression is **completely factored**. If each term of an expression is divisible by the same monomial, this monomial is called a **common factor** of the expression. Thus, a is a common factor of $ax + ay - aw$ since by the distributive law $ax + ay - aw = a(x + y - w)$; furthermore, $3ab$ is a common factor of $6a^3b + 3a^2b^2 + 18ab^3$ since

$$6a^3b + 3a^2b^2 + 18ab^3 = 3ab(2a^2 + ab + 6b^2)$$

The concept of a common factor is also applied to cases in which the term that is a factor of each other term or each group of terms is not a monomial. For example, $a + b$ is a common factor of $(a + b)x - (a + b)y = (a + b)(x - y)$.

EXAMPLE 1

$$x(x + 3) + y(x + 3) + (x + 3) = (x + 3)(x + y + 1)$$

since $x + 3 = 1(x + 3)$.

EXAMPLE 2

$$(x - 2y)3a + (x - 2y)2b - (x - 2y)(a + b)$$
$$= (x - 2y)(3a + 2b - a - b)$$
$$= (x - 2y)(2a + b)$$

EXAMPLE 3 Factor $8xw - 12yw - 2x + 3y$ after grouping the terms.

Solution We begin by noticing that the first two terms have the common factor $4w$. We then write

$$(8xw - 12yw) - (2x - 3y)$$
$$= 4w(2x - 3y) - 1(2x - 3y)$$
$$= (4w - 1)(2x - 3y)$$

STUDENT'S NOTES

EXERCISE 3.3

Factor each of the following expressions.

1. $3x + 6 =$ $3(x+2)$
2. $5x - 15 =$ $5(x-3)$
3. $8x + 2 =$ $2(4x+1)$
4. $9x - 6 =$ $3(3x-2)$
5. $5ax - 10ay =$ $5a(x-2y)$
6. $3bx + 6by =$ $3b(x+y)$
7. $6rt - 9rs =$ $3r(t-3s)$
8. $4xt + 8xs =$ $2x(2t+4s)$
9. $3xy^2 + 6x^2y =$ $3xy(y+2x)$
10. $5a^2b + 10ab^2 =$ $5ab(a+2b)$
11. $9a^2b^3 - 12a^3b^2 =$ $3a^2b^2(3b-4a)$
12. $6x^3y^3 + 9x^4y^2 =$ $3x^3y^2(2y+3x)$
13. $3a + 6b - 9c =$ $3(a+2b-3c)$
14. $5a^{13} - 10a^{12} + 15a^{11} =$ $5a^{11}(a^2-2a+3)$
15. $2x^2y^5z - 8xy^6 + 12y^5z^2x =$ $2xy^5(xz-4y+6z^2)$
16. $7xy^2z + 14x^2z - 21y^2z^2 =$ $7z(xy^2+2x^2-3y^2z)$
17. $2p^{24}qr + 3p^{23}q^2r + 5p^{23}q =$ $p^{23}q(2pr+3qr+5)$
18. $4x^2y^2z - 6xy^3 + 8xy^2z =$ $2xy^2(2xz-3y+4z)$
19. $9a^2b^8c^2 - 6a^2b^7 - 3a^2b^7c^2 =$ $3a^2b^7(3bc^2-2-c^2)$
20. $5a^3bc^2 - 10a^2c + 15abc^2 =$ $5ac(a^2bc-2a+3bc)$
21. $6x^3yz^6 + 15x^2yz^6 + 9xyz^6 =$ $3xyz^6(2x^2+5x+3)$
22. $9x^2y^2z^8 - 15xy^2z^8 - 6y^2z^8 =$
23. $12x^2yz^2 - 2x^2yz - 4x^2y =$ $2x^2y(6z^2-z-2)$
24. $18xy^2z + 21xyz - 9xz =$
25. $3a^{12} + 2a^{11}b - 5a^{11}c + 4a^{11}d =$ $a^{11}(3a+2b-5c+4d)$
26. $3x^5 - 6x^4w + 9x^4y - 12x^4z =$
27. $3a^2 - 6ax + 9ay - 12az =$ $3a(a-2x+3y-4z)$
28. $5b^2y - 10by + 20bx - 15bw =$

Chapter 3: Special Products and Factoring

29 $3x^2y^6z - 7xy^8z - 2xy^6z^2 + 5xy^6z =$ $xy^6z(3x - 7y^2 - 2z + 5)$

30 $2a^2bc^8 - 6a^2b^2c^7 - 4ab^2c^8 + 8abc^7 =$ $2abc^7(ac - 3ab - 2bc + 4)$

31 $6a^{13}b^2c^7 - 9a^{12}b^3c^7 - 15a^{12}b^2c^8 + 12a^{12}b^2c^7 =$ $3a^{12}b^2c^7(2a - 3b - 5c + 4)$

32 $7x^2y^3z^2 + 21x^3y^2z^2 - 14x^2y^2z^3 + 28x^2y^2z^2 =$ $7x^2y^2z^2(y + 3x - 2z + 4)$

33 $(x + 3)y + (x + 3)z =$ $(x+3)(y+z)$

34 $x(y + 1) - 2(y + 1) =$ $(x-2)(y+1)$

35 $x(y + w) - 3s(y + w) =$ $(y+w)(x-3s)$

36 $p(d + q) - 2a(d + q) =$ $(d+q)(p-2a)$

37 $(a + b)c + (a + b)(a - b) =$ $(a+b)c + (a-b) = (a+b)(c+a-b)$

38 $(m - n)p + (m - n)(m - 2n) =$ $(m-n)(p+m-2n)$

39 $(x - y)4 - (x - y)(y + 2) =$ $(x-y)(4 - (y+2)) = (x-y)(2-y)$

40 $(x + 2y)5 + (x + 2y)(x - 3) =$ $(x+2y)(5+x-3) = (x+2y)(2+x)$

Factor each of the following expressions after grouping.

41 $ab + ac + bd + cd =$ $a(b+c) + d(b+c)$
    $(b+c)(a+d)$

42 $xy - y^2 + zx - zy =$ $x(y+z) + y(y+z)$
    [likely: $x(y-y) ...$] $(y+z)(x+y)$

43 $ac - ad - bc + bd =$ $a(c-d) - b(c-d)$
    $(c-d)(a-b)$

44 $x^2 - xz - yx + yz =$ $x(x-z) - y(x-z) = (x-z)(x-y)$

45 $2a^2 - 2ab - 3ac + 3bc =$ $2a(a-b) - 3c(a-b) =$
    $(a-b)(2a-3c)$

46 $6x^2 - 4xy - 9xw + 6yw =$ $2x(3x-2y) - 3w(3x-2y) = (3x-2y)(2x-3w)$

47 $5wx - 15xz + 3wy - 9yz =$ $5x(w-3z) + 3y(w-3z) = (w-3z)(5x+3y)$

48 $6xs - 10xt + 9ys - 15yt =$ $2x(3s-5t) + 3y(3s-5t) =$
    $(2x+3y)(3s-5t)$

3.6 THE DIFFERENCE OF TWO SQUARES

We obtain the rule for factoring the difference of the squares of two numbers by restating the rule of Sec. 3.3 in the following form:

The **difference of the squares** of two numbers is equal to the product of the sum and the difference of the two numbers.

$$a^2 - b^2 = (a + b)(a - b) \qquad (3.4)$$

EXAMPLE 1

$16x^2 - 25y^2 = (4x)^2 - (5y)^2 = (4x + 5y)(4x - 5y)$

EXAMPLE 2

$9x^6 - 4y^4 = (3x^3)^2 - (2y^2)^2$
$= (3x^3 + 2y^2)(3x^3 - 2y^2)$

We also use this method for factoring expressions like $(x + 2y)^2 - z^2$. Since in this expression the two numbers that are squared are $(x + 2y)$ and z, we have

$(x + 2y)^2 - z^2 = [(x + 2y) + z][(x + 2y) - z]$
$= (x + 2y + z)(x + 2y - z)$

EXAMPLE 3

$(x + 2y)^2 - (2x - y)^2$
$= [(x + 2y) + (2x - y)][(x + 2y) - (2x - y)]$
$= (x + 2y + 2x - y)(x + 2y - 2x + y)$
$= (3x + y)(-x + 3y)$

Frequently, one or both of the factors obtained by this method can be factored further either by this method or by one of the other methods of factoring.

EXAMPLE 4

$x^4 - y^4 = (x^2)^2 - (y^2)^2$
$= (x^2 - y^2)(x^2 + y^2)$
$= (x - y)(x + y)(x^2 + y^2)$ \quad factoring $x^2 - y^2$

EXAMPLE 5

$x^4 + 2x^2y^2 + 9y^4$
$= (x^4 + 6x^2y^2 + 9y^4) - 4x^2y^2$
$= (x^2 + 3y^2)^2 - (2xy)^2$
$= (x^2 + 3y^2 + 2xy)(x^2 + 3y^2 - 2xy)$

Problems 1 to 32 of Exercise 3.4 may be worked now.

3.7 THE SUM OR DIFFERENCE OF TWO CUBES

If we divide $x^3 + y^3$ by $x + y$, we have

$$\begin{array}{r}
x^2 - xy + y^2 \\
x + y \overline{\smash{\big)} x^3 + y^3} \\
\underline{x^3 + x^2y } \\
-x^2y + y^3 \\
\underline{-x^2y - xy^2 } \\
xy^2 + y^3 \\
\underline{xy^2 + y^3}
\end{array}$$

Note: Although all signs of the dividend and divisor are plus, the sign of the middle term in the quotient is minus. Hence,

$x^3 + y^3 = (x + y)(x^2 - xy + y^2)$ \quad **sum of two cubes** (3.5)

Similarly,

$x^3 - y^3 = (x - y)(x^2 + xy + y^2)$ \quad **difference of two cubes** (3.6)

Now, since x and y can represent any two numbers whatsoever, we have the following conclusions:

One factor of the **sum of the cubes** of two numbers is the sum of the numbers. The other factor is the square of the first number minus the product of the first and the second plus the square of the second.

One factor of the **difference of the cubes** of two numbers is the difference of the numbers. The other factor is the square of the first number plus the product of the first and the second plus the square of the second.

These two statements should be memorized and thoroughly understood; otherwise, you must resort to long division to factor the sum or the difference of two cubes.

EXAMPLE 1

$8a^3 + 27b^3 = (2a)^3 + (3b)^3$
$= (2a + 3b)[(2a)^2 - (2a)(3b) + (3b)^2]$
$= (2a + 3b)(4a^2 - 6ab + 9b^2)$

EXAMPLE 2

$64x^3 - 125y^3 = (4x)^3 - (5y)^3$
$= (4x - 5y)[(4x)^2 + (4x)(5y) + (5y)^2]$
$= (4x - 5y)(16x^2 + 20xy + 25y^2)$

If we apply (3.5) to $35 = 2^3 + 3^3$, we have

$$35 = 2^3 + 3^3 = (2 + 3)(2^2 - 2(3) + 3^2)$$
$$= 5(4 - 6 + 9) = 5(7)$$

Note the similarity between (3.5), (3.6), and

$$x^5 + y^5 = (x + y)(x^4 - x^3y + x^2y^2 - xy^3 + y^4)$$
$$x^7 - y^7 = (x - y)(x^6 + x^5y + x^4y^2 + x^3y^3 + x^2y^4 + xy^5 + y^6)$$

EXERCISE 3.4

Factor each of the following expressions.

1. $x^2 - y^2 =$

2. $a^2 - b^2 = (a-b)(a+b)$ $\quad a^2+ab-ab+b^2$

3. $w^2 - 9 =$

4. $p^2 - 16 = (p-4)(p+4)$ $\quad p^2+4p-4p-16$

5. $x^2 - 16y^2 =$

6. $b^2 - 4a^2 = (b-2a)(b+2a)$ $\quad b^2+2ab-2ab-4a^2$

7. $9a^2 - b^2 =$

8. $x^2 - 25y^2 = (x-5y)(x+5y)$ $\quad x^2-5xy+5xy-25y^2$

9. $a^2 - 36b^2 =$

10. $q^2 - 49r^2 =$

11. $r^2 - 9s^2 =$

12. $s^2 - 64t^2 =$

13. $4x^2 - 25y^2 =$

14. $9y^2 - 16t^2 =$

15. $16t^2 - 49x^2 =$

16. $25a^2 - 9b^2 =$

17. $(x + y)^2 - w^2 =$

18. $a^2 - (b + c)^2 =$

19. $t^2 - (3x - a)^2 =$

20. $(y + w)^2 - 9x^2 =$

21. $(x + 2y)^2 - (2x - y)^2 =$

22. $(3a - b)^2 - (2a + b)^2 =$

23. $(5p - 3q)^2 - (2p + 7q)^2 =$

24. $(2q - 3r)^2 - (3q - 2r)^2 =$

25. $x^4 - y^2 =$

26. $(2x^2)^2 - 9y^2 =$

27. $36x^4 - 25y^6 =$

28. $49x^6 - 16x^4 =$

29. $x^4 - y^4 =$

30. $x^4 - 16y^4 =$

31. $81a^4 - 256b^4 =$

32. $625x^4 - 81y^4 =$

33 $a^3 + c^3 =$

34 $b^3 + d^3 =$

35 $x^3 - w^3 =$

36 $s^3 - t^3 =$

37 $x^3 - 8y^3 =$

38 $x^3 - 27y^3 =$

39 $125a^3 + b^3 =$

40 $64a^3 + b^3 =$

41 $a^3 + (b - c)^3 =$

42 $(w + y)^3 + x^3 =$

43 $a^3 - (b - 2c)^3 =$

44 $x^3 - (2y - 3w)^3 =$

45 $a^9 - b^9 =$

46 $(a^2)^3 - (b^4)^3 =$

47 $27x^{12} + 8y^3 =$

48 $64x^9 + 27y^6 =$

Factor each of the following after grouping terms.

49 $x - y - x^2 + y^2 =$

50 $x + y + x^3 + y^3 =$

51 $a^3 + b^3 + a^2 - b^2 =$

52 $a^3 - b^3 + a^2 - b^2 =$

53 $x + 2 + x^2 - 4 =$

54 $2x - 3y + 4x^2 - 9y^2 =$

55 $3w - 4y - 9w^2 + 16y^2 =$

56 $2a - 5b + 4a^2 - 25b^2 =$

57 $x^3 + 27y^3 - 2x - 6y =$

58 $2a^3 + 16b^3 + 5a + 10b =$

59 $2a^2 - 10ab - a^2b + 25b^3 =$

60 $6a^2 + 9ab + 4a^2b - 9b^3 =$

3.8 QUADRATIC TRINOMIALS

The general meaning of **quadratic** is "pertaining to or resembling a square." A trinomial in the form $px^2 + qxy + ry^2$ or $px^2 + qx + r$ is a quadratic trinomial.

FACTORS OF A QUADRATIC TRINOMIAL The process of factoring such a trinomial is the inverse of the method discussed in Sec. 3.1. By the method of that section, we have the relation

$$acx^2 + (ad + bc)(xy) + bdy^2 = (ax + by)(cx + dy) \quad (3.7)$$

which we shall use as a basis for our method. The expression $(ad + bc)(xy)$ in (3.7) is obtained by multiplying ax by dy and by by cx and then adding the products. These products are called the **cross products**. We shall use a trial-and-error process, as the following example will illustrate. To express $3x^2 + 5xy + 2y^2$ as the product of two binomials, we know that the first terms in the binomials must be factors of $3x^2$ and thus are $3x$ and x. The second terms must be factors of $2y^2$ and are therefore $2y$ and y. However, the second terms must be so placed in the binomials that the sum of the cross products is $5xy$. Obviously, our possibilities are $(3x + y)(x + 2y)$ and $(3x + 2y)(x + y)$. The sum of the cross products in the first of the above possibilities is $7xy$; that in the second is $5xy$, which is the result desired. Hence, $3x^2 + 5xy + 2y^2 = (3x + 2y)(x + y)$.

As illustrated by the above example, if the first and last terms of a quadratic trinomial can be factored in only one way, we first write the factors of the first term as the first terms of the binomial and the factors of the last term as the second terms of the binomial. Then if the sum of the cross products is not the one we seek, we interchange the last terms in the binomials.

If the first term or the last term of the trinomial can be factored in more than one way, several trials may have to be made before the correct combination of the binomial factors is found. We shall illustrate a systematic approach in Example 1.

EXAMPLE 1 To get the factors of $5x^2 - 13xy - 6y^2$, we note that the only factors of $5x^2$ are $5x$ and x. Therefore, as a first step, we write the factors in the form $(5x + by)(x + dy)$. Now referring to (3.7), we see that $a = 5$, $c = 1$ and the coefficient of the sum of the cross products is $5d + b$. Therefore, we must so choose b and d that $bd = -6$ and $5d + b = -13$. Therefore, we need only try the factors of -6 in $5d + b$ to see which combination gives -13. If $d = -3$ and $b = 2$, the conditions are satisfied. Therefore, $5x^2 - 13xy - 6y^2 = (5x + 2y)(x - 3y)$.

If both the first and last terms of the trinomial have two or more sets of factors, the number of possible combinations is increased considerably. In Example 2, we present in detail a systematic approach, but after some experience, you will usually be able to find the correct combination after a few trials.

EXAMPLE 2 Factor $4x^2 + 3xy - 27y^2$.

Solution Here the possibilities for the first terms of the factors are $2x$ and $2x$ and also $4x$ and x; hence, the possibilities for the factors with by and dy standing for the second terms are $(2x + by)(2x + dy)$ and $(4x + by)(x + dy)$. We start with the first of these and see if there exist values of b and d that give the correct factors. We have

$$4x^2 + 3xy - 27y^2 = (2x + by)(2x + dy)$$
$$= 4x^2 + xy(2d + 2b) + bdy^2$$

Hence,

$$bd = -27 \quad (1)$$

and

$$2d + 2b = 3 \quad (2)$$

We can rule this possibility out at once by considering (2). Since the left member of (2) is $2d + 2b = 2(d + b)$, for any integers assigned to b and d, $2(d + b)$ is equal to a number divisible by 2, and therefore can never be equal to 3.

We therefore try the second possibility for factors and have

$$4x^2 + 3xy - 27y^2 = (4x + by)(x + dy)$$
$$= 4x^2 + (4d + b)(xy) + bdy^2$$

Therefore,

$$bd = -27 \quad (3)$$

and

$$4d + b = 3 \quad (4)$$

From (3) we have the following possible pairs of corresponding values of b and d:

b	1	3	9	27	−1	−3	−9	−27
d	−27	−9	−3	−1	27	9	3	1

If we try them successively in (4), we find that $b = -9$ and $d = 3$ satisfy the condition. Therefore, $4x^2 + 3xy - 27y^2 = (4x - 9y)(x + 3y)$.

By methods described in Chap. 11, we can show that if p, q, and r are integers, the trinomial $px^2 + qx + r$ **can be factored** into $(ax + b)(cx + d)$ with integer coefficients if and only if

$$q^2 - 4pr \text{ is a perfect square} \tag{5}$$

Also $px^2 + qx + r$ **is a perfect square** $(ax + b)^2$ with integer coefficients if and only if

$$q^2 - 4pr = 0 \tag{6}$$

These rules only tell when there are factors; they do not tell how to find them.

In Example 2, $p = 4$, $q = 3y$, and $r = -27y^2$, and so $q^2 - 4pr = 9y^2 - 4(4)(-27y^2) = 9y^2 + 432y^2 = 441y^2 = (21y)^2$, which is a perfect square. Thus by (5), we know that $4x^2 + 3xy - 27y^2$ is factorable.

In Example 3, below, $q^2 - 4pr = (12y)^2 - 4(4)(9y^2) = 144y^2 - 144y^2 = 0$, and we thus find by (6) that $4x^2 + 12xy + 9y^2$ is a perfect square.

By using the method of Sec. 3.2, we have

$(a + b)^2 = a^2 + 2ab + b^2$

$(a - b)^2 = a^2 - 2ab + b^2$

where a and b stand for any two numbers. Hence we see that if a trinomial is a perfect square, the terms can be so arranged that the first and last terms are perfect squares and hence are positive, and the second term is twice the product of the square roots of the first and last terms. If the second term is positive, then the trinomial is the square of the sum of the square roots; if the second term is negative, then the trinomial is the square of the difference of the square roots.

In Examples 3 to 5, test the given trinomial to see if it is a perfect square. If it is, express it as the square of a binomial.

EXAMPLE 3

$4x^2 + 12xy + 9y^2$

Solution The first term is $(2x)^2$, the third is $(3y)^2$, and the second is $2(2x)(3y) = 12xy$. Hence, the given expression is a perfect square, and

$4x^2 + 12xy + 9y^2 = (2x + 3y)^2$

EXAMPLE 4

$9c^2 - 12cd + 16d^2$

$9c^2 = (3c)^2, \quad 16d^2 = (4d)^2,$
$\quad\quad\quad\quad\quad\quad\quad\text{and}\quad 2(3c)(4d) = 24cd$

Thus, since the middle term of the given trinomial is $-12cd$, the trinomial is not a perfect square.

EXAMPLE 5

$16a^2 - 40ab + 25b^2$

Solution

$16a^2 = (4a)^2, \quad 25b^2 = (5b)^2$
$\quad\quad\quad\quad\quad\quad\text{and}\quad 2(4a)(5b) = 40ab$

Hence,

$16a^2 - 40ab + 25b^2 = (4a - 5b)^2$

EXAMPLE 6

$a^2 + 2ab - b^2$

Solution This trinomial has only one term a^2 that is a positive perfect square; therefore, the trinomial is not a perfect square.

EXAMPLE 7

$9x^2 + 15xy + 25y^2$

Solution The first term is the square of $3x$, and the third term is the square of $5y$. However, $2(3x)(5y) = 30xy$, and this is not equal to the middle term. Therefore, the trinomial is not a perfect square.

EXERCISE 3.5

Put the correct sign between the terms of each binomial.

1. $2a^2 - 5ab + 3b^2 = (2a\ -\ 3b)(a\ +\ b)$

2. $3a^2 - 10ab + 3b^2 = (3a\ -\ b)(a\ -\ 3b)$
   $(3a-b)(a-3b)$

3. $2x^2 - 7xy + 6y^2 = (2x\quad 3y)(x\quad 2y)$

4. $5x^2 - 11xy + 2y^2 = (5x\quad y)(x\quad 2y)$

5. $3p^2 + 8pq + 4q^2 = (p\ +\ 2q)(3p\quad 2q)$

6. $4p^2 + 12pq + 5q^2 = (2p\quad 5q)(2p\quad q)$

7. $3r^2 + 11rs + 10s^2 = (3r\quad 5s)(r\quad 2s)$

8. $5r^2 + 8rs + 3s^2 = (5r\quad 3s)(r\quad s)$

9. $2w^2 + 5wt - 3t^2 = (2w\quad t)(w\quad 3t)$

10. $3w^2 + 2wt - 8t^2 = (3w\quad 4t)(w\quad 2t)$

11. $4a^2 + ab - 3b^2 = (4a\quad 3b)(a\quad b)$

12. $5a^2 + 7ab - 6b^2 = (5a\quad 3b)(a\quad 2b)$

Complete the factoring.

13. $9x^2 + 18xy + 8y^2 = (3x+\quad)(3x+\quad)$

14. $5x^2 - 17xy - 12y^2 = (5x+\quad)(x-\quad)$

15. $6x^2 + 7xy - 20y^2 = (2x+\quad)(3x-\quad)$

16. $15x^2 - 22xy + 8y^2 = (3x-\quad)(5x-\quad)$

Factor each trinomial.

17. $a^2 + 5a + 6 = (a+3)(a+2)$

18. $a^2 + 5a + 4 = (a+4)(a+1)$

19. $a^2 + 8a + 15 = (a+3)(a+5)$

20. $a^2 + 7a + 12 = (a+3)(a+4)$

21. $b^2 - 7b + 10 = (b\quad)(b\quad)$

22. $b^2 - 6b + 8 = (b-2)(b-4)$

23. $b^2 - 9b + 14 = (b-2)(b-7)$

24. $b^2 - 9b + 18 = (b-3)(b-6)$

25. $c^2 + c - 12 = (c+4)(c-3)$

26. $c^2 - 3c - 10 = (c-5)(c+2)$

27. $c^2 + 4c - 21 = (c-3)(c+7)$

28. $c^2 - c - 20 = (c-5)(c+4)$

Chapter 3: Special Products and Factoring

29 $8x^2 + 22xy + 15y^2 =$

$(4x + 5y)(2x + 3y)$

30 $6x^2 + 19xy + 15y^2 =$

31 $6x^2 + 23xy + 20y^2 =$

$(2x + 5y)(3x + 4y)$

32 $12x^2 + 23xy + 10y^2 =$

33 $12x^2 + xy - 6y^2 =$

$(3x - 2y)(4x + 3y)$

34 $12x^2 + xy - 35y^2 =$

35 $36x^2 + 7xy - 15y^2 =$

$(4y + 3y)(9x - 5y)$

36 $18x^2 - 11xy - 24y^2 =$

37 $6a^2 - 17ab + 12b^2 =$

$(2a - 3b)(3a - 4b)$

38 $6a^2 - 23ab + 20b^2 =$

39 $15a^2 - 23ab + 4b^2 =$

40 $18a^2 - 27ab + 10b^2 =$

Classify the trinomial as perfect squares or not perfect squares and factor the perfect squares.

41 $4x^2 + 20xy + 25y^2 =$

42 $16x^2 - 56xy + 49y^2 =$

43 $25x^2 - 20xy + 8y^2 =$

44 $64x^2 - 40xy + 9y^2 =$

45 $9p^2 + 6pq - q^2 =$

46 $9p^2 - 24pq + 16q^2 =$

47 $25p^2 + 60pq + 36q^2 =$

48 $16p^2 - 20pq + 25q^2 =$

3.9
CHAPTER SUMMARY

We began by finding the product of $ax + by$ and $cx + dy$ and then considering two special cases and the product of selected polynomials. The last half of the chapter was devoted to removing common factors, factoring the difference of two squares, the sum and difference of two cubes, and quadratic trinomials. The essential equations of the chapter are

$$(ax + by)(cx + dy) = acx^2 + (ad + bc)(xy) + bdy^2 \tag{3.1}$$
$$(a + b)^2 = a^2 + 2ab + b^2 \tag{3.2}$$
$$(a - b)^2 = a^2 - 2ab + b^2 \tag{3.3}$$
$$(a + b)(a - b) = a^2 - b^2 \tag{3.4}$$
$$x^3 + y^3 = (x + y)(x^2 - xy + y^2) \tag{3.5}$$
$$x^3 - y^3 = (x - y)(x^2 + xy + y^2) \tag{3.6}$$

STUDENT'S NOTES

1. common factor
2. grouping (4 or more terms)
3. general trinomial (3 terms)

$2x^2 + 7x + 6$

$(2x+3)(x+2)$

EXERCISE 3.6 REVIEW

Find the indicated product.

1. $(x + 4)(x + 7) =$

2. $(x - 3)(x + 5) =$

3. $(x - 2)(x - 6) =$

4. $(3x + y)(2x + y) =$

5. $(4x + y)(x + 3y) =$

6. $(5x - y)(x + 2y) =$

7. $(2x + 5y)(3x - 2y) =$

8. $(4x - 3y)(3x + 5y) =$

9. $(x + 3)^2 =$

10. $(2x + 1)^2 =$

11. $(3x + 5y)^2 =$

12. $(x + 7)(x - 7) =$

13. $(3x - 1)(3x + 1) =$

14. $(2x + 9y)(2x - 9y) =$

15. $(a - 2b + 3c)^2 =$

16. $(3a + 2b - c + 4d)^2 =$

17. $(2a - b + 2c)(2a + b - 2c) =$

18. $(p - 2d + 3q + 4r)(p - 2d - 3q - 4r) =$

19. $(p + 2q - r + 3s - t)(p + 2q - r - 3s + t) =$

Factor the expression.

20. $2a - 8b =$

21. $3xy + 6xy^2 =$

22. $5x - 10y + 20w =$

23. $3a^2bc - 12ab^2 + 18bc^2 =$

24 $15a^2bc + 9abc - 6bc =$

25 $6x^3w + 13x^2yw + 6xy^2w =$

26 $5abc - 8ab^2c - 2abc^2 + 7a^2bc =$

27 $6x^2y^2w^2 - 18x^3yw^2 - 12x^2y^2w^3 + 30x^3y^2w^2 =$

28 $(x + 2)a - (x + 2)b =$

29 $(x + y)(x - y) + (x + y)(x + 2y - w)$

30 $ab + ac - b^2 - bc =$

31 $6rx + 4ry - 10sy - 15sx =$

32 $a^2 - 25 =$

33 $9x^2 - 16y^2 =$

34 $(x - 2y)^2 - y^2 =$

35 $(3a - 4b)^2 - (2a - 5b)^2 =$

36 $4x^4 - 9y^4 =$

37 $16a^4 - 81b^4 =$

38 $a^3 + 27b^3 =$

39 $(x + 2y)^3 - w^3 =$

40 $8a^9 - 27b^{12} =$

41 $64b^6 - 125c^{15} =$

42 $x^2 - y^2 + x^3 + y^3 =$

43 $6a^3 - 15a^2b + 4a^2 - 25b^2 =$

44 $6x^2 + 11xy - 10y^2 =$

45 $6x^2 + 5xy - 21y^2 =$

Classify the trinomials as perfect squares or not perfect squares and factor the perfect squares.

46 $4x^2 + 12xy + 9y^2 =$

47 $9x^2 + 15xy + 25y^2 =$

48 $4x^2 + 30xy + 25y^2 =$

49 $9x^2 - 24xy + 16y^2 =$

50 Verify that $(x + y)^3 = x^3 + 3x^2y + 3xy^2 + y^3$.

NAME _____ DATE _____ SCORE _____

EXERCISE 3.7 CHAPTER TEST

Find the product indicated.

1 $(3x + 5y)(2x - 3y) =$

2 $(5x - 4)^2 =$

3 $(a - 3b + 2c + 5d)^2 =$

4 $(3p + 2q - 3r + s)(3p - 2q + 3r + s) =$

Factor the expression.

5 $6ax - 15ay =$

6 $3x^3 + 5x^2y + 2xy^2 =$

7 $9a^2 - 16b^4 =$

8 $8x^3 + 125y^3 =$

9 $(2x - 3)^2 - (x - 4)^2 =$

10 $81a^4 - 16b^4 =$

11 $4x^2 - 9y^2 + 8x^3 - 27y^3 =$

12 $6x^2 + xy - 35y^2 =$

13 $3x^3 + 4x^2y - 15xy^2 =$

Algebraic Fractions

4.1 DEFINITIONS AND FUNDAMENTAL PRINCIPLES

Skill in performing operations on fractions is essential in order to use mathematics as an art, a science, or a tool.

Any number of the form n/d, where n and d are the numbers of arithmetic or algebraic expressions, is called a **fraction**. The symbol n is called the **numerator** and d the **denominator** of the fraction. Together they are called the **members** of the fraction. Thus in $\tfrac{2}{3}$ and $(2x + 1)/(5x - 4)$, the number 2 and the expression $2x + 1$ are the numerators, whereas 3 and $5x - 4$ are the denominators.

We shall now prove a theorem that enables us to say whether two fractions are equal. In the theorem, we shall use a/b and c/d as the fractions where a, b, c, and d stand for arithmetic numbers or algebraic expressions and $b \neq 0$, $d \neq 0$. The theorem can then be stated as

$$\frac{a}{b} = \frac{c}{d} \quad \text{if and only if} \quad ad = bc \qquad (4.1)$$

PROOF We first assume that $ad = bc$ and then prove that $a/b = c/d$. If we multiply each member of $ad = bc$ by $1/bd$, we get

$$\frac{1}{bd} ad = \frac{1}{bd} bc$$

$$\frac{ad}{bd} = \frac{bc}{bd}$$

since

$$\frac{1}{bd} ad = \frac{ad}{bd} \quad \text{and} \quad \frac{1}{bd} bc = \frac{bc}{bd}$$

and then

$$\frac{a}{b}\frac{d}{d} = \frac{b}{b}\frac{c}{d}$$

since

$$\frac{ad}{bd} = \frac{a}{b}\frac{d}{d} \quad \text{and} \quad \frac{bc}{bd} = \frac{b}{b}\frac{c}{d}$$

so that

$$\frac{a}{b}1 = 1\frac{c}{d}$$

since

$$\frac{d}{d} = \frac{b}{b} = 1$$

and

$$\frac{a}{b} = \frac{c}{d}$$

since

$$1x = x$$

To prove that $ad = bc$ if $a/b = c/d$, we assume that $a/b = c/d$ and show that $ad = bc$. Thus multiplying each member of $a/b = c/d$ by bd gives

$$bd\frac{a}{b} = bd\frac{c}{d}$$

Now

$$\begin{aligned} bd\frac{a}{b} &= bd\left(\frac{1}{b}\right)a & \text{since } \frac{a}{b} = \frac{1}{b}a \\ &= b\frac{1}{b}ad & \text{by the commutative axiom} \\ &= 1(ad) & \text{since } b\frac{1}{b} = 1 \\ &= ad \end{aligned}$$

Similarly

$$bd\frac{c}{d} = bc$$

Consequently, $ad = bc$.

EXAMPLE 1 Prove that

$$\frac{2(x-y)}{3(x+y)} = \frac{6(x^2-y^2)}{9(x+y)^2}$$

Proof We shall use (4.1) with $a = 2(x-y)$, $b = 3(x+y)$, $c = 6(x^2-y^2)$, and $d = 9(x+y)^2$. Then

$$\begin{aligned} ad &= 2(x-y)(9)(x+y)^2 \\ &= 18(x-y)(x+y)(x+y) \\ &= 18(x^2-y^2)(x+y) \end{aligned}$$

since

$$(x-y)(x+y) = x^2 - y^2$$

and

$$\begin{aligned} bc &= 3(x+y)(6)(x^2-y^2) \\ &= 18(x+y)(x^2-y^2) \\ &= 18(x^2-y^2)(x+y) \end{aligned}$$

Hence, $ad = bc$, since each is equal to $18(x^2 - y^2) \times (x + y)$, and therefore the two given fractions are equal.

Frequently in computations involving fractions, it is necessary to convert a given fraction to an equal fraction in which the members differ from the members of the given fraction. This conversion is accomplished by use of the **fundamental principle of fractions**:

If the numerator and denominator of a given fraction are multiplied (or divided) by the same nonzero number and the products (or quotients) are used as the numerator and denominator, respectively, of a second fraction, then the two fractions are equal.

Symbolically, this statement is

$$\frac{na}{da} = \frac{n}{d} \qquad a \neq 0 \quad d \neq 0 \qquad (4.2)$$

and

$$\frac{n/b}{d/b} = \frac{n \div b}{d \div b} = \frac{n}{d} \qquad b \neq 0 \quad d \neq 0 \qquad (4.3)$$

PROOF By the commutative axiom, $nad = dan$. It then follows by (4.1) that $na/da = n/d$.

To prove (4.3), we replace a in (4.2) by $1/b$, $b \neq 0$, and get

$$\frac{n(1/b)}{d(1/b)} = \frac{n}{d}$$

We have

$$n\frac{1}{b} = \frac{n}{b} \quad \text{and} \quad d\frac{1}{b} = \frac{d}{b}$$

Hence,

$$\frac{n(1/b)}{d(1/b)} = \frac{n/b}{d/b}$$

This completes the proof.

SIGNS OF A FRACTION There are three signs associated with any fraction. They are the sign that precedes the numerator, the sign that precedes the denominator, and the sign that precedes the fraction. We remind the reader that if no sign appears in any one of the above positions, the sign is understood to be plus. We shall now prove that *if in a given fraction two of these signs are changed, the fraction obtained is equal to the given fraction*. In the proof of this statement, we shall use

$$\frac{a}{c}\frac{b}{d} = \frac{ab}{cd}$$

We also use the following equalities which follow from the law of signs for division:

$$-1a = -a$$

$$\frac{-1}{-1} = 1 \qquad \frac{-1}{1} = -1 \quad \text{and} \quad \frac{1}{-1} = -1$$

If the given fraction is n/d, we have

$$\frac{n}{d} = 1\frac{n}{d}$$

$$= \frac{-1}{-1}\frac{n}{d} \qquad \text{replacing 1 by } -1/-1$$

$$= \frac{-n}{-d}$$

Furthermore,

$$-\frac{n}{d} = -1\frac{n}{d}$$

$$= \frac{-1}{1}\frac{n}{d} \qquad \text{replacing } -1 \text{ by } -1/1$$

$$= \frac{-n}{d}$$

Similarly,

$$-\frac{n}{d} = -1\frac{n}{d} = \frac{1}{-1}\frac{n}{d} = \frac{n}{-d}$$

Therefore, we have

$$\frac{n}{d} = \frac{-n}{-d} = -\frac{-n}{d} = -\frac{n}{-d} \qquad (4.4)$$

Also

$$-\frac{n}{d} = \frac{-n}{d} = \frac{n}{-d} = -\frac{-n}{-d}$$

EXAMPLE 2

$$\frac{-a}{b-a} = \frac{-(-a)}{-(b-a)} = \frac{a}{a-b}$$

EXAMPLE 3

$$\frac{a^2+b^2}{b^2-a^2} = -\frac{a^2+b^2}{-(b^2-a^2)} = -\frac{a^2+b^2}{a^2-b^2}$$

Problems 1 to 12 in Exercise 4.1 may be done now.

4.2 REDUCTION TO LOWEST TERMS

If the members of a given fraction have a common factor other than the numeral 1, it is equal to a fraction that is simpler in form than the given fraction. For example, by (4.2)

$$\frac{a^2 - ab - 2b^2}{a^2 + 2ab + b^2} = \frac{(a+b)(a-2b)}{(a+b)(a+b)} = \frac{a-2b}{a+b}$$

If the members of a fraction have no common factor, the fraction is said to be in **lowest terms.** If a fraction is not in lowest terms, we convert it to an equal fraction in lowest terms by dividing the numerator and denominator by the product of the irreducible factors that are common to both. If the common factors of the members of a fraction are not evident after a brief inspection, it is advisable to factor the numerator and denominator as the first step in the reduction.

EXAMPLE 1

$$\frac{8a^2b}{12abc} = \frac{2a}{3c} \qquad \text{dividing each member by } 4ab$$

EXAMPLE 2

$$\frac{2a^3 - 4a^2b - 6ab^2}{4a^4 - 4a^2b^2} = \frac{2a(a+b)(a-3b)}{4a^2(a+b)(a-b)}$$

factoring each member

$$= \frac{a-3b}{2a(a-b)} \qquad \text{dividing each member by } 2a(a+b)$$

Problems 13 to 24 in Exercise 4.1 may be done now.

4.3 MULTIPLICATION OF FRACTIONS

By referring to (2.34) we see that

$$\frac{a}{c}\frac{b}{d} = \frac{ab}{cd}$$

PRODUCT OF FRACTIONS Hence, we have the following rule:

To get the product of two or more fractions, we first obtain the product of the numerators to get the numerator of the product; next, we obtain the product of the denominators to get the denominator of the product, and finally we convert the product to lowest terms.

We shall illustrate the process with several examples and make some suggestions for possible short cuts.

EXAMPLE 1

$$\frac{3x^2}{2y} \frac{4y^2(x-y)}{9(x+y)} \frac{x^2-y^2}{x^2y^2} = \frac{3x^2(4y^2)(x-y)(x^2-y^2)}{2y(9)(x+y)(x^2y^2)}$$
$$= \frac{12x^2y^2(x-y)(x^2-y^2)}{18x^2y^3(x+y)}$$
$$= \frac{2(x-y)(x-y)}{3y}$$

dividing numerator and denominator by $6x^2y^2(x+y)$

$$= \frac{2(x-y)^2}{3y}$$

Usually, in obtaining the product of two or more fractions, it is advisable to factor the numerators and denominators of the fractions involved before multiplying. If this is done, the factors that will be common to the members of the product can readily be detected. The procedure used in Example 2 is recommended.

EXAMPLE 2 Find the product of

$$\frac{a-2b}{a^2-b^2} \quad \frac{2a^2+3ab+b^2}{a^2-5ab+6b^2}$$

and

$$\frac{a^2+2ab-3b^2}{2a+b}$$

Solution We first factor the numerators and denominators of the fractions and get

$$\frac{a-2b}{a^2-b^2} \frac{2a^2+3ab+b^2}{a^2-5ab+6b^2} \frac{a^2+2ab-3b^2}{2a+b}$$
$$= \frac{a-2b}{(a-b)(a+b)} \frac{(2a+b)(a+b)}{(a-2b)(a-3b)}$$
$$\times \frac{(a+3b)(a-b)}{2a+b}$$

By the commutative axiom, we can arrange the factors in the numerator and denominator of the product in any order that is convenient. Hence, in writing the product, we select the factors that appear in both a numerator and a denominator of the factored form of the given fractions and write these factors first. At the same time, we arrange them so that each factor in the denominator will be directly below its equal factor in the numerator. Thus, we obtain

$$\frac{(a-2b)(2a+b)(a+b)(a-b)(a+3b)}{(a-2b)(2a+b)(a+b)(a-b)(a-3b)} \quad (1)$$

Now it is evident that each of $a-2b$, $2a+b$, $a+b$, and $a-b$ is a factor of the numerator and denominator of (1). Therefore, if we divide each member of (1) by the product of these factors, we get $(a+3b)/(a-3b)$, and this is the required product in lowest terms.

A serious error may occur if you attempt to obtain the product of two or more fractions before the members of the fractions are factored. For example, if you are careless in reading the problem

$$\frac{(x+1)x-2}{x+1} \frac{x-1}{x-2}$$

you may conclude that $x-2$ and $x+1$ will be common factors of the members of the product, and thus decide that the answer is $x-1$. This is quite incorrect, because the numerator $(x+1)x-2$ is not in factored form and is actually equal to $x^2+x-2 = (x+2)(x-1)$ and is divisible by neither $x+1$ nor $x-2$. If we express the first numerator in this form and complete the multiplication, we get

$$\frac{(x+1)x-2}{x+1} \frac{x-1}{x-2} = \frac{(x+2)(x-1)}{x+1} \frac{(x-1)}{x-2}$$
$$= \frac{(x-1)^2(x+2)}{(x+1)(x-2)}$$

which is the correct answer.

Recall from Sec. 1.10 that

$$\frac{a/b}{c/d} = \frac{a}{b}\frac{d}{c}$$

Hence we see that to divide two fractions we invert the denominator (write its reciprocal) and multiply the result by the numerator.

EXAMPLE 3

$$\frac{\frac{4}{5}}{\frac{8}{15}} = \frac{4}{5}\frac{15}{8} = \frac{4(5)(3)}{5(4)(2)} = \frac{3}{2}$$

EXAMPLE 4

$$\frac{4x^2}{9y^2} \div \frac{8x^3}{21y} = \frac{4x^2(21y)}{9y^2(8x^3)} = \frac{7}{6xy}$$

EXERCISE 4.1

Fill in the blank so as to justify the use of the equality sign.

1. $\dfrac{ab}{c} = \dfrac{}{3cab}$

2. $\dfrac{4x^2y^3}{6xy^4} = \dfrac{2xy^2}{}$

3. $\dfrac{-12p^3r^2s}{3pr^2s^2} = \dfrac{-4p^2}{}$

4. $\dfrac{9u^2v^2w}{-12u^3vw^2} = \dfrac{}{4uw}$

5. $\dfrac{a}{a-b} = \dfrac{}{b-a}$

6. $\dfrac{x-y}{x+y} = -\dfrac{}{x+y}$

7. $\dfrac{x-y}{p-r} = \dfrac{y-x}{}$

8. $\dfrac{p+q}{p-q} = \dfrac{}{q-p}$

9. $\dfrac{x-2y}{x-y} = \dfrac{}{(x-y)^2}$

10. $\dfrac{x+3y}{x-y} = \dfrac{}{x^2-y^2}$

11. $\dfrac{x^2-y^2}{x^2-2xy+y^2} = \dfrac{x+y}{}$

12. $\dfrac{2x^2+5xy-3y^2}{2x^2-3xy+y^2} = \dfrac{x+3y}{}$

Reduce the fractions to lowest terms.

13. $\dfrac{x^2y}{xy^2} =$

14. $\dfrac{a^2bc^3}{ab^2c^2} =$

15. $\dfrac{xy^4}{x^2y^2w^2} =$

16. $\dfrac{x^2y^5}{xyw} =$

17. $\dfrac{3x-3y}{x^2-y^2} =$

18. $\dfrac{x^3-x^2}{x-1} =$

19. $\dfrac{3xy-6xy^2}{1-4y^2} =$

20. $\dfrac{1+4x+4x^2}{3y+6yx} =$

21. $\dfrac{x^2 - x - 2}{2x^2 - 3x - 2} =$

22. $\dfrac{2x^2 - xy - 6y^2}{4x^2 + 8xy + 3y^2} =$

23. $\dfrac{(x - 2y)(x^2 + 4xy + 3y^2)}{(x + 3y)(2x^2 + xy - y^2)} =$

24. $\dfrac{(2x + y)(3x^2 + 7xy - 6y^2)}{(x + 3y)(9x^2 - 9xy + 2y^2)} =$

Find the indicated products.

25. $\dfrac{2}{3} \cdot \dfrac{3}{7} = \dfrac{2}{7}$

26. $\dfrac{4}{5} \cdot \dfrac{10}{17} = \dfrac{8}{17}$

27. $\dfrac{11}{5} \cdot \dfrac{10}{33} = \dfrac{2}{3}$

28. $\dfrac{3}{8} \cdot \dfrac{16}{9} =$

29. $\dfrac{a}{b^3} \cdot \dfrac{b^2}{a^4} = \dfrac{1}{a^3 b}$

30. $\dfrac{x^4}{y^3} \cdot \dfrac{y^5}{x^7} =$

31. $\dfrac{3x^6}{y^5} \cdot \dfrac{y^2}{6x^3} = \dfrac{x^3}{2y}$

32. $\dfrac{x^7}{7y^6} \cdot \dfrac{21y^2}{x^5} =$

33. $\dfrac{p^3 q^4}{qr^2} \cdot \dfrac{rs^3}{s^2 p^4} = \dfrac{qs}{rp}$

34. $\dfrac{a^7 b^3}{bc^2} \cdot \dfrac{c^3 d}{d^2 a^4} =$

35. $\dfrac{v^5 w^2}{x^3 y^4} \cdot \dfrac{x^2 y}{v^4 w^3} =$

36. $\dfrac{r^9 s^5}{t^2 u^4} \cdot \dfrac{tu^5}{r^8 s^6} =$

37. $\dfrac{5a^2bc^3}{7b^2c^4d^3} \cdot \dfrac{21c^4d^5a^2}{10d^3a^5b^3} =$

38. $\dfrac{3x^4y^2z^3}{11y^3z^2w^5} \cdot \dfrac{33z^4w^4x}{27w^2x^5y^2} =$

39. $\dfrac{13p^4d^2q^3}{6d^3q^2r^5} \cdot \dfrac{21qr^2p}{39rp^3d^2} =$

40. $\dfrac{24s^5t^4u^2}{35t^2u^5v^4} \cdot \dfrac{14u^2v^5s}{8v^2s^4t} =$

41. $\dfrac{2(a+b)}{x(a-b)} \cdot \dfrac{x(a-b)}{4y(a+b)} = \dfrac{1}{2y}$

42. $\dfrac{3(x+3y)}{y(3x+y)} \cdot \dfrac{x(3x+y)}{6(x+3y)} =$

43. $\dfrac{15(c-2d)}{x(d-c)} \cdot \dfrac{2x^2(d-c)}{3y(2d-c)} = \dfrac{-10y}{y}$

44. $\dfrac{17x(x-3y)}{2y(3x-y)} \cdot \dfrac{6(y-3x)}{51x(x-3y)} =$

45. $\dfrac{14x^2(y-x)}{13y(y+x)} \cdot \dfrac{39y^2(x+y)^2}{28x^4(y-x)} =$

46. $\dfrac{7a(a-b)^2}{3b(a+b)} \cdot \dfrac{9b^2(a+b)}{21a^3(a-b)} = \dfrac{b(a-b)}{3a^2}$

47. $\dfrac{17x^3(x+2y)}{52y(x-2y)^3} \cdot \dfrac{39y^3(x-2y)}{34x^4(x+2y)} =$

48. $\dfrac{19p^4(p-3q)^3}{34q^2(p+2q)} \cdot \dfrac{51q^3(p+2q)}{57p^5(3q-p)} =$

140 | Chapter 4: Algebraic Fractions

49 $\dfrac{2x^2 + 6xy}{3xy - 6y^2} \cdot \dfrac{x^3 - 2x^2y}{xy + 3y^2} =$

50 $\dfrac{2p^3 - 3p^2q}{9pq + 6q^2} \cdot \dfrac{6pq^2 + 4q^3}{8p^2 - 12pq} =$

51 $\dfrac{a^4 + 2a^3b}{a^2 + 2ab} \cdot \dfrac{4a - 8b}{2ab - 4b^2} =$

52 $\dfrac{6a^3 + 12a^2b}{ab - 2b^2} \cdot \dfrac{3ab^2 - 6b^3}{a^2 + 2ab} =$

53 $\dfrac{3v^3}{t^2(t - u)} \cdot \dfrac{2(t^2 - u^2)}{v^2(3t + u)} \cdot \dfrac{t^3(3t + u)}{6(t + u)} =$

54 $\dfrac{10(a^2 - b^2)}{a^3(a + 2b)} \cdot \dfrac{6(a + 2b)}{5(a - b)} \cdot \dfrac{a^2b}{3(a + 2b)} =$

55 $\dfrac{9x^2 - y^2}{x - 3y} \cdot \dfrac{x - 3y}{y(3x + y)} \cdot \dfrac{xy^3}{3x - y} =$

56 $\dfrac{x^2 - 4y^2}{2x + y} \cdot \dfrac{x^2y}{x + 2y} \cdot \dfrac{2x + y}{x(x - 2y)} =$

57 $\dfrac{(x + y)(x - 3y)}{(x + 2y)(x + 3y)} \cdot \dfrac{x^2 + xy - 2y^2}{x^2 - 4xy + 3y^2} =$

58 $\dfrac{6x^2 - 5xy - 6y^2}{(3x + 2y)(x - 2y)} \cdot \dfrac{(2x - y)(x + 2y)}{4x^2 - 8xy + 3y^2} =$

59 $\dfrac{3p^2 - 7pq - 6q^2}{(3p + 2q)(4p + 3q)} \dfrac{(2p + 5q)(3p - 7q)}{2p^2 - pq - 15q^2} =$

60 $\dfrac{2x^2 - 7xy + 3y^2}{(2x - y)(x + 2y)} \dfrac{3x^2 + 7xy + 2y^2}{(x - 3y)(x + 3y)} =$

61 $\dfrac{3a^2 - 7ab + 2b^2}{2a + 3b} \dfrac{a + 2b}{a^3 - 8b^3} \dfrac{a^2 + 2ab + 4b^2}{3a - b} =$

62 $\dfrac{2a - b}{a^3 - b^3} \dfrac{a^2 + ab + b^2}{6a^2 + ab - b^2} \dfrac{3a^2 - 10ab + 3b^2}{2a - b} =$

63 $\dfrac{x + 5y}{3x(x + 3y)} \dfrac{x^4 + 5x^3y}{xy + 5y^2} \dfrac{x^2 + 7xy + 12y^2}{x^2 + 2xy - 15y^2} =$

64 $\dfrac{x^2 + 2xy}{2xy^2 - y^3} \dfrac{2x^2 - 3xy + y^2}{x^2 + 4xy + 3y^2} \dfrac{x^2 - 9y^2}{x^2 + xy - 2y^2} =$

Perform the indicated multiplications and divisions.

65 $\dfrac{2}{3} \div \dfrac{5}{9} =$

66 $\dfrac{3}{7} \div \dfrac{12}{14} =$

67 $\dfrac{5}{13} \div \dfrac{15}{26} =$

68 $\dfrac{2}{11} \div \dfrac{10}{33} =$

69 $\dfrac{4b}{3y^2} \div \dfrac{5b^2}{6y^3} =$

70 $\dfrac{7x^4}{5y^3} \div \dfrac{21x^3}{10y^2} =$

71. $\dfrac{x^2 - y^2}{x + 2y} \div \dfrac{x + y}{x^2 + 2xy} =$

72. $\dfrac{3x - 9y}{8x + 4y} \div \dfrac{12x + 6y}{2x^2 + xy} =$

73. $\dfrac{xy}{x^2 - y^2} \cdot \dfrac{x^2 - 3xy + 2y^2}{x^3 - y^3} \div \dfrac{x^2 - xy - 2y^2}{x^2 + xy + y^2} =$

74. $\dfrac{2a - b}{a - b} \cdot \dfrac{(a - b)^3}{2ab} \div \dfrac{4a^2 - b^2}{4a + 2b} =$

75. $\dfrac{x^2 - y^2}{2x^2 - 3xy + y^2} \cdot \dfrac{2x^2 + 5xy - 3y^2}{x^2 + 4xy + 3y^2} \div \dfrac{x^2 - 4xy + 3y^2}{x^2 - 2xy - 3y^2} =$

76. $\dfrac{a^2 + 7ab + 12b^2}{a^2 - 16b^2} \cdot \dfrac{a^2 - 2ab - 8b^2}{a^2 - 9b^2} \div \dfrac{a^2 - 4b^2}{a^2 - 6ab + 9b^2} =$

77. $\dfrac{3a^2 + ab}{3a^2 + 10ab + 3b^2} \cdot \dfrac{2ab + b^2}{2a^2 + 5ab + 2b^2} \div \dfrac{2a^2b - ab^2}{2a^2 + 3ab - 2b^2} =$

78. $\dfrac{a^2 - 9b^2}{2a^2 - 5ab - 3b^2} \cdot \dfrac{4a^2 - b^2}{2a^2 + 5ab - 3b^2} \div \dfrac{a^3 - b^3}{a^2 - ab} =$

79. $\dfrac{(x + 2)x - 3}{(x - 3)x + 2} \cdot \dfrac{(x - 1)x - 2}{(x + 4)x + 3} \div \dfrac{(x + 4)x + 4}{(x - 1)x - 6} =$

80. $\dfrac{(x - 2)x + x - 2}{(x - 4)x + 2(x - 4)} \cdot \dfrac{(x + 1)x - 2}{(x + 1)x - 6} \div \dfrac{(x - 2)x - 3}{(x - 3)x - 4} =$

4.4 LOWEST COMMON DENOMINATOR

Like arithmetic fractions, two or more algebraic fractions cannot be added unless they have the same denominator. Any expression that has the denominator of each of several fractions as a factor is called a **common denominator**. The common denominator of lowest degree is called the **lowest common denominator** (lcd). The product of all denominators is a common denominator, but it is not necessarily the lowest one; as a matter of fact, it is not the lowest one if two or more denominators have a factor in common. All denominators should be factored as a preliminary step before finding the lcd.

> The lcd of a set of fractions must have as a factor the highest power of each prime factor that appears in any denominator and must have no other factors.

EXAMPLE 1 If the denominators of several fractions are $(x + 2)(3x + 2)^2$, $(x + 2)^4(2x - 1)$, and $(3x + 2)(2x - 1)$, then the lcd is $(x + 2)^4(3x + 2)^2 \times (2x - 1)$ since that is the product of all distinct factors that occur in any denominator and each is raised to the highest power to which it enters in any denominator.

EXAMPLE 2 If the denominators of four fractions are $(x - 1)^3$, $3(x - 1)(x + 2)^2$, $2(x + 2)(x - 3)$, and $5(x - 1)(x - 3)$, then the lcd is

$$30(x - 1)^3(x + 2)^2(x - 3)$$

since it contains each distinct factor of the denominators, each factor is to the highest power to which it enters any denominator, and no other factors are contained.

EXAMPLE 3 Convert

$$\frac{3x - 2}{x^2 - 2x - 3} \qquad \frac{x + 2}{3(x - 3)} \quad \text{and} \quad \frac{x - 1}{6(x + 1)}$$

into a set of fractions such that each has the lcd as a denominator.

Solution We begin by finding that the first denominator $x^2 - 2x - 3 = (x + 1)(x - 3)$. Consequently, the lcd is $6(x + 1)(x - 3)$. If we use this as the denominator of each of the new fractions, we get

$$\frac{3x - 2}{(x + 1)(x - 3)} = \frac{6}{6} \frac{3x - 2}{(x + 1)(x - 3)}$$
$$= \frac{18x - 12}{6(x + 1)(x - 3)}$$
$$\frac{x + 2}{3(x - 3)} = \frac{2(x + 1)}{2(x + 1)} \frac{x + 2}{3(x - 3)}$$
$$= \frac{2x^2 + 6x + 4}{6(x + 1)(x - 3)}$$

and

$$\frac{x - 1}{6(x + 1)} = \frac{x - 3}{x - 3} \frac{x - 1}{6(x + 1)}$$
$$= \frac{x^2 - 4x + 3}{6(x + 1)(x - 3)}$$

as the desired set of fractions.

Problems 1 to 16 of Exercise 4.2 may be done now.

4.5 ADDITION OF FRACTIONS

The distributive law states that

$$\frac{a}{n} + \frac{b}{n} + \frac{c}{n} + \cdots + \frac{m}{n}$$
$$= \frac{a + b + c + \cdots + m}{n}$$

Therefore:

> The sum of two or more fractions with identical denominators is equal to a fraction whose numerator is the sum of the numerators and whose denominator is the common denominator.

EXAMPLE 1

$$\frac{3x}{2ac} + \frac{5x}{2ac} - \frac{y}{2ac} = \frac{3x + 5x - y}{2ac} = \frac{8x - y}{2ac}$$

EXAMPLE 2

$$\frac{a + b}{a + 3b} + \frac{a - b}{a + 3b} - \frac{2a + b}{a + 3b}$$
$$= \frac{(a + b) + (a - b) - (2a + b)}{a + 3b}$$
$$= \frac{a + b + a - b - 2a - b}{a + 3b}$$
$$= \frac{-b}{a + 3b}$$

If the denominators of fractions are different, we convert the fractions to equal fractions with a common denominator and proceed as in the above examples.

EXAMPLE 3 Express the following sum as a single fraction:

$$\frac{1}{6a} + \frac{1}{3b} - \frac{3a + 2b}{12ab}$$

We first convert the fractions to equal fractions with the lcd, $12ab$, and then proceed as in Examples 1 and 2. The details of the solution follow:

$$\frac{1}{6a} + \frac{1}{3b} - \frac{3a + 2b}{12ab} = \frac{2b}{12ab} + \frac{4a}{12ab} - \frac{3a + 2b}{12ab}$$

$$= \frac{2b + 4a - (3a + 2b)}{12ab}$$

$$= \frac{2b + 4a - 3a - 2b}{12ab}$$

$$= \frac{a}{12ab} = \frac{1}{12b}$$

EXAMPLE 4 Combine

$$\frac{3a + b}{a^2 - b^2} - \frac{2b}{a(a - b)} - \frac{1}{a + b}$$

into a single fraction.

Solution The first step in the solution is to obtain the lcd, $a(a^2 - b^2)$. Then we proceed as follows:

$$\frac{3a + b}{a^2 - b^2} - \frac{2b}{a(a - b)} - \frac{1}{a + b}$$

$$= \frac{a(3a + b)}{a(a^2 - b^2)} - \frac{(a + b)2b}{a(a^2 - b^2)} - \frac{a(a - b)}{a(a^2 - b^2)}$$

$$= \frac{a(3a + b) - (a + b)2b - a(a - b)}{a(a^2 - b^2)}$$

$$= \frac{3a^2 + ab - 2ab - 2b^2 - a^2 + ab}{a(a^2 - b^2)}$$

$$= \frac{2a^2 - 2b^2}{a(a^2 - b^2)} = \frac{2(a^2 - b^2)}{a(a^2 - b^2)} = \frac{2}{a}$$

After some practice, the first and second steps in Examples 3 and 4 can be omitted and the result in the third step written at once. If this is done, care should be taken to change the signs of all terms in this result contributed by a fraction preceded by a minus sign.

Most of the work in adding fractions is in finding the numerator of the sum, but you should check each step and make sure that you have not omitted the denominator.

EXERCISE 4.2

Convert the set of fractions into an equal set such that each fraction of a set has the lcd of the set as denominator.

1. $\dfrac{1}{2}, \dfrac{2}{3}, \dfrac{4}{5}$

2. $\dfrac{1}{2}, \dfrac{3}{7}, \dfrac{8}{11}$

3. $\dfrac{1}{2}, \dfrac{3}{4}, \dfrac{5}{6}$

4. $\dfrac{2}{3}, \dfrac{4}{5}, \dfrac{7}{10}$

5. $\dfrac{1}{a}, \dfrac{1}{b}, \dfrac{1}{ab}$

6. $\dfrac{2}{a}, \dfrac{3}{a^2}, \dfrac{1}{ab}$

7. $\dfrac{u}{v^2}, \dfrac{2}{u^2v}, \dfrac{v}{u}$

8. $\dfrac{x}{y}, \dfrac{y}{x^2}, \dfrac{1}{xy^2}$

9. $\dfrac{a}{a-b}, \dfrac{a-b}{a+b}, \dfrac{a-2b}{b-a}$

10. $\dfrac{x-y}{x+y}, \dfrac{x+2y}{2x-y}, \dfrac{x+3y}{4x-2y}$

11. $\dfrac{x-2y}{x-y}, \dfrac{x+y}{x+2y}, \dfrac{x+2y}{3x+6y}$

12. $\dfrac{x-3y}{x+3y}, \dfrac{x+3y}{x-3y}, \dfrac{x-4y}{3x+9y}$

13. $\dfrac{a-3b}{(a+b)(a-2b)}, \dfrac{a+2b}{(a+b)(a+3b)}, \dfrac{a-b}{(a-2b)(a+3b)}$

14. $\dfrac{a-2b}{(a-b)(a+3b)}, \dfrac{a+b}{(a+3b)(a+2b)}, \dfrac{a-3b}{(a-b)(a+2b)}$

15. $\dfrac{a-3b}{(a-4b)(a+4b)}, \dfrac{a+4b}{(a-3b)(a-4b)}, \dfrac{a-4b}{(a-3b)(a+4b)}$

16. $\dfrac{a-4b}{(a-2b)(a-3b)}, \dfrac{a-2b}{(a-3b)(a-4b)}, \dfrac{a-3b}{(a-4b)(a-2b)}$

146 | Chapter 4: Algebraic Fractions

Complete the solution and reduce to lowest terms.

17. $\dfrac{2}{3} + \dfrac{5}{6} - \dfrac{7}{8} - \dfrac{1}{4} = \dfrac{16 + }{24} =$

18. $\dfrac{5}{12} - \dfrac{1}{6} - \dfrac{2}{9} + \dfrac{3}{4} = \dfrac{15 - }{36} =$

19. $\dfrac{1}{3} + \dfrac{3}{4} - \dfrac{5}{6} + \dfrac{7}{9} = \dfrac{12 + }{36} =$

20. $\dfrac{2}{5} + \dfrac{5}{6} - \dfrac{7}{10} - \dfrac{8}{15} = \dfrac{12 + }{30} =$

21. $\dfrac{b}{a} - \dfrac{c}{b} + \dfrac{a}{c} = \dfrac{}{abc} =$

22. $\dfrac{y}{x} - \dfrac{w}{3y} + \dfrac{x}{2w} = \dfrac{}{6xyw} =$

23. $\dfrac{t}{2xy} - \dfrac{y}{5tx} + \dfrac{x}{3ty} = \dfrac{}{30txy} =$

24. $\dfrac{5t}{6sr} - \dfrac{s}{2rt} + \dfrac{2r}{3st} = \dfrac{}{6rst} =$

25. $\dfrac{x^2 + 3xy + y^2}{x(x + y)} - \dfrac{x + 3y}{x + y} = \dfrac{}{x(x + y)} =$

26. $\dfrac{x^2 - 6xy + 9y^2}{3x + 6y} - \dfrac{x^2 - 2xy}{x + 2y} = \dfrac{}{3(x + 2y)} =$

27. $\dfrac{x + 3y}{x + y} - \dfrac{x^2 + 3xy - y^2}{x^2 - y^2} = \dfrac{}{x^2 - y^2} =$

28. $\dfrac{x + y}{x + 2y} - \dfrac{x + y}{x - 3y} = \dfrac{}{(x + 2y)(x - 3y)} =$

Perform the indicated additions.

29 $\dfrac{1}{2} + \dfrac{1}{3} - \dfrac{1}{5} =$

30 $\dfrac{2}{3} + \dfrac{1}{4} + \dfrac{5}{6} - \dfrac{1}{12} =$

31 $\dfrac{1}{2} + \dfrac{1}{3} + \dfrac{1}{4} - \dfrac{5}{6} =$

32 $\dfrac{2}{3} + \dfrac{3}{5} - \dfrac{1}{10} + \dfrac{1}{6} =$

33 $\dfrac{a}{bc} + \dfrac{2b}{ac} + \dfrac{c}{ab} =$

34 $\dfrac{2p}{qr} - \dfrac{3q}{2pr} + \dfrac{r}{3pq} =$

35 $\dfrac{t}{2rs} + \dfrac{2s}{3rt} + \dfrac{3r}{5st} =$

36 $\dfrac{2a}{3bc} - \dfrac{3b}{5ac} + \dfrac{5c}{2ab} =$

37 $\dfrac{2-a}{ab} - \dfrac{b+2a}{ba^2} + \dfrac{b+1}{b} =$

38 $\dfrac{x+2}{2x} - \dfrac{x-3}{3y} - \dfrac{x+y}{xy} =$

4.2

39 $\dfrac{5x + 2y}{xy} - \dfrac{2y^2 - x}{xy^2} - \dfrac{5x^2 - y}{x^2 y} =$

40 $\dfrac{3x + 4y}{xy^2} - \dfrac{y + 2}{xy} - \dfrac{2x - xy}{x^2 y} =$

41 $\dfrac{a + 3b}{3a + b} - \dfrac{b}{a} =$

42 $\dfrac{2x - y}{x + 2y} + \dfrac{y}{x} =$

43 $\dfrac{2x + 5y}{x - 2y} - \dfrac{5y}{x} =$

44 $\dfrac{x + y}{x + 3y} - \dfrac{x}{3y} =$

45 $\dfrac{x + y}{2x - y} - \dfrac{x - y}{2x + y} =$

46 $\dfrac{3x - 2y}{x + 3y} - \dfrac{3x + 2y}{x - 3y} =$

47 $\dfrac{2x^2}{(y - 2x)(y - x)} + \dfrac{y + x}{y - x} =$

48 $\dfrac{x^2 + 3xy}{x^2 + 5xy + 6y^2} - \dfrac{2x - y}{x + 2y} =$

49 $\dfrac{x}{y} + \dfrac{y + 3x}{x - 3y} - \dfrac{y}{x} =$

50 $\dfrac{3x}{y} + \dfrac{2y}{x} - \dfrac{6x - 6y}{2y - 3x} =$

51 $-\dfrac{3x}{2y} - \dfrac{2y}{x} + \dfrac{3x - 6y}{2y - 3x} =$

52 $\dfrac{2x}{y} - \dfrac{3y}{2x} - \dfrac{2x - 3y}{2x + y} =$

53 $\dfrac{x + 2y}{x + 3y} - \dfrac{x - y}{x - 3y} + \dfrac{6xy + 3y^2}{x^2 - 9y^2} =$

54 $\dfrac{xy - 4y^2}{x^2 - 16y^2} - \dfrac{x + 3y}{x + 4y} - \dfrac{x - 2y}{x - 4y} =$

55 $\dfrac{9xy + y^2}{x^2 - y^2} + \dfrac{x - 3y}{x + y} - \dfrac{x + 5y}{x - y} =$

56 $\dfrac{x + 3y}{x + 2y} - \dfrac{x - 3y}{x - 2y} - \dfrac{2xy}{x^2 - 4y^2} =$

57 $\dfrac{x - 2y}{x - 3y} - \dfrac{x - y}{x + 5y} + \dfrac{x^2 + 23y^2}{x^2 + 2xy - 15y^2} =$

58 $\dfrac{x + 3}{x + 4} - \dfrac{x - 2}{x - 1} + \dfrac{x - 6}{x^2 + 3x - 4} =$

59 $\dfrac{x-3}{x+2} - \dfrac{x-1}{x+4} + \dfrac{x^2+6}{x^2+6x+8} =$

60 $\dfrac{x+5}{x+6} - \dfrac{x-1}{x+1} - \dfrac{10}{x^2+7x+6} =$

61 $\dfrac{3}{x+1} + \dfrac{1}{x-1} - \dfrac{8}{2x-1} =$

62 $\dfrac{3}{x-3} + \dfrac{1}{x+3} - \dfrac{2}{2x-3} =$

63 $\dfrac{1}{x+y} - \dfrac{4}{2x-y} + \dfrac{1}{x-2y} =$

64 $\dfrac{2}{x+2y} - \dfrac{3}{2x-y} + \dfrac{1}{x-y} =$

65 $\dfrac{r+3s}{(r-s)(r+2s)} + \dfrac{r+3s}{(r+s)(r+2s)} - \dfrac{6s}{r^2-s^2} =$

66 $\dfrac{2r+4s}{(r+s)(r-3s)} - \dfrac{r}{(r-s)(r+s)} - \dfrac{5s}{(r-3s)(r-s)} =$

67 $\dfrac{3a^2}{(2a+b)(a-b)} + \dfrac{a^2+ab-b^2}{(2a-b)(a-b)} - \dfrac{4a^2}{(2a+b)(2a-b)} =$

68 $\dfrac{11a+2b}{2(3a+b)(2a-b)} - \dfrac{7a}{2(3a+b)(2a+3b)} - \dfrac{6b}{(2a+3b)(2a-b)} =$

4.6
COMPLEX FRACTIONS

A complex fraction is a fraction in which the numerator or denominator contains fractions. The following are examples of complex fractions:

$$\frac{1-\frac{1}{2}}{3} \qquad \frac{2a-b}{a+\frac{2}{b}} \qquad \frac{\frac{x-y}{x+y}+\frac{x}{y}}{1-\frac{x^2}{x-y}}$$

SIMPLIFYING A COMPLEX FRACTION A complex fraction can be simplified by expressing it as an equivalent fraction in which neither the numerator nor the denominator contains fractions. The simplification of a complex fraction is fundamentally a problem in division, and it can be accomplished by simplifying the numerator and the denominator of the complex fraction and then finding the quotient of the results. Usually, however, the most efficient method consists of the following steps:

1. Find the lcm of the denominators of the fractions that appear in the complex fraction.
2. Multiply the members of the complex fraction by the above lcm.
3. Simplify the result obtained in step 2.

EXAMPLE 1 Simplify

$$\frac{3+\frac{1}{a}}{9-\frac{1}{a^2}}$$

Solution The fractions appearing in the complex fraction are $1/a$ and $1/a^2$, and the lcm of the denominators is a^2. Therefore, we multiply each member of the complex fraction by a^2 and obtain

$$\frac{3+\frac{1}{a}}{9-\frac{1}{a^2}} = \frac{a^2\left(3+\frac{1}{a}\right)}{a^2\left(9-\frac{1}{a^2}\right)}$$

$$= \frac{3a^2+a}{9a^2-1}$$

$$= \frac{a(3a+1)}{(3a+1)(3a-1)} \qquad \text{factoring members of fraction}$$

$$= \frac{a}{3a-1} \qquad \text{dividing each member by } 3a+1$$

It is sometimes advisable to simplify the numerator and the denominator before multiplying by the lcm of the denominators. This is illustrated in Example 2.

EXAMPLE 2 Simplify

$$\frac{3+\frac{4}{a-1}}{\frac{a}{a+1}-\frac{a+1}{a-1}}$$

Finding sum of fractions in numerator and denominator gives

$$\frac{\frac{3a-3+4}{a-1}}{\frac{a^2-a-a^2-2a-1}{a^2-1}}$$

Combining similar terms we have

$$\frac{\frac{3a+1}{a-1}}{\frac{-3a-1}{a^2-1}}$$

Multiplying members by the lcm of denominators gives

$$\frac{(a^2-1)\frac{3a+1}{a-1}}{(a^2-1)\frac{-3a-1}{a^2-1}}$$

Factoring leads to

$$\frac{\frac{(a-1)(a+1)(3a+1)}{a-1}}{\frac{(a^2-1)(-1)(3a+1)}{a^2-1}} =$$

$$\frac{(a+1)(3a+1)}{(-1)(3a+1)} = \frac{a+1}{-1} = -a-1$$

STUDENT'S NOTES

EXERCISE 4.3

Change the following complex fractions to simple fractions and reduce to lowest terms.

1. $\dfrac{1}{1 - \frac{3}{4}} =$

2. $\dfrac{3}{2 - \frac{5}{6}} =$

3. $\dfrac{2}{1 + \frac{3}{7}} =$

4. $\dfrac{3}{1 - \frac{3}{5}} =$

5. $\dfrac{5 - \frac{5}{6}}{4 + \frac{4}{9}} =$

6. $\dfrac{2 - \frac{1}{2}}{3 + \frac{6}{7}} =$

7. $\dfrac{\frac{2}{3} + \frac{1}{2}}{\frac{2}{9} - \frac{1}{6}} =$

8. $\dfrac{\frac{1}{2} + \frac{3}{4}}{\frac{5}{6} + \frac{1}{4}} =$

9. $\dfrac{a - \frac{1}{a}}{1 - \frac{1}{a^2}} =$

10. $\dfrac{c - \frac{9}{c}}{1 - \frac{3}{c}} =$

11. $\dfrac{2 + \frac{1}{a}}{4 - \frac{1}{a^2}} =$

12. $\dfrac{2 - \frac{8}{a^2}}{a - 4 + \frac{4}{a}} =$

13. $\dfrac{\dfrac{3}{x} - \dfrac{x}{3}}{\dfrac{1}{x} + \dfrac{2}{3x}} =$

14. $\dfrac{\dfrac{x}{6} - \dfrac{1}{3}}{\dfrac{1}{2} - \dfrac{1}{x}} =$

15. $\dfrac{\dfrac{x}{2} + \dfrac{y}{3}}{\dfrac{3}{y} + \dfrac{2}{x}} =$

16. $\dfrac{\dfrac{x}{2y} - \dfrac{1}{2}}{\dfrac{x}{3y} - \dfrac{1}{3}} =$

17. $\dfrac{1 + \dfrac{y}{x+y}}{1 + \dfrac{3y}{x-y}} =$

18. $\dfrac{2 - \dfrac{x}{2x+y}}{2 - \dfrac{5x}{x-y}} =$

19. $\dfrac{1 + \dfrac{5}{x-y}}{1 + \dfrac{3}{x+y}} =$

20. $\dfrac{3 + \dfrac{5x}{x - 3y}}{4 - \dfrac{5y}{2x - y}} =$

21. $\dfrac{x + 4 + \dfrac{1}{x + 2}}{x + 1 + \dfrac{4}{x + 5}} =$

22. $\dfrac{\dfrac{x - 1}{x + 2} - \dfrac{x + 3}{x + 1}}{5 - \dfrac{3}{x + 2}} =$

23. $\dfrac{\dfrac{x - 2}{x + 3} - \dfrac{x}{x - 1}}{\dfrac{10}{x + 3} - 3} =$

24. $\dfrac{\dfrac{x + 2}{x - 2} - \dfrac{x}{x + 2}}{3 + \dfrac{2}{x}} =$

25. $\dfrac{\dfrac{x}{x + 1} - \dfrac{x^2}{x^2 - 1}}{\dfrac{1}{x - 1} + 1} =$

26. $\dfrac{\dfrac{x}{x + 2} - 1}{\dfrac{2x + 3}{x + 2} - \dfrac{2x}{x + 1}} =$

27 $\dfrac{\dfrac{3}{2x+3} - \dfrac{5}{x+2}}{\dfrac{3x+1}{x+2} + 4} =$

28 $\dfrac{2 - \dfrac{3}{x+1}}{\dfrac{5}{x+2} - 2} =$

Show that the first number is smaller than the second.

29 $2 + \dfrac{1}{2 + \frac{1}{2}},\ 2 + \dfrac{1}{2 + \dfrac{1}{2 + \frac{1}{2}}}$

30 $3 + \dfrac{1}{3 + \frac{1}{3}},\ 3 + \dfrac{1}{3 + \dfrac{1}{3 + \frac{1}{3}}}$

31 $3 + \dfrac{1}{2 + \dfrac{1}{6 + \frac{1}{4}}},\ 3 + \dfrac{1}{2}$

32 $3 + \dfrac{1}{2 + \frac{1}{6}},\ 3 + \dfrac{1}{2}$

4.7
CHAPTER SUMMARY

This chapter was devoted entirely to algebraic fractions, including complex fractions, and operations on them. We began with some definitions and the fundamental principle and followed with reduction to lowest terms, multiplication, division, and addition of fractions. The final section dealt with complex fractions. Several fundamental formulas were presented:

$$\frac{a}{b} = \frac{c}{d} \quad \text{if and only if} \quad ad = bc \quad \begin{array}{l} b \neq 0 \\ d \neq 0 \end{array} \tag{4.1}$$

$$\frac{na}{da} = \frac{n}{d} \quad \begin{array}{l} a \neq 0 \\ d \neq 0 \end{array} \tag{4.2}$$

$$\frac{n/b}{d/b} = \frac{n \div b}{d \div b} = \frac{n}{d} \quad \begin{array}{l} b \neq 0 \\ d \neq 0 \end{array} \tag{4.3}$$

$$\frac{n}{d} = -\frac{-n}{d} = -\frac{n}{-d} = \frac{-n}{-d} \quad d \neq 0 \tag{4.4}$$

STUDENT'S NOTES

EXERCISE 4.4 REVIEW

Fill in the blanks so as to justify the use of the equality sign.

1. $\dfrac{ax}{by} = \dfrac{}{2aby}$

2. $\dfrac{3x^3y^2}{6xy^4} = \dfrac{}{2y^2}$

3. $\dfrac{a-b}{c+d} = -\dfrac{}{c+d}$

4. $\dfrac{a+b}{a-b} = \dfrac{}{a^2 - b^2}$

Reduce the fraction to lowest terms.

5. $\dfrac{6x^2y^3}{4x^3y^5} =$

6. $\dfrac{x^3 - x}{x - 1} =$

7. $\dfrac{(x-y)(x^2 - xy - 6y^2)}{(x+2y)(x^2 - 5xy + 4y^2)} =$

Find the indicated products and quotients.

8. $\dfrac{2(x-y)}{3(x+y)} \cdot \dfrac{6x(x+y)}{y(x-y)} =$

9. $\dfrac{2x^2 + xy - 3y^2}{6x^2 + xy - y^2} \cdot \dfrac{2x^2 + 3xy - 2y^2}{4x^2 + 4xy - 3y^2} =$

10. $\dfrac{x^2 - y^2}{x - y} \div \dfrac{(x+y)x}{x + 2y} =$

11. $\dfrac{x^2 - y^2}{2x^2 - 3xy - 2y^2} \cdot \dfrac{2x^2 + 7xy + 3y^2}{3x^2 - 2xy - y^2} \div \dfrac{x^2 + 5xy + 6y^2}{3x^2 - 5xy - 2y^2} =$

Convert the set of fractions to an equal set with the lcd as each denominator.

12. $\dfrac{x + 2y}{x - y} \qquad \dfrac{2x - y}{x + 3y} \qquad \dfrac{4x + y}{2x + y}$

13. $\dfrac{a + 2b}{(a - 3b)(a + 3b)} \qquad \dfrac{a + 3b}{(a - 2b)(a - 3b)} \qquad \dfrac{a - 3b}{(a - 2b)(a + 3b)}$

Chapter 4: Algebraic Fractions

Perform the additions indicated.

14 $\dfrac{x-2}{x+3} - \dfrac{2x-1}{x+1} + \dfrac{x^2+9x+2}{x^2+4x+3} =$

15 $\dfrac{5x+2y}{2x+5y} - \dfrac{3x+y}{x+3y} - \dfrac{2x-y}{x-2y} =$

16 $\dfrac{3x}{2x^2+3xy-2y^2} - \dfrac{2x+y}{2x^2-5xy+2y^2} + \dfrac{y}{x^2-4y^2}$

Reduce the complex fraction to a simple fraction.

17 $\dfrac{\frac{2}{3}+\frac{1}{2}}{\frac{1}{2}+\frac{5}{6}} =$

18 $\dfrac{\frac{x}{6}-\frac{1}{3}}{\frac{1}{2}-\frac{1}{x}} =$

19 $\dfrac{3x+\frac{x-5}{x-1}}{x-\frac{5}{3x-2}} =$

20 $\dfrac{\frac{5}{2-x}-\frac{3}{1+2x}}{1-\frac{1+12x}{2-x}} =$

21 $2 + \dfrac{1}{3+\dfrac{1}{4+\frac{1}{2}}}$

EXERCISE 4.5 CHAPTER TEST

Reduce to lowest terms.

1. $\dfrac{9x^3y^4z}{6xy^2z^3} =$

2. $\dfrac{(x-y)(2x^2+7xy+3y^2)}{(2x+y)(3x^2-2xy-y^2)} =$

Find the indicated products and quotients.

3. $\dfrac{x^2-xy-2y^2}{4x^2-y^2} \cdot \dfrac{2x^2+5xy-3y^2}{x^2+xy-6y^2} =$

4. $\dfrac{x(x+3)+4(x+3)}{x(x+1)+4(x+1)} \cdot \dfrac{x^2-(2x+3)}{x^2-9} =$

5. $\dfrac{(x-2)x+1}{(x-1)x-2} \cdot \dfrac{(x-2)x+x-2}{x^2-1} \div \dfrac{x+6}{x+1} =$

6. $\dfrac{3x^2-2xy-y^2}{2xy+4x^2} \cdot \dfrac{2x^2-3xy-2y^2}{3x^2+4xy+y^2} \div \dfrac{2x^2-xy-y^2}{xy^2+y^3} =$

Chapter 4: Algebraic Fractions

Perform the indicated additions.

7. $\dfrac{5x}{3y} - \dfrac{2x + 5y}{x - 2y} =$

8. $\dfrac{x + 3y}{2x - y} + \dfrac{3x + y}{2x + y} - \dfrac{8x^2 + 2y^2}{4x^2 - y^2} =$

Simplify the complex fraction.

9. $\dfrac{\dfrac{a + 2}{a - 2} - \dfrac{a}{a + 2}}{3 - \dfrac{4}{a + 2}} =$

10. $\dfrac{\dfrac{3}{2a + 3} - \dfrac{5}{a + 2}}{\dfrac{3a + 1}{a + 2} + 4} =$

CHAPTER 5

Relations, Functions, and Graphs

5.1 FORMULAS AND CHARTS

It is not unusual for us to want the value of some quantity and to be able to find it only by making use of one or more others. This is ordinarily done by making use of a **formula** that contains the desired quantity and one or more known quantities. For example, the volume V of a right circular cylinder of radius r and height h is given by $V = \pi r^2 h$. We can find the value of V if we know corresponding values of r and h by substituting them in the formula. Another formula we often need to use is $I = Prt$, where I, P, r, and t are simple interest, principal, rate, and time, respectively.

EXAMPLE 1 Find the volume of a right circular cylinder if its radius is 4 in and its height is 6 in.

Solution If we substitute the given values of r and h in the formula for the volume, we have

$$V = \pi(4^2)(6) = 96\pi = 301.632$$

using 3.142 as an approximation to π.

Another way of finding the value of one quantity when the value of another related one is known is to use a **chart**. A chart can be constructed if we know several pairs of corresponding values of two variables.

We shall illustrate the use of a chart by considering the following table, which gives corre-

Height	Weight	Height	Weight
60	126	66	144
61	128	68	152
62	130	70	161
63	133	72	172
64	136	74	184

sponding heights in inches and weights in pounds for a 30-year-old man. We display the data by showing height along a vertical and weight along a horizontal line. Thus after drawing a vertical line to represent the height above each corresponding weight, we have the vertical lines in Fig. 5.1. This is called a **bar graph.**

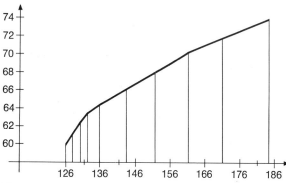

Fig. 5.1

The relation can be shown in another way by putting a dot or cross above each weight and across from the corresponding height and drawing a smooth curve connecting the points thus determined. In this manner, we get the connected segments in Fig. 5.1.

If we now want to determine the weight that corresponds to a specified value of the height, we draw a horizontal line from the given height value until it intersects the curve and then estimate the corresponding value of weight by dropping a perpendicular to the weight line.

EXAMPLE 2 Find the weight that corresponds to a height of 67 in.

Solution We draw a horizontal line from the height of 67 until it intersects the curve, then drop a perpendicular to the weight line, and read the weight. We thus are led to a weight of 148 lb.

Problems 1 to 20 in Exercise 5.1 may be done now.

5.2
NUMERICAL GEOMETRY

In this section we shall present the usual formulas dealing with the plane figures and solids illustrated and described below. We shall use the following general abbreviations for plane figures:

p = perimeter, or length of boundary of plane figure bounded by straight-line segments
C = length of circumference of circle
A = area of plane figure

The important formulas associated with each geometrical figure are given beside the figure that illustrates it.

TRIANGLES

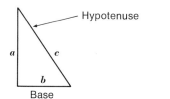

Right triangle
$p = a + b + c$
$A = \frac{1}{2}ab$
$c^2 = a^2 + b^2$

Other triangles
$p = a + b + c$
$A = \frac{1}{2}(\text{base} \times \text{altitude})$
$\quad = \frac{1}{2}bh$

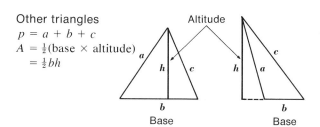

QUADRILATERALS

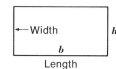

Rectangle: opposite sides parallel and all angles right angles
$p = 2h + 2b$
$A = bh$

Parallelogram: opposite sides parallel
$p = 2a + 2b$
$A = bh$

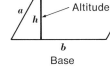

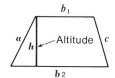

Trapezoid: two sides parallel
$p = b_1 + b_2 + a + c$
$A = \frac{1}{2}(b_1 + b_2)h$

REGULAR POLYGONS

Regular polygon: all sides equal and all angles equal
$A = \frac{1}{2}(\text{perimeter} \times \text{apothem})$

CIRCLES

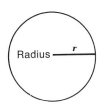

Circle: all points on perimeter equidistant from center
$C = 2\pi r$
$A = \pi r^2$
$\pi = 3.1416$ (approximately)

SOLIDS

The **lateral area** of a prism or a pyramid is the sum of the areas of all faces other than the base or bases; the lateral area of either a cylinder or a cone is the area of the curved surface. The **total surface area** is the sum of the lateral area and the area of the base (or bases). We shall use the following abbreviations in connection with the solids illustrated below:

L = lateral area V = volume
S = total surface area B = area of base

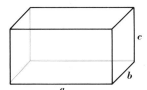

Parallelepiped: all faces are rectangles
$S = 2ab + 2ac + 2bc$
$V = abc$

Right prism: bases in parallel planes, other faces rectangles
L = perimeter of base × altitude
$\quad = (a + b + c)(h)$
$S = L + 2B$
V = area of base × altitude
$\quad = Bh$

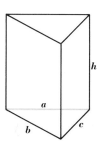

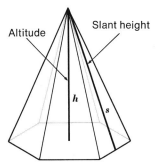

Regular pyramid: all faces congruent triangles
$L = \frac{1}{2}$(perimeter of base × slant height)
$\quad = \frac{1}{2}ps$
$S = L + B$
$V = \frac{1}{3}$(area of base × altitude)
$\quad = \frac{1}{3}Bh$

Right circular cylinder: bases are circles in parallel planes
L = circumference of base × altitude
$\quad = 2\pi rh$
$V = \pi r^2 h$
$S = L + 2B$
$\quad = 2\pi rh + 2\pi r^2$
$\quad = 2\pi r(h + r)$

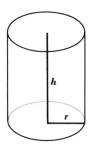

Right circular cone: base is a circle, and line joining vertex to center of base is perpendicular to base
$L = \frac{1}{2}$(circumference of base × slant height)
$\quad = \pi rs$
$S = L + B$
$\quad = \pi(rs + r^2)$
$V = \frac{1}{3}$(area of base × altitude)
$\quad = \frac{1}{3}\pi r^2 h$

Sphere: all points equidistant from center
$S = 4\pi r^2$
$V = \frac{4}{3}\pi r^3$

EXAMPLE 1 Find the number of acres in a rectangular field $\frac{1}{2}$ mi long and 600 yd wide.

Solution

1 acre = 4840 yd², $\frac{1}{2}$ mi = $\frac{1}{2} \times 1760$ = 880 yd

Area of field in square yards = 880(600)

$$\text{Area in acres} = \frac{880(600)}{4840}$$

$$= \frac{880}{4840} \cdot 600$$

$$= \frac{2}{11} \cdot 600 \quad \text{dividing 880 and 4840 by 440}$$

$$= \frac{1200}{11} = 109\tfrac{1}{11}$$

EXAMPLE 2 Find the capacity in gallons of an oil drum in the shape of a right circular cylinder 4.0 ft high and 30 in in diameter.

Solution

1 gal = 231 in³   4.0 ft = 48 in
Volume of cylinder = $\pi \times$ radius² $\times$ altitude
Radius = $\frac{1}{2}$(30 in) = 15 in
π = 3.14 to three significant digits

$$\text{Volume in gallons} = \frac{3.14(15)(15)(48)}{231}$$

$\qquad\qquad\qquad = \dfrac{33{,}900}{231}\quad$ to three significant digits

$\qquad\qquad\qquad = 146.8$

$\qquad\qquad\qquad = 150 \qquad$ to two significant digits

Note: The measurements in the data are given to two significant figures, so we round the result off to two significant figures. The zero in 150 is not significant.

EXERCISE 5.1

1. The time in seconds required to make a photographic enlargement of area A from a given plate is $t = 0.33A$. Find the time required to make a 5 by 7 enlargement and an 11 by 14 enlargement.

2. If the tax liability of a person with a taxable income between \$4000 and \$6000 is given by $T = \$840 + 0.26E$, where E is the excess of income over \$4000, find the tax due by a person with an income of \$5726.37.

3. If the normal blood pressure of a person is $P = 110 + 0.5A$, where A is the person's age, find the normal pressure for a person of age 30, 50, 70.

4. The distance in feet that a freely falling body will travel in t seconds is $s = 16t^2$. Find the distance that such a body will fall in 1, 4, and 10 s.

5. The principal P invested at simple interest that is needed to accumulate to A dollars in t years at rate r per year is $P = A/(1 + rt)$. Find the principal that would accumulate to \$500 in 3 years at 8%; to \$500 in 3 years at 6%.

6. The number of bushels that a rectangular container will hold is $B = LWH/1846$, where L, W, and H are the length, width, and height of the container in inches. How many bushels will a box hold if it is 66 by 21 by 36 in; 44 by 28 by 45 in?

7. If $l = a + nd - d$, find l for $a = 3$, $n = 7$, and $d = 4$; for $a = 7$, $n = 4$, and $d = 7$.

8. If $s = (a - rl)/(1 - r)$, find s for $a = 1$, $r = 2$, and $l = 64$; for $a = 243$; $r = \frac{2}{3}$, and $l = 32$.

Show the data in the following problems graphically by using a bar graph and a continuous curve. Graph paper is suggested.

9 This table shows the normal weight for 20-year-old women of various heights.

Height, in	Weight, lb	Height, in	Weight, lb
60	106	65	125
61	110	66	128
62	114	67	131
63	117	68	135
64	121	69	139

10 The minimum number of hundreds of calories required by boys of various ages is as follows:

Age	Calories	Age	Calories
8	17	13	25
9	19	14	26
10	21	15	27
11	21	16	27
12	23	17	28

11 The following table shows the percent of single men who marry within a year after reaching the specified age.

Age	Percent	Age	Percent
21	9.4	26	17.3
22	12.5	27	17.3
23	15.3	28	17.1
24	15.9	29	16.8
25	17.0	30	15.9

12 The following table shows the percent of single men who eventually marry after having reached a specified age.

Age	Percent	Age	Percent
21	92.3	26	85.9
22	91.8	27	83.4
23	90.9	28	80.3
24	89.6	29	76.6
25	88.0	30	72.3

13 The following table shows the average weight of 40-year-old men of various heights.

Height	Weight	Height	Weight
60	134	66	156
61	137	68	165
62	140	70	174
63	144	72	183
64	148	74	192

14 The following table shows the average weight of 40-year-old women of various heights.

Height	Weight	Height	Weight
60	127	65	143
61	130	66	147
62	133	67	151
63	136	68	155
64	140	69	159

15 The following table shows the median height for boys of different ages.

Age	Height	Age	Height
1	30.2	6	45.9
2	34.6	7	48.1
3	37.8	8	50.5
4	40.8	9	52.8
5	43.4	10	54.3

16 The following table shows the median height for girls of different ages.

Age	Height	Age	Height
1	29.4	6	45.9
2	33.8	7	47.8
3	37.5	8	50.0
4	40.7	9	52.2
5	43.4	10	54.5

17 The crude oil production in the United States in billions of barrels for several years is given below:

Year	Production	Year	Production
1945	1.7	1970	3.5
1950	2.0	1972	3.5
1955	2.5	1973	3.4
1960	2.6	1974	3.2
1965	2.8	1975	3.1

18 The liquid natural gas production in the United States in 10^{15} cubic feet for several years is shown below:

Year	Production	Year	Production
1945	3.9	1970	21.9
1950	6.3	1972	22.5
1955	9.4	1973	22.6
1960	12.8	1974	21.6
1965	16.0	1975	20.1

19 The average hourly earnings of manufacturing production workers in the United States for several years is given below.

Year	Earnings	Year	Earnings
1955	$1.86	1972	$3.81
1960	2.26	1973	4.08
1965	2.61	1974	4.41
1970	3.36	1975	4.81
1971	3.57	1976	5.15

20 The cost of medical care in billions of dollars for the United States for several years is given below.

Year	Cost	Year	Cost
1950	$ 8.8	1970	$47.4
1955	12.8	1973	68.3
1960	19.1	1974	76.1
1965	28.1	1975	86.4

21 Find the area of a circle whose radius is 13 in.

22 Find the area of a regular hexagon (a regular polygon with six sides) with each side equal to 9.0 in and apothem equal to 7.6 in.

23 Find the lateral area of a regular pyramid whose slant height is 8.0 ft and whose base has a perimeter of 16 ft.

24 Find the circumference of a circle with a radius of 28 yd.

25 Find the surface area of a sphere whose radius is 14 in.

26 Find the lateral area of a right circular cylinder with a base circumference of 28 in and an altitude of 35 in.

27 Find the volume of a sphere whose radius is 6.0 yd.

28 Find the lateral area of a right prism with an altitude of 12 in and a base perimeter of 65 in.

29 Find the volume of a parallelepiped with an altitude of 6.0 ft and a base area of 48 ft².

30 Find the area of a right triangle with a base of 4.0 ft and a hypotenuse of 5.0 ft.

31 Find the total surface area of a right circular cone that has a slant height of 11 in and a base of radius 3.0 in.

32 Find the volume of a right circular cylinder 14 ft high whose base is 3.0 ft in diameter.

33 Find the volume of a parallelepiped with an altitude of 90 mm and a base 60 mm long and 40 mm wide.

34 Find the volume of a right prism whose altitude is 12 in and whose base is a triangle with a base of 8.0 in and an altitude of 6.0 in.

35 Find the area in square feet of a rectangular flower bed 15 ft long and 3.0 yd wide.

36 Find the number of acres in a trapezoidal field whose 900- and 600-yd parallel sides are 450 yd apart.

37 Find the area in square yards of a parallelogram with a base of 14 ft and an altitude of 10 ft.

38 A conical vessel has an altitude of 6.0 in and a base with a diameter of 4.0 in. How many pints will it hold?

39 How many yards of trimming would be necessary to trim the edge of a circular tablecloth 4.0 ft in diameter?

40 How many square yards of cloth are there in the tablecloth mentioned in Prob. 39?

41 A right circular cylindrical can with a diameter of 26 in is 3.0 ft high. How many gallons will it hold?

42 If mercury weighs 13.6 g/cm^3, how many grams of mercury will a hollow glass ball with an internal diameter of 2.0 in hold?

43 How many 50-lb bags of fertilizer will be required to fertilize a rectangular lawn 65 by 85 ft if 5 lb is applied to each 100 ft^2?

44 If 1 gal of paint will cover 165 ft^2, how many gallons will be needed to cover four cylindrical porch columns each of which is 8.0 ft tall and 12 in in diameter?

5.3 RELATIONS

The formulas of Secs. 5.1 and 5.2 expressed one quantity in terms of one or more others in such a way that we could find the one if corresponding values of the others were given. Some of the quantities involved are fixed, and others vary.

A symbol that takes on any element of a set of values is called a **variable**. If there is only one element in the set, the symbol is a **constant**. If V represents the number of gallons of water in a barrel that is being filled with a hose, it may be any number from 0 to the capacity of the barrel. If V is the amount of water in the barrel when full, then V is a constant. The set of values that a variable may take on is called the **domain** of the variable.

If two variables are so related that one or more values of the second are determined when a value in its domain is assigned to the first, we say there is a **relation** between them. The set of values taken on by the second variable is called its **range**.

EXAMPLE 1 Find the range of the volume of water that can be put in a right circular cylinder of radius 4 in and height 15 in.

Solution Since the volume of a right circular cylinder of radius r and height h is $V = \pi r^2 h$, the volume of water at any time in the one under consideration is $V = \pi 4^2 h = 16\pi h$, where h is the height of the liquid. Therefore, the range is all numbers from 0 to $16\pi(15) = 240\pi = 754$, approximately.

If the meaning or significance of each of a pair of numbers a and b written as (a,b) or (b,a) depends on its position in the pair, we say we have an **ordered pair**. Any set of ordered pairs is called a **relation**.

EXAMPLE 2 The sets $\{(3,2), (4,1), (4,3), (5,6)\}$ and $\{(3,2), (4,1), (6,3), (5,6)\}$ are both relations.

The set of first numbers in a relation is the domain, and the set of second numbers is the range. Thus, in the first set in Example 2 the domain is $\{3,4,5\}$ and the range is $\{2,1,3,6\}$.

The relation between two variables can be determined by various means including an equation which uses the two variables.

EXAMPLE 3 Find the values of y for $x = 0, 1, 2$ in the relation defined by $y = 3x - 1$.

Solution We need only replace x in $3x - 1$ by the values in the domain. Thus, we get $y = 3(0) - 1 = -1$, $y = 3(1) - 1 = 2$, and $y = 3(2) - 2 = 4$. Therefore, the range is $\{-1,2,4\}$.

EXAMPLE 4 If $y^2 = 16 - x^2$, find the range that corresponds to the domain $x = 0, 1, 2, 3, 4$.

Solution We can either solve the defining equation for y and then substitute each value of x, or we can substitute each value of x and then solve each resulting equation for y. If we do the former, we get $y = \pm\sqrt{16 - x^2}$ and can construct the following table:

x	0	1	2	3	4
$\pm\sqrt{16 - x^2}$	± 4	$\pm\sqrt{15}$	$\pm\sqrt{12}$	$\pm\sqrt{7}$	0

Consequently, the relation is $\{(0,\pm 4), (1, \pm\sqrt{15}), (2, \pm\sqrt{12}), (3, \pm\sqrt{7}), (4,0)\}$.

5.4 FUNCTIONS

If we examine the relation in Example 4 of Sec. 5.3, we find that there are two values of y for most values of x. In Example 3, however, there is exactly one value of y for each value of x. The difference between these two situations is essential. If a relation is such that there is exactly one value of y for each value of x, we say that y is a **function** of x.

EXAMPLE 1 The equation $y = x^2 + 3$ defines a function, and the equation $y^2 = x^2 + 3$ defines a relation that is not a function.

We can see from the definition of a function given above and the definition of a relation as a set of ordered number pairs that a function is a set of ordered pairs in which there is exactly one second number corresponding to each first number in a pair. The definition of a function is often put in the following form.

If D is a set of numbers, if there exists a rule such that for each element x of D exactly one number y is determined, and if R is the set of all numbers y, then the set of ordered pairs $\{(x,y) | x \in D$ and $y \in R\}$ is a function with domain D and range R.

The essential difference between a function and a relation is that for a function there is exactly one value of the second number for a given first number whereas for a relation there can be more than one second number for a given first number. The rule that determines the set of ordered pairs is often an equation and may be indicated by $y = f(x)$.

EXAMPLE 2 Find the set of ordered pairs that make up the function determined by the equation $y = x^2 + 2x - 1$ for $D = \{x | x$ is an integer from -3 to 2, inclusive$\}$.

Solution The numbers in the domain are $x = -3, -2, -1, 0, 1, 2$, and we must find each corresponding value y of the range. If $x = -3$, then $y = (-3)^2 + 2(-3) - 1 = 2$. Similarly, $y(-2) = -1$, $y(-1) = -2$, $y(0) = -1$, $y(1) = 2$, $y(2) = 7$.

Therefore, the function is $\{(-3,2), (-2,-1), (-1,-2), (0,-1), (1,2), (2,7)\}$.

EXAMPLE 3 If a function is defined by $\{(x,f(x)) | f(x) = 3x - 2\}$, $D - \{-3, -1, 0, 1, 2\}$, find the range.

Solution To find the range we need only substitute each number in the domain in the equation that defines the function. Thus we get $f(-3) = 3(-3) - 2 = -11$, $f(-1) = -5$, $f(0) = -2$, $f(1) = 1$, $f(2) = 4$. Hence, the range is $R = \{-11, -5, -2, 1, 4\}$.

At times we shall refer to x in $y = f(x)$ as the **independent variable** and to y as the **dependent variable.** Thus, in a function the first member in each ordered pair is the independent variable and the second is the dependent variable.

EXERCISE 5.2

1. Is {(1,2), (3,4), (4,3), (5,6)} a function? Why?

2. Is {(1,2), (3,4), (3,5), (5,6)} a function? Why?

3. Is {(1,2), (3,2), (2,3), (3,9)} a function? Why?

4. Is {(1,2), (3,2), (2,3), (4,9)} a function? Why?

5. Find the set of ordered pairs obtained by pairing the nth element of {1,3,5,7} with the $(n + 1)$th element of {−1,1,3,5,7}, $n = 1, 2, 3, 4$. Is this a function?

6. Find the set of ordered pairs obtained by pairing the nth letter of foot with the nth letter of ball. Is this a function?

7. Find the set of ordered pairs obtained by pairing the nth letter of ball with the nth letter of park. Is this a function?

8. Find the set of ordered pairs obtained by pairing the nth letter of shame with the $(6 − n)$th letter of faced. Is this a function?

9. Is the set of ordered pairs $\{(x,y) | y = \sqrt{x}, x = 0, 1, 4, 9\}$ a function?

10. Is the set of ordered pairs $\{(x,y) | y = x, x = 1, 2, 3, 4\}$ a function?

11. Is the set of ordered pairs $\{(x,y) | y = x^2, x = 1, 2, 4, 8\}$ a function?

12. Is the set of ordered pairs $\{(x,y) | y^2 = x^2, x = 1, 4, 9, 16\}$ a function?

13. If $f(x) = 3x + 1$, find $f(−1), f(0)$, and $f(3)$.

14. If $f(x) = 2 − 5x$, find $f(−2), f(1)$, and $f(2)$.

15. If $f(x) = \sqrt{5x + 4}$, find $f(0), f(1)$, and $f(9)$.

16. If $f(x) = 2x^2 − 7$, find $f(−2), f(0)$, and $f(3)$.

Find the range for the given function f and the specified domain.

17. $f = \{(x, 2x − 1) | x = −2, −1, 1, 3\}$

18. $f = \{(x, 3x + 2) | x = −3, −1, 0, 2\}$

178 | Chapter 5: Relations, Functions, and Graphs

19 $f = \{(x, \sqrt{17 - x}) | x = 1, 8, 16, -8\}$

20 $f = \{(x, \sqrt{x^2 - 11}) | x = \sqrt{47}, 6, 3\sqrt{3}, 2\sqrt{5}\}$

Write out the ordered pairs determined and state whether they are a function.

21 $\{(x, 4x - 3) | x = -1, 0, 1, 2\}$

22 $\{(x, 1 - 3x) | x = -2, -1, 0, 1\}$

23 $\{(x, \sqrt{26 - x}) | x = 1, 10, 17, 22\}$

24 $\{(x, x^2 - 5) | x = -4, -1, 2, 5\}$

25 $\{(x,y) | y = 2x + 3, D = \{-2, 0, 2, 3\}\}$

26 $\{(x,y) | y = 5 - 2x, D = \{-2, -1, 1, 4\}\}$

27 $\{(x,y) | y = x^2 - x + 1, D = \{-3, -1, 1, 3\}\}$

28 $\{(x,y) | y = 2x^2 - 3x + 5, D = \{-2, 0, 2, 3\}\}$

Find $[f(x + h) - f(x)]/h$.

29 $f(x) = 2x - 1$

30 $f(x) = 3x + 2$

31 $f(x) = x^2 + 2x - 1$

32 $f(x) = x^2 + 4x - 3$

Find $f \cap g$.

33 $f = \{(x, x - 1)\}, g = \{(x, 2x + 1)\}, x = -2, 0, 1, 2$

34 $f = \{(x, 3x + 4)\}, g = \{(x, 2x + 3)\}, x = -2, -1, 1, 3$

35 $f = \{(x, 2x^2 + x + 1)\}, g = \{(x, x^2 - 2x - 1)\}, x = -3, -2, -1, 2$

36 $f = \{(x, 3x^2 - 3x + 5)\}, g = \{(x, 2x^2 + x + 2)\}, x = -3, -1, 1, 3$

5.5 THE RECTANGULAR COORDINATE SYSTEM

In Chap. 1 we explained a method for associating a real number with a point on the number line. In this section we shall discuss a device for associating an ordered pair of numbers with a point in a plane. The device is called the **rectangular coordinate system.**

In order to set up this system, we construct a vertical and a horizontal line in a plane and choose a suitable scale on each, as in Fig. 5.2. The two

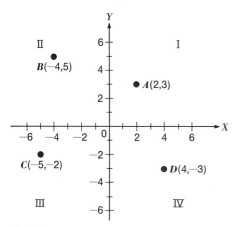

Fig. 5.2

perpendicular lines are called the **coordinate axes.** The horizontal line is the X axis, the vertical line is the Y axis, and their intersection is the **origin.** These lines divide the plane into four sections called **quadrants** that are numbered I, II, III, and IV counterclockwise, as indicated in Fig. 5.2. We agree that horizontal distances measured to the right from the Y axis are positive and those measured to the left are negative. Similarly, vertical distances measured upward from the X axis are positive, and those measured downward are negative. These distances are called **directed distances.** Finally, we agree that the first number in an ordered pair represents the directed distance from the Y axis to a point and the second number in the pair represents the directed distance from the X axis to the point. Consequently, an ordered pair of numbers uniquely determines the position of a point in the plane. For example, (2,3) determines a point that is 2 units to the right of the Y axis and 3 units above the X axis and thus locates the point A in Fig. 5.2. Similarly, $(-5, -2)$ determines the point C that is 5 units to the left of the Y axis and 2 units below the X axis. Conversely, each point in the plane determines a unique pair of numbers. For example, in Fig. 5.2 the point B is 4 units to the left of the Y axis and 5 units above the X axis and thus determines the pair $(-4,5)$. Similarly the point D determines the pair $(4,-3)$. The first number in an ordered pair that is associated with a point is called the **abscissa,** and the second number is the **ordinate.** The two numbers are called the **coordinates** of the point. The procedure for locating a point by means of its coordinates is called **plotting the point.** The notation $P(a,b)$ indicates the point whose coordinates are (a,b).

EXAMPLE 1 Plot the points $Q(4,-2)$, $R(-5,-6)$, $S(-5,3)$, and $T(6,4)$.

Solution We first construct the coordinate axes as shown in Fig. 5.3. Then, to plot $Q(4,-2)$, we count 4 units to the right of 0 on the X axis, and then from this point, we count 2 units downward and thus obtain the point $Q(4,-2)$. The positions of the other points are shown in the figure and are obtained in a similar manner.

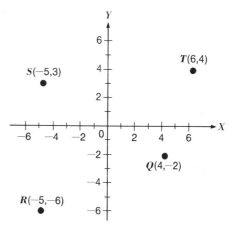

Fig. 5.3

5.6 THE GRAPH OF A FUNCTION AND A RELATION

By use of the rectangular coordinate system, we can obtain a geometric representation, or a geometric "picture," of a function or a relation. For this purpose, we require that each ordered pair (x,y) be the coordinates of a point in the plane, with x as the abscissa and y as the ordinate. Now we define the graph as follows:

The **graph** of a function (relation) is the totality of points (x,y) whose coordinates constitute the set of ordered pairs of the function (relation) with x a number in the domain D and y the corresponding number in the range R.

If the domain D of a function is the set of all real numbers, or if D is the set of all real numbers between two given limits, the graph is usually a straight line or a smooth continuous curve.* In some cases, however, the graph is a curve and a finite number of isolated points, and in others the graph is two or more disjoint curves. Furthermore, if a curve is the graph of a function, the coordinates of each point on the graph are the numbers in an ordered pair of a function, and the numbers in each pair of the function are the coordinates of a point on the curve. Hence, if $P(x,y)$ is a point on the graph, its coordinates (x,y) satisfy the equation $y = f(x)$.

We obtain the graph of a function by the following steps:

1. Find a sufficient number of pairs of the function to determine the nature and direction of the graph.
2. Plot the points determined by the pairs obtained in step 1.
3. Draw a smooth curve through the points obtained in step 2.

We shall illustrate the method with several examples.

EXAMPLE 1 Construct the graph of the function $\{(x,y) | y = 2x - 1\}$.

Solution We first assign the consecutive integers from -3 to 3 to x, calculate the corresponding values of y, and tabulate the associated number pairs as below:

x	-3	-2	-1	0	1	2	3
y	-7	-5	-3	-1	1	3	5

Next we plot the points as shown in Fig. 5.4. These points appear to lie along a straight line. In fact, they do, since it is proved in analytic geometry that the graph of the function determined by an

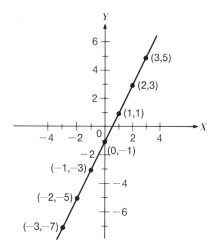

Fig. 5.4

equation of the type $y = ax + b$ is a straight line. Therefore we draw the line shown in the figure. Since the domain is the set of all real numbers, the line is of unlimited length, so the figure shows only a portion of the graph.

It is usually advisable first to assign zero and the first few consecutive positive integers to x, compute each corresponding value of y, and plot each point as its coordinates are found. Continue this process until either a definite trend for the curve is found or the coordinates become too large to be plotted in the space allotted. Repeat this procedure with negative integers.

EXAMPLE 2 Construct the graph of the function $\{(x,y) | y = \frac{1}{2}x^2 - 2\}$.

Solution We start by assigning the numbers 0, 1, 2, 3, 4, and 5 to x, and then we compute each value of y and get the pairs of corresponding numbers in the right side of the table below. For example, for $x = 4$, $y = \frac{1}{2}(4^2) - 2 = \frac{16}{2} - 2 = 8 - 2 = 6$. When these points are plotted, we get the points on and to the right of the Y axis in Fig. 5.5. Next, we assign the negative integers from -1 to -5 to x and proceed as above. Thus, we get the following table:

x	-5	-4	-3	-2	-1	0	1	2	3	4	5
y	10.5	6	2.5	0	-1.5	-2	-1.5	0	2.5	6	10.5

Finally, we draw a smooth curve through the points thus obtained and get the curve in Fig. 5.6. Since y increases very rapidly as x increases from 5 or decreases from -5, the curve extends upward and slightly to the right from (5,10.5), and upward and slightly to the left from $(-5,10.5)$.

---

* At present we are not in a position to give a rigorous definition of a smooth, continuous curve. For our purposes, however, the following description will suffice: a smooth, continuous curve contains no breaks or gaps, and there are no sudden or abrupt changes in its direction.

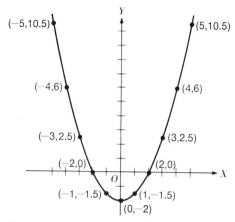

Fig. 5.5

EXAMPLE 3 Construct the graph of $\{(x,y)|y = x^3 - x - 1)\}$.

Solution We follow the above directions and obtain the following table of corresponding values:

x	-2	-1	0	1	2	3
y	-7	-1	-1	-1	5	23

We discard the point (3,23), because the ordinate is too large for the space allotted. Values of x greater than 3 and less than -2 also yield values of y that are numerically too large for our use. If we plot the points listed in the table, we obtain Fig. 5.6. Clearly, we need additional points in order to draw the curve.

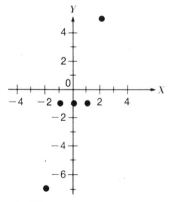

Fig. 5.6

Hence, we assign the values $-1\frac{1}{2}$, $-\frac{1}{2}$, $\frac{1}{2}$, and $1\frac{1}{2}$ to x and, after calculating y in each case, we obtain the following table:

x	$-1\frac{1}{2}$	$-\frac{1}{2}$	$\frac{1}{2}$	$1\frac{1}{2}$
y	$-2\frac{7}{8}$	$-\frac{5}{8}$	$-1\frac{3}{8}$	$\frac{7}{8}$

After these points and those previously obtained are plotted, we have a definite pattern and can draw the curve as shown in Fig. 5.7.

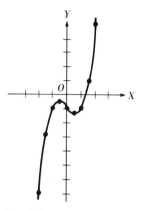

Fig. 5.7

EXAMPLE 4 Construct the graph of $\{(x,y)|y = 2\sqrt{25 - x^2}\}$.

Solution In this equation the domain of x is limited to values from -5 to 5, inclusive, since if x is greater than 5 or less than -5, the radicand is negative, and a real value of y does not exist. By using the integral values of x in this domain, we obtain the following table of corresponding values:

x	-5	-4	-3	-2	-1	0	1	2	3	4	5
y	0	6	8	9.2	9.8	10	9.8	9.2	8	6	0

We then plot the points and connect them with a smooth curve, and obtain Fig. 5.8.

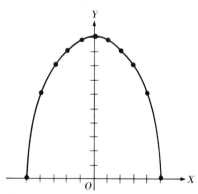

Fig. 5.8

EXAMPLE 5 Graph the relation $x = y^2 + 1$.

Solution A table of values is given below.

x	1	2	5	10
y	0	±1	±2	±3

The graph is shown in Fig. 5.9

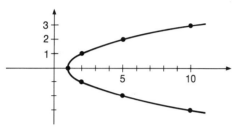

Fig. 5.9

EXAMPLE 6 Graph the relation $x^2 + 4y^2 = 36$.

Solution Without providing a table of values, we give the graph in Fig. 5.10.

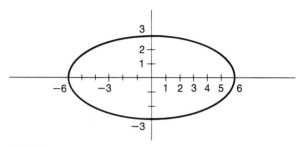

Fig. 5.10

Note that *in the graph of a function, every vertical line intersects the graph in at most one place.* With the graph of a relation, some vertical lines may intersect the graph in two or more points.

EXERCISE 5.3

Plot the points whose coordinates are given.

1. $(2,4), (3,-1), (-2,5), (-1,-2)$
2. $(-3,0), (-2,-3), (2,5), (4,-2)$
3. $(-3,-5), (1,6), (3,-2), (-2,3)$
4. $(4,-3), (0,3), (-3,-4), (-2,7)$

Sketch the lines described.

5. (a) The abscissa of each point is 4. (b) The ordinate of each point is -2.

6. The abscissa and ordinate of each point are equal and of the same sign.

7. The abscissa and ordinate of each point are numerically equal but of opposite signs.

8. (a) The abscissa and ordinate of each point are positive and equal. (b) The abscissa and ordinate of each point are numerically equal.

Construct the graph of the function defined by the equation.

9. $y = 2x + 1$
10. $y = 3x + 2$
11. $y = 0.5x - 3$
12. $y = 0.75x - 5$
13. $y = -3x + 1$
14. $y = -2x - 3$
15. $y = -4x + 5$
16. $y = -0.5x + 6$

17 $y = x^2$

18 $y = 2x^2 - x$

19 $y = 2x^2 + 5$

20 $y = 3x^2 + 2x - 4$

21 $y = -x^2 + 2$

22 $y = -x^2 - 3x$

23 $y = -3x^2 + 5x - 1$

24 $y = -2x^2 - 2x + 5$

25 $y = x^3 + x$

26 $y = x^3 + 3$

27 $y = x^3 - x^2 + 1$

28 $y = x^3 + 2x^2 - 3x$

29 $y = \sqrt{x}$

30 $y = \sqrt{2x + 5}$

31 $y = \sqrt{x^2 - x + 3}$

32 $y = \sqrt{2x^2 - 5x + 3}$

Graph the following relations.

33 $x = 2y^2 - 3$

34 $2x = 5y^2 + 3$

35 $x = y^2 + 2y + 3$

36 $x = 2y^2 - y + 4$

37 $x^2 + y^2 = 4$

38 $4x^2 + y^2 = 16$

39 $x^2 + 9y^2 = 81$

40 $4x^2 + 9y^2 = 36$

5.7 CHAPTER SUMMARY

We began this chapter with a section on formulas and charts and followed with work on areas and volumes of geometrical figures. We then defined and discussed the concept of a relation and followed that with a study of the special case of a relation that is known as a function. Finally we presented the rectangular coordinate system and showed how to sketch the graph of a function.

STUDENT'S NOTES

EXERCISE 5.4 REVIEW

1. The volume V of a pyramid of height h and square base of side s is given by $V = \frac{1}{3}s^2h$. Find the volume of a pyramid of height 27 in and square base of side 19 in.

2. The area of a trapezoid with bases b and B is given by $A = 0.5h(b + B)$, where h is the altitude. Find the area to the justifiable degree of accuracy if the altitude is measured to be 24.6 cm and the bases are calculated to be 42.3 and 84.7 cm after appropriate measurements.

3. The number of millions of pounds of wool produced in the United States for several years is shown in the table below. Show this information on a bar graph.

Year	Wool	Year	Wool
1940	434	1971	172
1950	249	1972	169
1960	299	1973	153
1965	225	1974	138
1970	177	1975	124

4. The world production of natural rubber in thousands of metric tons is given for several years in the table below. Show this information on a bar graph.

Year	Rubber	Year	Rubber
1940	1421	1965	2353
1945	276	1970	3103
1950	1854	1971	3085
1955	1924	1973	3513
1960	2002	1975	3293

5. Is $\{(2,1), (4,3), (6,5), (8,7)\}$ a function? Why?

6. Is $\{(2,1), (4,1), (6,5), (8,7)\}$ a function? Why?

7. Is $\{(2,1), (2,3), (6,5), (8,7)\}$ a function? Why?

8. Is the set of ordered pairs $\{(x, \sqrt{x}) | x = 0,1,4,9\}$ a function?

9. Is the set of ordered pairs $\{(x,y) | y^2 = x, x = 0, 1, 4, 9\}$ a function?

10. If $f(x) = 4x - 3$, find $f(-1)$, $f(0)$, and $f(2)$.

188 | Chapter 5: Relations, Functions, and Graphs

11 If $f(x) = \sqrt{11 - x}$, find $f(-5)$, $f(2)$, and $f(7)$.

12 Find the range if $f = \{(x, 3x - 2) | x = -2, 0, 3\}$.

13 Write out the ordered pairs determined by $\{(x, 4x - 1) | x = -1, 0, 1, 3\}$.

14 Write out the ordered pairs determined by $\{(x, x^2 + 2x + 3) | x = -3, 1, 4, 5\}$.

15 If $f(x) = 3x - 4$, find $[f(x + h) - f(x)]/h$.

16 If $f = \{(x, 2x^2 - 3x + 1)\}$, $x = -1, 0, 1, 2, 3$ and $g = \{(x, x^2 + x - 2)\}$, $x = -1, 0, 1, 2, 3$, find $f \cap g$ and $f \cup g$.

17 Plot the points $(2,3)$, $(3,-1)$, $(-4,5)$, and $(-2,-6)$.

18 Sketch the line on which the abscissa is two less than the ordinate for each point.

Construct the graph defined by the equation.

19 $y = 1.5x + 4$

20 $y = x^2 - 3x + 1$

21 $y = -\sqrt{3x + 1}$

22 $y = x^3 - 5x^2 - 4x + 2$

23 $x = -y^2 + 2y + 4$

24 $x^2 + 16y^2 = 64$

25 $|x| + |y| = 1$

NAME _____ DATE _____ SCORE _____

EXERCISE 5.5 CHAPTER TEST

1. The volume V of a cone of height h and radius r of the base is $V = \pi r^2 h/3$. Find the volume of a cone of altitude $17/\pi$ and base radius 6.

2. The total coal production in the United States in millions of tons for each of several years is given below. Show these data by use of a bar graph.

Year	Production	Year	Production
1945	633	1970	612
1950	566	1972	603
1955	491	1973	599
1960	434	1974	610
1965	527	1975	643

3. Is $\{(7,8), (8,9), (9,10), (10,11)\}$ a function? Why?

4. Is $\{(7,8), (8,9), (8,10), (10,11)\}$ a function? Why?

5. Is the set of ordered pairs $\{(x, -\sqrt{x}) \mid x = 0, 1, 4, 16\}$ a function? Why?

6 Is the set of ordered pairs $\{(x,y) | y = x^4, x = 0, 1, 16\}$ a function? Why?

7 If $f(x) = 5x - 6$, find $f(-1), f(1), f(2)$.

8 Write out the ordered pairs determined by $\{(x, 3x + 4)\}$ for $x = -2, 0, 1, 3$.

9 Sketch the graph of $y = 0.5x + 5$.

10 Sketch the graph of $y = x^2 - x + 2$.

11 Sketch the graph of $y = |x - 1| + |x - 3|$.

CHAPTER 6

Exponents and Radicals

6.1 POSITIVE INTEGRAL EXPONENTS

In Chap. 2 we defined nonnegative integral exponents and derived two laws for dealing with them in multiplication and division. In this chapter we shall extend the definition of an exponent to include negative and fractional exponents, and we shall do so in such a way that the previously derived laws will hold for the extended definition.

We next restate the definitions and laws previously given and illustrate how they are used:

$$a \cdot a \cdot a \cdot a \cdots a \text{ to } n \text{ } a\text{'s} = a^n \qquad (6.1)$$

where n is a positive integer. The expression a^n is called the nth power of a; the letter a is the **base** and n is the **exponent.**

$$a^0 = 1 \qquad a \neq 0 \qquad (6.2)$$

The laws for obtaining the product and quotient of two integral powers of the same variable are

$$a^m a^n = a^{m+n} \qquad \text{product} \qquad (6.3)$$

$$\frac{a^m}{a^n} = a^{m-n} \qquad \begin{array}{c} m > n \\ a \neq 0 \end{array} \qquad \text{quotient} \qquad (6.4)$$

We shall next develop laws for obtaining (1) the power of a product, (2) the power of a quotient, and (3) the power of a power.

1. The nth power of the product of the two numbers a and b is written $(ab)^n$. By (6.1)

$$(ab)^n = ab \cdot ab \cdot ab \cdots$$
$$\qquad \text{to } n \text{ factors each of which is } ab$$
$$= (a \cdot a \cdot a \cdots \text{to } n \text{ } a\text{'s})$$
$$\qquad \times (b \cdot b \cdot b \cdots \text{to } n \text{ } b\text{'s})$$
$$\qquad \text{by commutative axiom}$$
$$= a^n b^n$$

Hence, the power of a product is

$(ab)^n = a^n b^n$ (6.5)

2. The nth power of the quotient of the two numbers a and b is written $(a/b)^n$, provided that $b \neq 0$. By again using (6.1), we have

$$\left(\frac{a}{b}\right)^n = \frac{a}{b} \frac{a}{b} \frac{a}{b} \cdots$$

to n factors each of which is $\frac{a}{b}$

$$= \frac{a \cdot a \cdot a \cdots \text{ to } n \text{ } a\text{'s}}{b \cdot b \cdot b \cdots \text{ to } n \text{ } b\text{'s}}$$

by definition of product of fractions

$$= \frac{a^n}{b^n} \quad b \neq 0$$

Hence, the power of a quotient is

$$\left(\frac{a}{b}\right)^n = \frac{a^n}{b^n} \quad b \neq 0 \quad (6.6)$$

3. The nth power of a^m is written $(a^m)^n$. By the use of (6.1) we have

$(a^m)^n = a^m \cdot a^m \cdot a^m \cdots$ to n factors each of which is a^m

$= a^{m+m+m+\cdots}$ to n m's   by (6.3)
$= a^{mn}$

Hence, the power of a power is

$(a^m)^n = a^{mn}$ (6.7)

EXAMPLE 1

$y^5 \cdot y^3 = y^{5+3} = y^8$ by (6.3)

EXAMPLE 2

$z^2 \cdot z^7 \cdot z^4 = z^{2+7+4} = z^{13}$ by (6.3)

EXAMPLE 3

$\dfrac{x^9}{x^5} = x^{9-5} = x^4$ by (6.4)

EXAMPLE 4

$\dfrac{a^8 a^4}{a^3} = \dfrac{a^{12}}{a^3} = a^9$ by (6.3) and (6.4)

EXAMPLE 5

$(a^2 b^3)^4 = (a^2)^4 (b^3)^4$ by (6.5)
$= a^8 b^{12}$ by (6.7)

EXAMPLE 6

$\left(\dfrac{x^4}{y^5}\right)^2 = \dfrac{(x^4)^2}{(y^5)^2}$ by (6.6)

$= \dfrac{x^8}{y^{10}}$ by (6.7)

In order to obtain the product of two monomials involving powers of the same variable we *multiply* the coefficients and *add* the exponents.

EXAMPLE 7 Simplify $(3a^2 b^4)(5a^3 b^6)$.

Solution

$(3a^2 b^4)(5a^3 b^6) = 3(5)(a^2)(a^3)(b^4)(b^6)$
                       by commutative axiom
$= 15 a^5 b^{10}$ by (6.3)

In order to simplify an expression involving positive integral exponents, we perform all possible combinations by use of laws (6.2) to (6.7).

EXAMPLE 8 Simplify $(3x^3 y^4)(4x^5 y^2)$.

Solution

$(3x^3 y^4)(4x^5 y^2) = 3(4)(x^3)(x^5)(y^4)(y^2)$
                        by commutative axiom
$= 12 x^8 y^6$ by (6.3)

EXAMPLE 9 Simplify

$$\frac{12 a^7 b^5 c^2}{8 a^3 b^5 c^6}$$

Solution

$$\frac{12 a^7 b^5 c^2}{8 a^3 b^5 c^6} = \frac{12}{8} \frac{a^7}{a^3} \frac{b^5}{b^5} \frac{c^2}{c^6}$$

$$= \frac{3}{2} a^{7-3} (b^{5-5}) \frac{c^{2-2}}{c^{6-2}} \quad \text{by (6.4)}$$

$$= \frac{3}{2} (a^4)(b^0) \frac{c^0}{c^4}$$

$$= \frac{3 a^4}{2 c^4} \quad \text{by (6.2) and } b^0 = c^0 = 1$$

EXAMPLE 10 Simplify

$$\left(\frac{4x^3 y^2}{5z^4}\right)^3$$

Solution

$$\left(\frac{4x^3 y^2}{5z^4}\right)^3 = \frac{(4x^3 y^2)^3}{(5z^4)^3} \quad \text{by (6.6)}$$

$$= \frac{4^3 (x^3)^3 (y^2)^3}{5^3 (z^4)^3} \quad \text{by (6.5)}$$

$$= \frac{64 x^9 y^6}{125 z^{12}} \quad \text{by (6.7)}$$

EXAMPLE 11 Simplify

$$\left(\frac{4a^3 b^4}{3c^2 d^3}\right)^4 \left(\frac{9c^5 d}{2a^2 b^5}\right)^2$$

Solution

$$\left(\frac{4a^3b^4}{3c^2d^3}\right)^4 \left(\frac{9c^5d}{2a^2b^5}\right)^2$$

$$= \frac{4^4(a^3)^4(b^4)^4}{3^4(c^2)^4(d^3)^4} \frac{9^2(c^5)^2(d)^2}{2^2(a^2)^2(b^5)^2} \quad \text{by (6.5) and (6.6)}$$

$$= \frac{256a^{12}b^{16}}{81c^8d^{12}} \frac{81c^{10}d^2}{4a^4b^{10}} \quad \text{by (6.7)}$$

$$= \frac{256}{81} \frac{81}{4} \frac{a^{12}}{a^4} \frac{b^{16}}{b^{10}} \frac{c^{10}}{c^8} \frac{d^2}{d^{12}} \quad \text{by commutative axiom}$$

$$= \frac{64a^8b^6c^2}{d^{10}} \quad \text{by (6.4)}$$

Note: In each of the above examples every step in the simplification process is shown. After some practice, you should be able to perform two or more of these steps simultaneously and thereby abbreviate the process.

EXAMPLE 12 Simplify

$$\left(\frac{2a^2b^3}{3cd^4}\right)^3$$

Solution

$$\left(\frac{2a^2b^3}{3cd^4}\right)^3 = \frac{8a^6b^9}{27c^3d^{12}} \quad \text{by (6.5) to (6.7)}$$

STUDENT'S NOTES

EXERCISE 6.1

Circle the correct expressions.

1. $4a^3(3a^4) =$ $7a^7$ $7a^{12}$ $12a^7$ $12a^{12}$

2. $2a^2(3a^5) =$ $6a^7$ $5a^7$ $5a^{10}$ $6a^{10}$

3. $\dfrac{8a^4}{2a} =$ $4a^4$ $4a^3$ $6a^3$ $6a^4$

4. $\dfrac{9a^6}{3a^2} =$ $3a^3$ $3a^4$ $6a^4$ $6a^3$

5. $(2a^2b^3)^3 =$ $8a^5b^9$ $8a^6b^6$ $6a^6b^9$ $8a^6b^9$

6. $(3a^4b^2)^2 =$ $6a^6b^4$ $9a^6b^4$ $9a^8b^4$ $5a^8b^4$

7. $\left(\dfrac{12a^3}{3a}\right)^3 =$ $12a^4$ $7a^7$ $12a^8$ $64a^6$

8. $\left(\dfrac{8a^5}{2a^3}\right)^2 =$ $6a^6$ $16a^4$ $16a^5$ $6a^5$

The table shows the values of powers of 2 and 3 through the twelfth. Use it to solve the following problems.

n	1	2	3	4	5	6	7	8	9	10	11	12
2^n	2	4	8	16	32	64	128	256	512	1,024	2,048	4,096
3^n	3	9	27	81	243	729	2,187	6,561	19,683	59,049	177,147	531,441

9. $32(64) = 2^5(2^6) = 2^{11} =$

10. $16(256) =$

11. $32(128) =$

12. $8(512) =$

13 $27(6561) =$

14 $81(243) =$

15 $243(729) =$

16 $729(27) =$

17 $\dfrac{4096}{512} =$

18 $\dfrac{2048}{64} =$

19 $\dfrac{531{,}441}{19{,}683} =$

20 $\dfrac{177{,}147}{2187} =$

Simplify the following expressions.

21 $3^2 3^3 =$ **22** $5(5^2) =$

23 $6^0 6^3 =$ **24** $2^5 2^4 =$

25 $\dfrac{5^7}{5^2} =$ **26** $\dfrac{7^8}{7^6} =$

27 $\dfrac{4^5}{4^3} =$ **28** $\dfrac{6^6}{6^3} =$

29 $(2^2)^2 =$ **30** $(3^3)^3 =$

31 $(7^1)^4 =$ **32** $(4^3)^0 =$

33. $(3^2 2^3)^2 =$

34. $(2^4 4^2)^2 =$

35. $(2^3 5^2)^0 =$

36. $(5^2 2^3)^3 =$

37. $\left(\dfrac{3^2}{2^3}\right)^1 =$

38. $\left(\dfrac{2^0}{5^2}\right)^3 =$

39. $\left(\dfrac{7^2}{2^4}\right)^2 =$

40. $\left(\dfrac{3^1}{2^0}\right)^0 =$

41. $\left(\dfrac{2a^2 b^3}{c}\right)^3 =$

42. $\left(\dfrac{3ab^4}{c^2}\right)^2 =$

43. $\left(\dfrac{5^0 a^3 b}{c^3}\right)^4 =$

44. $\left(\dfrac{2^2 a^4 b^0}{c^3}\right)^3 =$

45. $(3a^2 b^3)2a^3 =$

46. $(2a^5 b^0)5b^2 =$

47. $5a^3(3a^2 b^4) =$

48. $2b^2(7a^3 b) =$

49. $\dfrac{15b^3 c^5}{3b^2 c^3} =$

50. $\dfrac{18c^4 d}{6c^0 d^3} =$

51. $\dfrac{24b^4 d^5}{8b^5 d} =$

52. $\dfrac{12a^6 b^3}{6a^5 b^4} =$

53. $2ab^2 c^0 (a^2 b)^2 =$

54. $3a^2 xy^3 (ax^2)^3 =$

55. $a^3 b(ab^2 c^3)^2 =$

56. $a^0 c^2 (ab^3 c^2)^3 =$

57. $\dfrac{(2w^2 xy^3)^2}{(3wxy^2)^3} =$

58. $\dfrac{(3w^3 x^0 y)^4}{(2w^0 x^2 y)^5} =$

59 $\dfrac{(12w^0xy^2)^3}{(6w^2xy^0)^5} =$

60 $\dfrac{(15w^3x^2y^4)^3}{(5w^5x^4y^5)^2} =$

61 $\left(\dfrac{w^3x^3}{y^4}\right)^2 \left(\dfrac{y^3}{wx^2}\right)^3 =$

62 $\left(\dfrac{3x^3y^2}{w^3}\right)^3 \left(\dfrac{w^2}{x^3y}\right)^2 =$

63 $\left(\dfrac{6a^2b}{c^3}\right)^2 \left(\dfrac{c^2}{3ab^2}\right)^3 =$

64 $\left(\dfrac{8a^3}{b^2c^4}\right)^3 \left(\dfrac{bc^3}{4a^2}\right)^4 =$

65 $\left(\dfrac{x^2y^3}{w^4}\right)^3 \div \left(\dfrac{xy^2}{w^3}\right)^4 =$

66 $\left(\dfrac{x^2y^5}{w^3}\right)^3 \div \left(\dfrac{x^3y^2}{w^2}\right)^2 =$

67 $\left(\dfrac{x^4y^3}{w^3}\right)^3 \div \left(\dfrac{x^3y^4}{w}\right)^2 =$

68 $\left(\dfrac{x^5y^0}{w}\right)^3 \div \left(\dfrac{x^3y}{w^0}\right)^5 =$

69 $\left[\left(\dfrac{2x^4}{3y^2}\right)^4 \left(\dfrac{9y^2}{4x^2}\right)^3\right]^2 =$

70 $\left[\left(\dfrac{3x}{2y^2}\right)^3 \left(\dfrac{4y}{9x^2}\right)^2\right]^3 =$

71 $\left[\left(\dfrac{2x^4}{3y^2}\right)^3 \left(\dfrac{3y^3}{x^5}\right)^2\right]^2 =$

72 $\left[\left(\dfrac{5x^3}{2y^2}\right)^4 \left(\dfrac{4y^4}{10x^5}\right)^2\right]^2 =$

6.2 NEGATIVE EXPONENTS

If we disregard the restriction $m > n$ in law (6.4) and apply the law to a^n/a^{2n}, we obtain

$$\frac{a^n}{a^{2n}} = a^{n-2n} = a^{-n} \tag{1}$$

and this gives a result a^{-n} that has no meaning according to our present definition of exponents. However, if we divide the numerator and denominator of the left member of (1) by a^n, we get

$$\frac{a^n}{a^{2n}} = \frac{1}{a^{2n-n}} = \frac{1}{a^n} \tag{2}$$

NEGATIVE EXPONENT Therefore, since the left members of (1) and (2) are the same, it is logical to define a^{-n}, for $n > 0$, as

$$a^{-n} = \frac{1}{a^n} \quad \text{provided that } a \neq 0 \tag{6.8}$$

EXAMPLE 1

$$3^{-2} = \frac{1}{3^2} = \frac{1}{9} \qquad \left(\frac{1}{2}\right)^{-3} = \frac{1}{(\frac{1}{2})^3} = \frac{1}{\frac{1}{8}} = 8$$

We can prove that laws (6.2) to (6.7) hold for this definition of exponents, and we shall indicate the method by using laws (6.3) and (6.4). In law (6.3) we shall assume that $a \neq 0$ and shall replace m and n by $-x$ and $-y$, respectively, and prove below that $a^{-x}a^{-y} = a^{-x-y}$.

$$a^{-x}a^{-y} = \frac{1}{a^x} \frac{1}{a^y} \qquad \text{by (6.8)}$$

$$= \frac{1}{a^x a^y} = \frac{1}{a^{x+y}} \qquad \text{by (6.3)}$$

$$= a^{-(x+y)} \qquad \text{by (6.8)}$$

$$= a^{-x-y}$$

We shall next replace n by $-x$ in law (6.5) and show that $(ab)^{-x} = a^{-x}b^{-x}$, provided that neither a nor b is zero.

$$(ab)^{-x} = \frac{1}{(ab)^x} \qquad \text{by (6.8)}$$

$$= \frac{1}{a^x b^x} \qquad \text{by (6.5)}$$

$$= \frac{1}{a^x} \frac{1}{b^x} = a^{-x}b^{-x} \qquad \text{by (6.8)}$$

It is frequently desirable to convert an expression involving negative exponents to an equal expression in which all exponents are positive. If the expression involves no fractions, we employ the definition of a negative exponent (6.8) for this purpose. For example, $ab^{-3} = a(1/b^3) = a/b^3$ and $xy^{-3} + x^{-2}y = x(1/y^3) + (1/x^2)(y) = x/y^3 + y/x^2$.

If an algebraic fraction involves negative exponents, we convert it to an equal fraction in which all exponents are positive by use of the fundamental principle of fractions. As an example of the process, we shall consider the fraction $x^n y^{-p}/z^{-r}w^t$. We first notice that by (6.3) and (6.2) $y^p y^{-p} = y^{p-p} = y^0 = 1$ and $z^{-r}z^r = z^{r-r} = z^0 = 1$. Hence, we multiply the given fraction by $y^p z^r / y^p z^r$ and get

$$\frac{x^n y^{-p}}{z^{-r} w^t} = \frac{x^n y^{-p}}{z^{-r} w^t} \frac{y^p z^r}{y^p z^r} = \frac{x^n y^{p-p} z^r}{z^{r-r} w^t y^p} = \frac{x^n y^0 z^r}{z^0 w^t y^p} = \frac{x^n z^r}{w^t y^p}$$

In the above fraction, no power of the same variable appeared in the numerator and denominator. If a power of the same variable does appear in both members of the fraction, we still use the procedure described above. However, some care should be exercised in choosing the product by which we multiply the numerator and denominator of the fraction. For example, in the fraction $a^{-4}b^{-1}c^{-5}/a^{-2}b^{-3}c^{-2}$ the negative powers which appear are a^{-2}, a^{-4}, b^{-1}, b^{-3}, c^{-2}, and c^{-5}. The negative powers of a, b, and c, however, which have exponents with the greater absolute value are a^{-4}, b^{-3}, and c^{-5}. Now we shall show that if we multiply the numerator and denominator of the fraction by $a^4 b^3 c^5$, no negative exponents will appear in the product:

$$\frac{a^{-4}b^{-1}c^{-5}}{a^{-2}b^{-3}c^{-2}} \frac{a^4 b^3 c^5}{a^4 b^3 c^5} = \frac{a^{-4+4}b^{-1+3}c^{-5+5}}{a^{-2+4}b^{-3+3}c^{-2+5}} = \frac{a^0 b^2 c^0}{a^2 b^0 c^3} = \frac{b^2}{a^2 c^3}$$

TO OBTAIN POSITIVE EXPONENTS The above example suggests the following procedure for converting a fraction in which the numerator or the denominator or both involve negative exponents to an equal fraction in which all exponents are positive.

1. Select the negative power of each number, or of each variable in which the exponent has the greater absolute value.
2. Form a product of the powers selected in step 1 with the sign of each exponent changed.
3. Multiply the numerator and denominator of the given fraction by the product obtained in step 2.
4. Simplify the result if possible.

EXAMPLE 2 Convert

$$\frac{2^{-3}a^{-2}bc^{-1}}{4^{-2}xy^{-3}z^4}$$

to an equal fraction that does not contain negative exponents.

Solution Here the same number or variable does not appear in both the numerator and denominator. Those with negative exponents, however, are 2^{-3}, 4^{-2}, a^{-2}, c^{-1}, and y^{-3}. Hence, we multiply the numerator and denominator of the fraction by $2^3 4^2 a^2 c y^3$ and get

$$\frac{2^{-3}a^{-2}bc^{-1}}{4^{-2}xy^{-3}z^4} = \frac{2^{-3}a^{-2}bc^{-1}}{4^{-2}xy^{-3}z^4} \cdot \frac{2^3 4^2 a^2 c y^3}{2^3 4^2 a^2 c y^3}$$

$$= \frac{2^{-3+3}4^2 a^{-2+2} bc^{-1+1}y^3}{2^3 4^{-2+2} a^2 c x y^{-3+3} z^4}$$

$$= \frac{2^0 4^2 a^0 bc^0 y^3}{2^3 4^0 a^2 c x y^0 z^4}$$

$$= \frac{16 b y^3}{8 a^2 c x z^4} \quad \text{since } 2^0 = 4^0 = a^0 = c^0$$
$$\quad = y^0 = 1$$

$$= \frac{2 b y^3}{a^2 c x z^4}$$

EXAMPLE 3 Express

$$\frac{2c^{-1}d^{-4}e^2}{3c^{-3}d^{-2}e^{-1}}$$

as an equal fraction without negative exponents.

Solution Here we have negative powers of c and d in both the numerator and denominator. We also have e^2 in the numerator and e^{-1} in the denominator. In the case of c^{-1} and c^{-3}, c^{-3} has the negative exponent with the greater absolute value since $|-3| > |-1|$. Similarly, for d^{-4} and d^{-2}, d^{-4} has the exponent with the greater absolute value. Furthermore, for the pair of numbers e^2 and e^{-1}, the only one with a negative exponent is e^{-1}. Consequently, according to step 3, the product we shall use is $c^3 d^4 e$, and we multiply the numerator and denominator of the given fraction by it and get

$$\frac{2c^{-1}d^{-4}e^2}{3c^{-3}d^{-2}e^{-1}} = \frac{2c^{-1}d^{-4}e^2}{3c^{-3}d^{-2}e^{-1}} \cdot \frac{c^3 d^4 e}{c^3 d^4 e}$$

$$= \frac{2c^2 d^0 e^3}{3c^0 d^2 e^0}$$

$$= \frac{2c^2 e^3}{3d^2} \quad \text{since } d^0 = c^0 = e^0 = 1$$

EXAMPLE 4 Express

$$\left(\frac{a^{-2}b^3}{2^{-2}a^{-1}b^{-2}}\right)^{-3}$$

without negative exponents.

Solution Here we have two alternatives for the first step. We can first convert the fraction in the parentheses to an equal fraction in which the exponents are positive and then apply (6.5), or we can apply (6.5) first and then proceed as in the above example. We shall work the problem by each method.

Method 1 First multiply the fraction in parentheses by $2^2 a^2 b^2 / 2^2 a^2 b^2$ and get

$$\left(\frac{a^{-2}b^3}{2^{-2}a^{-1}b^{-2}}\right)^{-3} = \left(\frac{a^{-2}b^3}{2^{-2}a^{-1}b^{-2}} \cdot \frac{2^2 a^2 b^2}{2^2 a^2 b^2}\right)^{-3}$$

$$= \left(\frac{2^2 a^0 b^5}{2^0 a b^0}\right)^{-3} = \left(\frac{4b^5}{a}\right)^{-3}$$

$$= \frac{1}{\left(\frac{4b^5}{a}\right)^3} = \frac{1}{\frac{64 b^{15}}{a^3}} = \frac{a^3}{64 b^{15}}$$

Method 2 If we apply (6.5) first, we obtain

$$\left(\frac{a^{-2}b^3}{2^{-2}a^{-1}b^{-2}}\right)^{-3} = \frac{(a^{-2})^{-3}(b^3)^{-3}}{(2^{-2})^{-3}(a^{-1})^{-3}(b^{-2})^{-3}}$$

$$= \frac{a^6 b^{-9}}{2^6 a^3 b^6} = \frac{a^3 b^{-9}}{64 b^6}$$

$$= \frac{a^3 b^{-9}}{64 b^6} \cdot \frac{b^9}{b^9} = \frac{a^3}{64 b^{15}}$$

EXAMPLE 5 Express

$$\frac{4x^{-2} - 9y^{-2}}{3x + 2y}$$

as an equal fraction without negative exponents.

Solution In this problem, we multiply the fraction by $x^2 y^2 / x^2 y^2$ and obtain

$$\frac{4x^{-2} - 9y^{-2}}{3x + 2y} = \frac{4x^{-2} - 9y^{-2}}{3x + 2y} \cdot \frac{x^2 y^2}{x^2 y^2}$$

$$= \frac{4y^2 - 9x^2}{x^2 y^2 (3x + 2y)}$$

$$= \frac{(2y - 3x)(2y + 3x)}{x^2 y^2 (3x + 2y)} \quad \text{factoring}$$

$$= \frac{2y - 3x}{x^2 y^2} \quad \text{dividing numerator and denominator by } 2y + 3x$$

EXAMPLE 6 Convert

$$\frac{16a^{-4} - 9b^{-4}}{4a^{-2} - 3b^{-2}}$$

to an equal fraction without negative exponents.

Solution The negative powers that occur here are a^{-4}, b^{-4}, a^{-2}, and b^{-2}. Hence, if we multiply the fraction by a^4b^4/a^4b^4, we obtain an equal fraction in which all exponents are positive. Accordingly, we have

$$\frac{16a^{-4} - 9b^{-4}}{4a^{-2} - 3b^{-2}} = \frac{16a^{-4} - 9b^{-4}}{4a^{-2} - 3b^{-2}} \cdot \frac{a^4b^4}{a^4b^4}$$

$$= \frac{16b^4 - 9a^4}{4a^2b^4 - 3a^4b^2}$$

$$= \frac{(4b^2 + 3a^2)(4b^2 - 3a^2)}{a^2b^2(4b^2 - 3a^2)}$$

$$= \frac{4b^2 + 3a^2}{a^2b^2} \qquad \text{dividing numerator and denominator by } 4b^2 - 3a^2$$

STUDENT'S NOTES

EXERCISE 6.2

Circle the correct expression.

1. $a^{-2} =$ $-a^2$ $\dfrac{1}{a^2}$ $\dfrac{-1}{a^2}$

2. $3a^{-4} =$ $\dfrac{1}{3a^4}$ $-3a^4$ $\dfrac{3}{a^4}$

3. $(2a)^{-2} =$ $\dfrac{1}{4a^2}$ $\dfrac{4}{a^4}$ $\dfrac{4}{a^2}$

4. $3a^3b^{-1} =$ $-3a^3b$ $\dfrac{3a^3}{b}$ $\dfrac{b}{3a^3}$

5. $a^{-1} + b^{-1} =$ $\dfrac{1}{a+b}$ $\dfrac{-1}{a}\dfrac{-1}{b}$ $\dfrac{a+b}{ab}$

6. $\left(\dfrac{-3a^{-1}}{b^2}\right)^2 =$ $\dfrac{-9}{a^2b^4}$ $\dfrac{-6}{a^2b^4}$ $\dfrac{9}{a^2b^4}$

7. $\left(\dfrac{-3a^2}{b^{-1}}\right)^{-1} =$ $\dfrac{3b}{a^2}$ $\dfrac{-1}{3a^2b}$ $\dfrac{-1}{3ab^0}$

8. $\dfrac{3a^{-1}}{a^{-1}+b^{-1}} =$ $\dfrac{a+b}{3a}$ $\dfrac{3b}{b+a}$ $\dfrac{3(a+b)}{a}$

Find the value of each of the following expressions.

9. $2^{-3} =$ 10. $4^{-1} =$

11. $5^{-2} =$ 12. $3^{-4} =$

13. $\dfrac{6^{-5}}{6^{-2}} =$ 14. $\dfrac{5^{-4}}{5^{-1}} =$

15 $\dfrac{3^{-2}}{3^{-3}} =$

16 $\dfrac{2^{-1}}{2^{-4}} =$

17 $(2^{-1})^{-1} =$

18 $(3^{-2})^{-3} =$

19 $(7^{-2})^1 =$

20 $(6^3)^{-2} =$

21 $(2^{-1}3^0)^{-2} =$

22 $(3^{-2}2^2)^{-2} =$

23 $(2^2 3^{-1})^{-1} =$

24 $(3^{-3}2^4)^{-1} =$

25 $\left(\dfrac{3}{2}\right)^{-1} =$

26 $\left(\dfrac{2}{5}\right)^{-2} =$

27 $\left(\dfrac{2^{-1}}{3^2}\right)^{-2} =$

28 $\left(\dfrac{3^{-2}}{2^3}\right)^{-3} =$

By use of negative exponents, if necessary, write the following expressions without a denominator.

29 $\dfrac{a^3}{b^2} =$

30 $\dfrac{4a^2}{b^4} =$

31 $\dfrac{a^3}{a^4} =$

32 $\dfrac{b^{-2}}{b} =$

33 $\dfrac{5a^7 b^3}{c^{-2} d} =$

34 $\dfrac{x^6 y^5 w^2}{3^0 a^{-3} b^2} =$

35 $\dfrac{a^5 b^2 c^4}{2^{-2} f^{-1} g} =$

36 $\dfrac{a^{-1} b^{-2} c^0}{2^{-1} g^{-2} h^{-3} k} =$

Write the following expressions without negative exponents.

37 $2a^{-3}b^{-1} =$

38 $3m^{-1}n^2 =$

39 $5^{-1}a^{-2}b^3 =$

40 $4^0 b^{-3} c =$

41 $\dfrac{a^{-1}}{a^{-3}} =$

42 $\dfrac{b^{-2}}{b^{-6}} =$

43 $\dfrac{c^{-4}}{c^{-2}} =$

44 $\dfrac{d^{-5}}{d^{-3}} =$

45 $\dfrac{a^{-2}b^{-3}c^{-4}}{a^{-4}b^{-3}c^{-2}} =$

46 $\dfrac{a^{-3}b^0 c^{-7}}{a^{-5}b^{-2}c^{-6}} =$

47 $\dfrac{a^{-9}b^3 c^{-2}}{a^{-6}b^0 c^3} =$

48 $\dfrac{a^{-3}b^2 c^{-1}}{a^{-4}b^{-1}c^{-2}} =$

49 $\dfrac{2^{-1}x^{-3}y^2 w^{-2}}{3^{-2}x^{-4}y^0 w^3} =$

50 $\dfrac{6a^{-3}b^{-1}y}{3^0 a^{-5}b^2 y^{-2}} =$

51 $\dfrac{2^{-3}r^3 s^4 t^3}{3^{-2}r^{-2}s^0 t^{-5}} =$

52 $\dfrac{5m^{-3}n^2 p^{-5}}{2^{-1}mn^{-3}p^2} =$

206 | Chapter 6: Exponents and Radicals

53 $\left(\dfrac{a^{-4}b^3}{a^{-3}b^{-1}}\right)^{-2} =$

54 $\left(\dfrac{w^{-3}z^{-6}}{w^{-5}z^2}\right)^{-3} =$

55 $\left(\dfrac{3x^{-5}y^{-4}z^3}{7^0x^2y^3z^{-2}}\right)^{-3} =$

56 $\left(\dfrac{2^{-1}x^{-2}y^{-3}z^{-4}}{4^{-1}x^{-3}y^{-2}z^{-1}}\right)^{-2} =$

57 $\left(\dfrac{a^3b^{-2}}{3^{-1}a^{-2}b^{-1}}\right)^{-4} =$

58 $\left(\dfrac{a^{-1}b^3}{2^{-1}a^{-2}b^0}\right)^{-1} =$

59 $\left(\dfrac{3^{-1}a^{-2}b^0}{a^{-4}b^3}\right)^{-2} =$

60 $\left(\dfrac{2^{-2}a^{-2}b^{-1}}{4^{-1}a^{-1}b^{-2}}\right)^{-5} =$

61 $\dfrac{9a^{-2} - 16b^{-2}}{4a + 3b} =$

62 $\dfrac{x^{-2} - 25y^{-2}}{5x - y} =$

63 $\dfrac{2xy^{-1} - 1 - x^{-1}y}{y + 2x} =$

64 $\dfrac{6xy^{-1} - 11 - 10x^{-1}y}{2y + 3x} =$

6.3 ROOTS OF NUMBERS

In arithmetic we say the square roots of 4 are 2 and -2, since $2^2 = (-2)^2 = 4$; a cube root of 27 is 3, since $3^3 = 27$; and fourth roots of 16 are 2 and -2, since $2^4 = (-2)^4 = 16$. In general, we say

If n is a positive integer, an nth root of a is b if and only if $b^n = a$.

As the above definition implies, and as the square roots of 4 and fourth roots of 16 illustrate, there may be more than one nth root of a. However, it is customary to define one of these roots as the **principal nth root,** and this is done in the following way:

If a positive nth root of a exists, then it is the principal nth root; if no positive nth root of a exists but there is a negative nth root, then the negative nth root is the principal nth root of a.

The principal nth root of a is designated by $\sqrt[n]{a}$. If n is a positive integer, the expression $\sqrt[n]{a}$ is called a **radical** of order n, the letter a is the **radicand,** and n is called the **index** of the radical.*

There is no real number which is $\sqrt{-8}$, since -8 is negative but $x^2 \geq 0$ for every real number x. However, $\sqrt[3]{-8} = -2$ since $(-2)^3 = -8$.

It follows from the definition of the principal nth root above that $\sqrt{x^2}$ is positive or 0. We must thus have $\sqrt{x^2} = |x|$, not just x. However if x is positive, then $\sqrt{x^2} = x$. Also $\sqrt[4]{x^4} = x$ if x is positive, and similarly for any even index.

Regardless of whether x is positive or negative, $\sqrt[3]{x^3} = x, \sqrt[5]{x^5} = x$, and similarly if the index is odd.

POSITIVE RADICAND WITH EVEN INDEX For the above reasons, whenever we write $\sqrt[n]{x^n}$ or $\sqrt[n]{a}$ with n even, we shall assume that x or a is positive.

6.4 FRACTIONAL EXPONENTS

In this section we shall extend the definition of exponents so that fractional exponents will have a meaning and the laws of Sec. 6.1 will still hold. By the definition of the principal nth root of a, we have

$$(\sqrt[n]{a})^n = a \qquad (1)$$

If (6.7) holds with $m = 1/n$, we have

$$(a^{1/n})^n = a^{n/n} = a \qquad (2)$$

Now if we compare (1) and (2), we see that statement (2) is true if we define $a^{1/n}$ as follows:

$$a^{1/n} = \sqrt[n]{a} \qquad (6.9)$$

Furthermore, if (6.7) is to hold when $m = 1/p$ and $n = q$, we must have

$$(a^{1/p})^q = a^{q/p} \qquad (3)$$

By (6.9), however, $a^{1/p} = \sqrt[p]{a}$. Consequently, it follows that $a^{q/p} = (a^{1/p})^q = (\sqrt[p]{a})^q$. Hence, we define $a^{q/p}$ to be $(\sqrt[p]{a})^q$. We can prove that if $\sqrt[p]{a}$ is real,† $(\sqrt[p]{a})^q = \sqrt[p]{a^q}$. Therefore we have

$$a^{q/p} = (\sqrt[p]{a})^q = \sqrt[p]{a^q} \qquad \text{if } \sqrt[p]{a} \text{ is real} \qquad (6.10)$$

We call attention to the fact that (6.10) holds only if $\sqrt[p]{a}$ is a real number. In order to see that this restriction is necessary, we shall consider the case for $a = -4$ and $p = q = 2$. Since the square of any real number is positive, $\sqrt{-4}$ is not a real number. If we attempt to apply (6.10) for $a = -4$ and $p = q = 2$, however, we have $(\sqrt{-4})^2 = -4$, by the definition of a square root. But, $\sqrt{(-4)^2} = \sqrt{16} = 4$, by the definition of the principal square root. Consequently, $(\sqrt{-4})^2$ is not equal to $\sqrt{(-4)^2}$.

We can prove that all the laws (6.2) to (6.7) hold for the above interpretation of fractional exponents. However, the general proofs are rather tedious, and we shall indicate the method for a special case by using law (6.5) with a and b positive and $n = \frac{1}{2}$. We wish to prove that

$$(ab)^{1/2} = a^{1/2}b^{1/2} \qquad \begin{array}{l} a > 0 \\ b > 0 \end{array} \qquad (4)$$

We first note that the principal square roots of ab, a, and b are positive; hence, both members of (4) are positive. Furthermore,

$$[(ab)^{1/2}]^2 = ab \qquad (a^{1/2}b^{1/2})^2 = a^{2/2}b^{2/2} = ab$$

Hence, each of the two members of (4) is the principal square root of ab, and hence the two members are equal.

---

* If no index appears in a radical, as in $\sqrt{a}$, the index is understood to be 2.

† This restriction excludes the case in which a is negative and p is an even integer.

EXAMPLE 1 Find the value of each of the following: $8^{1/3}$, $4^{3/2}$, and $(\frac{1}{27})^{1/3} + 3^{-1} + 3^0$.

Solution
$$8^{1/3} = \sqrt[3]{8} = 2$$
$$4^{3/2} = (\sqrt{4})^3 = 2^3 = 8$$
$$\left(\frac{1}{27}\right)^{1/3} + 3^{-1} + 3^0 = \sqrt[3]{\frac{1}{27}} + \frac{1}{3} + 1$$
$$= \frac{1}{3} + \frac{1}{3} + 1 = 1\tfrac{2}{3}$$

EXAMPLE 2 Find the product of $2x^{1/3}$ and $3x^{3/4}$.

Solution
$$2x^{1/3}(3x^{3/4}) = 2(3)(x^{1/3})(x^{3/4}) = 6x^{(1/3)+(3/4)}$$
$$= 6x^{(4+9)/12} = 6x^{13/12}$$

EXAMPLE 3 Divide $2a^{3/4}$ by $4a^{1/6}$.

Solution
$$2a^{3/4} \div 4a^{1/6} = \tfrac{2}{4}a^{(3/4)-(1/6)} = \tfrac{1}{2}a^{(9-2)/12} = \tfrac{1}{2}a^{7/12}$$

EXAMPLE 4 Perform the operations indicated in $(4a^4b^{3/4}c^2/9a^2b^{1/4}c)^{1/2}$.

Solution
$$\left(\frac{4a^4b^{3/4}c^2}{9a^2b^{1/4}c}\right)^{1/2} = \left(\frac{4}{9}\frac{a^4}{a^2}\frac{b^{3/4}}{b^{1/4}}\frac{c^2}{c}\right)^{1/2}$$
$$= (\tfrac{4}{9}a^2b^{2/4}c)^{1/2}$$
$$= \tfrac{2}{3}ab^{1/4}c^{1/2} \qquad \text{by (6.5) to (6.7)}$$

EXAMPLE 5 Express $2a^{2/3}$ and $(3c^2d^3)^{3/4}$ in radical form.

Solution
$$2a^{2/3} = 2\sqrt[3]{a^2} \qquad \text{by (6.10)}$$
$$(3c^2d^3)^{3/4} = \sqrt[4]{3^3(c^2)^3(d^3)^3} \qquad \text{by (6.10) and (6.5)}$$
$$= \sqrt[4]{27c^6d^9} \qquad \text{by (6.7)}$$

EXAMPLE 6 Express $\sqrt{3a^3}$ and $2\sqrt[4]{81a^8b^3c^2}$ without radicals.

Solution
$$\sqrt{3a^3} = 3^{1/2}a^{3/2}$$
$$2\sqrt[4]{81a^8b^3c^2} = 2(81^{1/4}a^{8/4}b^{3/4}c^{2/4})$$
$$= 2(3a^2b^{3/4}c^{1/2})$$
$$= 6a^2b^{3/4}c^{1/2}$$

An expression involving exponents is said to be **simplified** if all combinations using laws (6.2) to (6.7) have been made and the result is expressed without zero or negative exponents.

EXAMPLE 7 Simplify

$$\left(\frac{4a^2b^{3/4}c^{1/6}}{32a^{-1}b^0c^{-5/6}}\right)^{1/3}$$

Solution We first convert the fraction inside the parentheses to an equal fraction in which all exponents are positive by multiplying by $ac^{5/6}/ac^{5/6}$ and then proceed as indicated.

$$\left(\frac{4a^2b^{3/4}c^{1/6}}{32a^{-1}b^0c^{-5/6}}\right)^{1/3} = \left(\frac{4a^2b^{3/4}c^{1/6}}{32a^{-1}b^0c^{-5/6}}\frac{ac^{5/6}}{ac^{5/6}}\right)^{1/3}$$
$$= \left(\frac{4a^3b^{3/4}c^{1/6+5/6}}{32a^{-1+1}b^0c^{-(5/6)+(5/6)}}\right)^{1/3}$$
$$= \left(\frac{4a^3b^{3/4}c^{6/6}}{32a^0b^0c^0}\right)^{1/3} = \left(\frac{4a^3b^{3/4}c}{32}\right)^{1/3}$$
$$= (\tfrac{1}{8}a^3b^{3/4}c)^{1/3}$$
$$= (\tfrac{1}{8})^{1/3}(a^3)^{1/3}(b^{3/4})^{1/3}(c)^{1/3}$$
$$= \tfrac{1}{2}ab^{1/4}c^{1/3}$$

EXERCISE 6.3

Circle the correct expression.

1 $(36a^4b^{16})^{1/2} =$ $-6a^2b^8$ $6a^2b^8$ $18a^2b^8$

2 $(27a^6b^9)^{1/3} =$ $9a^2b^3$ $-3a^2b^3$ $3a^2b^3$

3 $\sqrt{4x^3y^4} =$ $2x^{3/2}y^2$ $x^{3/2}y^2$ $4x^{3/2}y^2$

4 $\sqrt[3]{27x^3y^9} =$ $9xy^3$ $-3xy^3$ $3xy^3$

5 $2a^{-1/2}b^{1/2} =$ $\dfrac{2b^{1/2}}{a^{1/2}}$ $2ab$ $\dfrac{2a^{1/2}}{b^{1/2}}$

6 $8^{1/3}a^{2/3}b^{-1/3} =$ $\dfrac{2a}{3b}$ $\dfrac{2a^{2/3}}{b^{1/3}}$ $\dfrac{8a^{2/3}}{3b^{1/3}}$

7 $(a^2 - b^2)^{1/2} =$ $\dfrac{a^2 - b^2}{2}$ $a - b$ $\sqrt{a^2 - b^2}$

8 $(a^3 + b^3)^{1/3} =$ $a + b$ $\dfrac{a^3 + b^3}{3}$ $\sqrt[3]{a^3 + b^3}$

Evaluate the following expressions without exponents or radicals.

9 $16^{1/2} =$

10 $16^{1/4} =$

11 $0.25^{1/2} =$

12 $0.008^{1/3} =$

13 $32^{2/5} =$

14 $9^{3/2} =$

15 $8^{2/3} =$

16 $64^{5/6} =$

Write the following expressions without radicals or fractional exponents.

17 $\sqrt{64} =$

18 $\sqrt[3]{64} =$

19 $\sqrt{0.01} =$

20 $\sqrt[3]{27^2} =$

21 $\sqrt[3]{8a^9} =$

22 $\sqrt[5]{32a^{10}} =$

23 $\sqrt[4]{81a^8} =$

24 $\sqrt{81a^8} =$

25 $\sqrt{25x^2y^4} =$

26 $\sqrt[3]{27x^6y^9} =$

27 $\sqrt[5]{243x^0y^{10}} =$

28 $\sqrt[4]{256a^8y^{12}} =$

29 $\sqrt[3]{\dfrac{8x^3}{27y^6}} =$

30 $\sqrt[4]{\dfrac{81x^8}{16y^{12}}} =$

31 $\sqrt{\dfrac{36x^4}{49y^{10}}} =$

32 $\sqrt[3]{\dfrac{125x^6}{27y^{12}}} =$

Write the following expressions by use of radicals.

33 $x^{2/3} =$

34 $y^{4/5} =$

35 $a^{2/5} =$

36 $b^{3/4} =$

37 $a^{-2/3} =$

38 $b^{-3/7} =$

39 $c^{-1/2} =$

40 $d^{-3/5} =$

41 $x^{2/5}x^{1/5} =$

42 $x^{-1/3}x^{2/3} =$

43 $y^{5/7}y^{-2/7} =$

44 $y^{4/5}y^{-2/5} =$

45 $8^{2/3}a^{1/3}b^{2/3} =$

46 $81^{3/4}a^{2/4}b^{1/4} =$

47 $9^{1/2}x^{3/2}y^{1/2} =$

48 $32^{3/5}a^{2/5}b^{4/5} =$

49 $x^{2/3}y^{1/4} = x^{8/12}y^{3/12} =$

50 $x^{1/3}y^{1/2} =$

51 $x^{1/3}y^{5/6}w^{2/9} =$

52 $x^{1/2}y^{3/4}w^{5/6} =$

Simplify the following expressions.

53 $3a^{1/3}y^{2/3}(4a^{-1/3}y^{1/3}) =$

54 $2a^{2/3}b^{-4/5}(5a^{-1/3}b^{2/5}) =$

55 $3x^{-4/7}y^{3/4}(4x^{5/7}y^{1/4}) =$

56 $2a^{-5/7}b^{2/3}(5a^{2/7}b^{-1/3}) =$

57 $\dfrac{12a^{4/5}b^{-2/3}c^{-2}}{6a^{2/5}b^{-1/3}c^0} =$

58 $\dfrac{30x^{6/7}y^{3/4}w^{-2/3}}{25x^{-1/7}y^{-1/4}w^{1/3}} =$

59 $\dfrac{6^{-1/2}xy^{-2/5}w^{3/2}}{36^{1/4}x^{-1/2}y^{3/5}w^{-1/2}} =$

60 $\dfrac{8^{1/3}x^{-2/3}y^{3/5}w^{-1/3}}{64^{1/6}x^{1/3}y^{-2/5}w^{1/3}} =$

61 $\left(\dfrac{27x^6}{64y^{3/5}}\right)^{1/3} =$

62 $\left(\dfrac{81y^4}{49w^{-4/7}}\right)^{-1/2} =$

63 $\left(\dfrac{243^{1/5}x^{-1/5}y^{2/5}}{3x^{2/5}y^{4/5}}\right)^5 =$

64 $\left(\dfrac{16^{1/4}x^{3/5}y^{2/3}}{625^{1/4}x^{-1/5}y^{-1/3}}\right)^{-5} =$

6.5 SIMPLIFICATION OF RADICALS

By use of (6.9) with $n = \frac{1}{2}$ and (6.5), we have
$\sqrt{128} = (128)^{1/2} = [64(2)]^{1/2} = 64^{1/2}2^{1/2} = \sqrt{64} \times \sqrt{2} = 8\sqrt{2}$. Similarly,

$$\begin{aligned}\sqrt[3]{54a^4b^7} &= [27(a^3)(b^6)(2ab)]^{1/3} \\ &= 27^{1/3}(a^3)^{1/3}(b^6)^{1/3}(2ab)^{1/3} \\ &= \sqrt[3]{27}\sqrt[3]{a^3}\sqrt[3]{b^6}\sqrt[3]{2ab} = 3ab^2\sqrt[3]{2ab}\end{aligned}$$

The above examples illustrate a law of radicals which we shall next derive.

The radical

$$\begin{aligned}\sqrt[n]{ab} &= (ab)^{1/n} &&\text{by (6.9)} \\ &= a^{1/n}b^{1/n} &&\text{by (6.5)} \\ &= \sqrt[n]{a}\sqrt[n]{b}\end{aligned}$$

Hence, we have

$$\sqrt[n]{ab} = \sqrt[n]{a}\sqrt[n]{b} \qquad (6.11)$$

If the radicand of a radical of order n has factors each of which is the nth power of a rational number or of a variable, we can simplify it by means of (6.11). The two examples at the beginning of this section and the following five examples illustrate the process.

EXAMPLE 1
$\sqrt{a^3} = \sqrt{a^2a} = \sqrt{a^2}\sqrt{a} = a\sqrt{a}$

EXAMPLE 2
$\sqrt{a^2b} = \sqrt{a^2}\sqrt{b} = a\sqrt{b}$

EXAMPLE 3
$\sqrt[3]{27a^6b^5} = \sqrt[3]{3^3}\sqrt[3]{(a^2)^3}\sqrt[3]{b^3b^2} = 3a^2b\sqrt[3]{b^2}$

EXAMPLE 4
$\sqrt{50} = \sqrt{5^2(2)} = 5\sqrt{2}$

EXAMPLE 5
$\sqrt[3]{81} = \sqrt[3]{3^3(3)} = 3\sqrt[3]{3}$

We can use (6.11) read from right to left to find the product of two or more radicals of the same order and for simplifying the result.

EXAMPLE 6
$$\begin{aligned}\sqrt{8a^3b}\sqrt{6a^2b^5} &= \sqrt{8a^3b(6a^2b^5)} &&\text{by (6.11)} \\ &= \sqrt{48a^5b^6} \\ &= \sqrt{4^2(3)(a^2)^2(a)(b^3)^2} \\ &= 4a^2b^3\sqrt{3a}\end{aligned}$$

A similar law for obtaining the quotient of two radicals of the same order is derived by use of (6.9) and (6.6). By (6.9) we have

$$\begin{aligned}\sqrt[n]{\frac{a}{b}} &= \left(\frac{a}{b}\right)^{1/n} \\ &= \frac{a^{1/n}}{b^{1/n}} &&\text{by (6.6)} \\ &= \frac{\sqrt[n]{a}}{\sqrt[n]{b}} &&\text{by (6.9)}\end{aligned}$$

Therefore,

$$\sqrt[n]{\frac{a}{b}} = \frac{\sqrt[n]{a}}{\sqrt[n]{b}} \qquad (6.12)$$

EXAMPLE 7
$$\begin{aligned}\frac{\sqrt{128x^3y^5}}{\sqrt{2xy^2}} &= \sqrt{\frac{128x^3y^5}{2xy^2}} &&\text{by (6.12)} \\ &= \sqrt{64x^2y^3} &&\text{simplifying radicand} \\ &= 8xy\sqrt{y}\end{aligned}$$

Problems 1 to 28 and 41 to 60 of Exercise 6.4 may now be worked.

6.6 RATIONALIZING MONOMIAL DENOMINATORS

In many situations a radical with a fractional radicand is easier to deal with if it is expressed in a form in which no radicals appear in the denominator. This process is called **rationalizing the denominator.**

In this section we shall consider the case in which the denominator of the radicand is a monomial and illustrate the rationalization process with three examples.

EXAMPLE 1 Rationalize the denominator of $\sqrt{3/2a}$.

Solution We first multiply the numerator and denominator of the radicand by a number that will change the radicand into a fraction whose denominator is a perfect square. If we use $2a$, we have

$$\begin{aligned}\sqrt{\frac{3}{2a}} &= \sqrt{\frac{6a}{4a^2}} &&\text{multiplying numerator and denominator of radicand by } 2a \\ &= \frac{\sqrt{6a}}{2a} &&\text{extracting square root of denominator}\end{aligned}$$

EXAMPLE 2 Rationalize the denominator of $\sqrt{3/8x^3y}$.

Solution The smallest perfect square that is a multiple of the denominator is $16x^4y^2$. Hence, we multiply the numerator and denominator of the radicand by $2xy$ and get

$$\sqrt{\frac{3}{8x^3y}} = \sqrt{\frac{6xy}{16x^4y^2}}$$
$$= \frac{\sqrt{6xy}}{4x^2y} \quad \text{extracting square root of denominator}$$

EXAMPLE 3 Rationalize the denominator of $\sqrt[3]{5d/6ab^5}$.

Solution In this problem we must convert the denominator into a perfect cube. The smallest perfect cube that is a multiple of $6ab^5$ is $216a^3b^6$. Hence, we multiply the numerator and denominator of the radicand by $36a^2b$ and obtain

$$\sqrt[3]{\frac{5d}{6ab^5}} = \sqrt[3]{\frac{180a^2bd}{216a^3b^6}}$$
$$= \frac{\sqrt[3]{180a^2bd}}{6ab^2} \quad \text{extracting cube root of denominator}$$

EXERCISE 6.4

Simplify the radical expression and rationalize all denominators.

1. $\sqrt{36} =$

2. $\sqrt{121} =$

3. $\sqrt{81} =$

4. $\sqrt{64} =$

5. $\sqrt[3]{27} =$

6. $\sqrt[4]{16} =$

7. $\sqrt[5]{3125} =$

8. $\sqrt[6]{4096} =$

9. $\sqrt{28} =$

10. $\sqrt{45} =$

11. $\sqrt[3]{24} =$

12. $\sqrt[5]{96} =$

13. $\sqrt[3]{54} =$

14. $\sqrt{162} =$

15. $\sqrt[4]{243} =$

16. $\sqrt[5]{128} =$

17. $\sqrt{3}\,\sqrt{27} =$

18. $\sqrt{128}\,\sqrt{8} =$

19. $\sqrt{18}\,\sqrt{8} =$

20. $\sqrt{5}\,\sqrt{45} =$

21. $\sqrt[3]{18}\,\sqrt[3]{12} =$

22. $\sqrt[3]{4}\,\sqrt[3]{128} =$

23. $\sqrt[4]{32}\,\sqrt[4]{8} =$

24. $\sqrt{27}\,\sqrt[5]{9} =$

25. $\sqrt{20}\,\sqrt{45} =$

26. $\sqrt{50}\,\sqrt{72} =$

27. $\sqrt[3]{32}\,\sqrt[3]{54} =$

28. $\sqrt[5]{36}\,\sqrt[5]{216} =$

29 $\dfrac{\sqrt{27}}{\sqrt{12}} =$

30 $\dfrac{\sqrt{48}}{\sqrt{27}} =$

31 $\dfrac{\sqrt[5]{972}}{\sqrt[5]{128}} =$

32 $\dfrac{\sqrt[3]{16}}{\sqrt[3]{54}} =$

33 $\sqrt{\dfrac{3}{7}} =$

34 $\sqrt{\dfrac{7}{11}} =$

35 $\sqrt{\dfrac{5}{32}} =$

36 $\sqrt{\dfrac{7}{18}} =$

37 $\sqrt[3]{\dfrac{2}{9}} =$

38 $\sqrt[3]{\dfrac{5}{4}} =$

39 $\sqrt[4]{\dfrac{17}{72}} =$

40 $\sqrt[5]{\dfrac{5}{162}} =$

41 $\sqrt{9a^2b^4} =$

42 $\sqrt{16a^6b^2} =$

43 $\sqrt[3]{8a^9b^6} =$

44 $\sqrt[4]{81a^8b^0} =$

45 $\sqrt{8a^4b^3} =$

46 $\sqrt{75a^6b^4} =$

47 $\sqrt[3]{54x^4y^3} =$

48 $\sqrt[3]{56x^6y^4} =$

49 $\sqrt[3]{x^9y^3w^4} =$

50 $\sqrt[3]{x^6y^8w^7} =$

51 $\sqrt[6]{x^7y^{14}w^9} =$

52 $\sqrt[7]{x^7y^{14}w^9} =$

53 $\sqrt{3x^2y}\ \sqrt{12x^4y^3} =$

54 $\sqrt{18x^6y^0}\ \sqrt{2x^4y^2} =$

55 $\sqrt{3x^3y^5}\ \sqrt{75xy^4} =$

56 $\sqrt{7x^4y^7}\ \sqrt{28x^3y} =$

57 $\sqrt[3]{16x^6y^3}\ \sqrt[3]{4x^4y^5w^6} =$

58 $\sqrt[5]{8x^4y^6}\ \sqrt[5]{4x^6y^2w^{10}} =$

59 $\sqrt[4]{3x^6y^4w^3}\ \sqrt[4]{27x^2y^5w^7} =$

60 $\sqrt[3]{48x^0yw^2}\ \sqrt[3]{108x^7y^4w^6} =$

61 $\dfrac{\sqrt{50a^7b^2}}{\sqrt{2a^3b^{-2}}} =$

62 $\dfrac{\sqrt{3a^4b^3c}}{\sqrt{27a^6b^{-1}c^3}} =$

63 $\dfrac{\sqrt{50a^{-2}b^{-1}c^0}}{\sqrt{18a^{-6}b^3c^2}} =$

64 $\dfrac{\sqrt{147a^{-4}b^{-7}c^3}}{\sqrt{48a^7b^{-3}c^2}} =$

65 $\sqrt{\dfrac{18a^3b^8c^5}{32a^4b^2c^3}} =$

66 $\sqrt{\dfrac{98a^6b^2w^6}{72a^3b^3w^3}} =$

67 $\sqrt{\dfrac{12a^3b^5c^4}{27a^7b^2c^3}} =$

68 $\sqrt{\dfrac{20a^7b^5c}{40a^3b^4c^5}} =$

69 $\sqrt[4]{\dfrac{48a^7b^3c^0}{162a^2b^5c^2}} =$

70 $\sqrt[5]{\dfrac{64a^9b^2c^6}{243a^4b^{-3}c^{11}}} =$

71 $\sqrt[3]{\dfrac{135a^4b^7c^2}{320a^6b^2c^{-1}}} =$

72 $\sqrt[3]{\dfrac{375a^7b^9c^{-2}}{192ab^2c^3}} =$

6.7 CHANGING THE ORDER OF A RADICAL

If it is possible to convert a radical into an equal radical with a smaller index, it is usually desirable to do so. This type of conversion is possible if the radical can be expressed as a power in which the exponent has a factor that is also a factor of the index of the radical. Thus, $\sqrt[10]{a^8}$ can be expressed as a radical with index less than 10 since 8 and 10 have a common factor. If it is possible to **reduce the order of a radical,** the reduction can be accomplished by writing the radical in fractional-exponent form, removing the common factors from numerator and denominator of the exponent, and then rewriting the expression in radical form. Thus

$$\sqrt[qn]{a^{pn}} = \sqrt[q]{a^p} \qquad (6.13)$$

since $\sqrt[qn]{a^{pn}} = a^{pn/qn} = a^{p/q} = \sqrt[q]{a^p}$.

EXAMPLE 1 Reduce $\sqrt[10]{a^8}$ to a radical of lower order.

Solution If we follow the procedure suggested above, we have

$$\sqrt[10]{a^8} = a^{8/10} = a^{4/5} = \sqrt[5]{a^4}$$

It is sometimes desirable to change the order of a radical so that the new index is an integer n times the original index. This can be done in essentially the same way as reducing the order of a radical. The two procedures differ only in that in one we remove a common factor from the terms of the fractional exponent and in the other we introduce a common factor.

EXAMPLE 2 Change $\sqrt[3]{a^2}$ to a radical of index 12.

Solution Since $\sqrt[3]{a^2} = a^{2/3}$ and we want a radical of index 12, we must multiply the denominator (and consequently the numerator) of the fractional exponent by 4. We thereby find that $a^{2/3} = a^{8/12} = \sqrt[12]{a^8}$ is the desired form.

By proper use of the procedure for changing the order of a radical we can express two radicals of different orders as radicals of the same order.

EXAMPLE 3 Change $\sqrt[3]{a}$ and $\sqrt[4]{a}$ to radicals of the same order.

Solution Since $\sqrt[3]{a} = a^{1/3}$ and $\sqrt[4]{a} = a^{1/4}$ and 3 and 4 are factors of 12, we shall express each of the given radicals as a radical of index 12. Thus we have

$$\sqrt[3]{a} = a^{1/3} = a^{4/12} = \sqrt[12]{a^4}$$
$$\sqrt[4]{a} = a^{1/4} = a^{3/12} = \sqrt[12]{a^3}$$

EXAMPLE 4 Convert $\sqrt{7ab}$, $\sqrt[3]{2a^2b}$, and $\sqrt[5]{3a^3b^4}$ to radicals of the same order.

Solution The indices of the radicals are 2, 3, and 5; their least common multiple is 30; hence, we shall convert each radical to an equal one of index 30. Thus, we get

$$\sqrt{7ab} = (7ab)^{1/2} = (7ab)^{15/30} = \sqrt[30]{7^{15}a^{15}b^{15}}$$
$$\sqrt[3]{2a^2b} = (2a^2b)^{1/3} = (2a^2b)^{10/30} = \sqrt[30]{2^{10}a^{20}b^{10}}$$
$$\sqrt[5]{3a^3b^4} = (3a^3b^4)^{1/5} = (3a^3b^4)^{6/30} = \sqrt[30]{3^6a^{18}b^{24}}$$

Problems 1 to 16 and 45 to 52 of Exercise 6.5 may be done now.

6.8 PRODUCTS OF POLYNOMIALS WHOSE TERMS CONTAIN RADICALS

The distributive law and the law of radicals (6.11) are used for obtaining the product of two polynomials whose terms involve radicals. We shall illustrate the method with two examples. In the first example the multiplier is a monomial, while in the second example, we obtain the product of two binomials.

EXAMPLE 1

$(\sqrt{2} + 2\sqrt{3} - \sqrt{5})\sqrt{6}$
$= \sqrt{2}\,\sqrt{6} + 2\sqrt{3}\,\sqrt{6} - \sqrt{5}\,\sqrt{6}$
  by distributive law
$= \sqrt{12} + 2\sqrt{18} - \sqrt{30}$ by (6.11)
$= 2\sqrt{3} + 6\sqrt{2} - \sqrt{30}$ by (6.11)

EXAMPLE 2 Obtain the product of $1 + 3\sqrt{2}$ and $2 - 5\sqrt{2}$.

Solution

$$1 + 3\sqrt{2}$$
$$2 - 5\sqrt{2}$$
$$\overline{2 + 6\sqrt{2}} \qquad \text{multiply by } 2$$
$$\underline{-5\sqrt{2} - 15\sqrt{4}} \qquad \text{multiply by } -5\sqrt{2}$$
$$2 + \sqrt{2} - 15\sqrt{4} = 2 + \sqrt{2} - 15(2)$$
$$= 2 + \sqrt{2} - 30$$
$$= -28 + \sqrt{2}$$

Problems 17 to 20 of Exercise 6.5 may be done now.

6.9 RATIONALIZING BINOMIAL DENOMINATORS

If we multiply the binomial $\sqrt{3} + \sqrt{2}$ by $\sqrt{3} - \sqrt{2}$, we obtain $(\sqrt{3})^2 - (\sqrt{2})^2 = 3 - 2 = 1$. Hence, if we multiply the two members of the fraction

$$\frac{\sqrt{3} - \sqrt{2}}{\sqrt{3} + \sqrt{2}}$$

by $\sqrt{3} - \sqrt{2}$, we get

$$\frac{\sqrt{3} - \sqrt{2}}{\sqrt{3} + \sqrt{2}} = \frac{(\sqrt{3} - \sqrt{2})(\sqrt{3} - \sqrt{2})}{(\sqrt{3} + \sqrt{2})(\sqrt{3} - \sqrt{2})}$$
$$= \frac{(\sqrt{3})^2 - 2\sqrt{3}\sqrt{2} + (\sqrt{2})^2}{(\sqrt{3})^2 - (\sqrt{2})^2}$$
$$= \frac{3 - 2\sqrt{6} + 2}{3 - 2} = \frac{5 - 2\sqrt{6}}{1}$$

and in the last fraction, the denominator is rational.

RATIONALIZING BINOMIAL DENOMINATORS The above example illustrates the following rule of procedure:

In order to rationalize the denominator of a fraction if it is a binomial that involves radicals of the second order, we multiply both members of the fraction by the binomial obtained by changing the sign between the terms of the denominator.

EXAMPLE 1

$$\frac{4}{\sqrt{5} - 1} = \frac{4(\sqrt{5} + 1)}{(\sqrt{5} - 1)(\sqrt{5} + 1)} = \frac{4(\sqrt{5} + 1)}{5 - 1}$$
$$= \frac{4(\sqrt{5} + 1)}{4} = \sqrt{5} + 1$$

EXAMPLE 2

$$\frac{2\sqrt{a} + \sqrt{b}}{\sqrt{a} - 2\sqrt{b}} = \frac{(2\sqrt{a} + \sqrt{b})(\sqrt{a} + 2\sqrt{b})}{(\sqrt{a} - 2\sqrt{b})(\sqrt{a} + 2\sqrt{b})}$$
$$= \frac{2a + 5\sqrt{ab} + 2b}{a - 4b}$$

Problems 21 to 28 and 53 to 56 of Exercise 6.5 may be done now.

6.10 ADDITION OF RADICALS

We obtain the sum of two or more radicals of the same order by first simplifying the radicals and then combining the coefficients of the radicals in which the radicands are the same.

EXAMPLE 1

$$\sqrt{8} + \sqrt{18} - \sqrt{32} = 2\sqrt{2} + 3\sqrt{2} - 4\sqrt{2}$$
$$= (2 + 3 - 4)\sqrt{2}$$
$$= \sqrt{2}$$

EXAMPLE 2

$$5\sqrt{3} + \sqrt{20} - \sqrt{27} - \sqrt{125}$$
$$= 5\sqrt{3} + 2\sqrt{5} - 3\sqrt{3} - 5\sqrt{5}$$
$$= (5 - 3)\sqrt{3} + (2 - 5)\sqrt{5}$$
$$= 2\sqrt{3} - 3\sqrt{5}$$

If the radicals appearing in a sum involve fractions, the denominators must be rationalized before combinations are attempted.

EXAMPLE 3 Simplify each term and then make all possible combinations in

$$\sqrt{\tfrac{1}{3}} + \sqrt{\tfrac{1}{8}} - 2\sqrt{27} + \sqrt{128}$$

Solution We rationalize the denominators of the first two terms and simplify the last two terms, obtaining

$$\sqrt{\tfrac{1}{3}} = \sqrt{\tfrac{3}{9}} = \frac{\sqrt{3}}{3}$$
$$\sqrt{\tfrac{1}{8}} = \sqrt{\tfrac{2}{16}} = \frac{\sqrt{2}}{4}$$
$$2\sqrt{27} = 2\sqrt{9(3)} = 6\sqrt{3}$$
$$\sqrt{128} = \sqrt{64(2)} = 8\sqrt{2}$$

Hence,

$$\sqrt{\frac{1}{3}} + \sqrt{\frac{1}{8}} - 2\sqrt{27} + \sqrt{128}$$
$$= \frac{\sqrt{3}}{3} + \frac{\sqrt{2}}{4} - 6\sqrt{3} + 8\sqrt{2}$$
$$= (\tfrac{1}{3} - 6)\sqrt{3} + (\tfrac{1}{4} + 8)\sqrt{2}$$
$$= -\tfrac{17}{3}\sqrt{3} + \tfrac{33}{4}\sqrt{2}$$

Problems 29 to 44 may be done now.

EXAMPLE 4

$$\frac{2\sqrt{a} - \sqrt{b}}{\sqrt{a} + \sqrt{b}} + \frac{\sqrt{a} + 2\sqrt{b}}{\sqrt{a} - \sqrt{b}}$$
$$= \frac{2\sqrt{a} - \sqrt{b}}{\sqrt{a} + \sqrt{b}} \cdot \frac{\sqrt{a} - \sqrt{b}}{\sqrt{a} - \sqrt{b}} + \frac{\sqrt{a} + 2\sqrt{b}}{\sqrt{a} - \sqrt{b}} \cdot \frac{\sqrt{a} + \sqrt{b}}{\sqrt{a} + \sqrt{b}}$$
$$= \frac{2a - 3\sqrt{ab} + b + a + 3\sqrt{ab} + 2b}{a - b}$$
$$= \frac{3(a + b)}{a - b}$$

Problems 57 to 64 may be done now.

STUDENT'S NOTES

EXERCISE 6.5

Reduce the order of the radical.

1 $\sqrt[6]{27} =$

2 $\sqrt[4]{9} =$

3 $\sqrt[8]{625} =$

4 $\sqrt[9]{8} =$

5 Change $\sqrt[5]{a^2}$ to a radical of order 10.

6 Change $\sqrt[4]{a^3}$ to a radical of order 8.

7 Change $\sqrt[3]{a}$ to a radical of order 9.

8 Change $\sqrt{a^3}$ to a radical of order 4.

Convert the radicals to radicals of the same order.

9 $\sqrt{a}, \sqrt[3]{a} =$

10 $\sqrt{a}, \sqrt[5]{a} =$

11 $\sqrt[3]{a}, \sqrt[6]{a} =$

12 $\sqrt[4]{a}, \sqrt[6]{a} =$

13 $\sqrt{5ab}, \sqrt[3]{2ab^2}, \sqrt[5]{3a^3b^2} =$

14 $\sqrt{3ab}, \sqrt[3]{3a^2b}, \sqrt[4]{3a^3b} =$

15 $\sqrt[3]{a^2b}, \sqrt[4]{ab^2}, \sqrt[5]{a^4b^3} =$

16 $\sqrt{ab}, \sqrt[3]{a^2b}, \sqrt[6]{a^4b^3} =$

Perform the multiplications and simplify the resulting radicals.

17 $(3\sqrt{2} + \sqrt{3} + \sqrt{5})\sqrt{6} =$

18 $(2\sqrt{2} + 3\sqrt{5} + \sqrt{3})\sqrt{10} =$

19 $(2 + 5\sqrt{2})(1 + 3\sqrt{2}) =$

20 $(\sqrt{2} - 2\sqrt{3})(3\sqrt{2} + \sqrt{3}) =$

Rationalize the denominator.

21 $\dfrac{2}{\sqrt{3} + 1} =$

22 $\dfrac{4}{\sqrt{5} - 1} =$

23 $\dfrac{\sqrt{7} - 2}{1 - \sqrt{7}} =$

24 $\dfrac{2\sqrt{3} + 5}{2 - \sqrt{3}} =$

25 $\dfrac{\sqrt{5} - \sqrt{3}}{\sqrt{5} + \sqrt{3}} =$

26 $\dfrac{2\sqrt{7} - \sqrt{2}}{\sqrt{7} - 3\sqrt{2}} =$

27 $\dfrac{\sqrt{5} + 3\sqrt{7}}{2\sqrt{5} - \sqrt{7}} =$

28 $\dfrac{2\sqrt{11} - 3\sqrt{10}}{3\sqrt{11} + 2\sqrt{10}} =$

Simplify the radicals and then perform the indicated additions and subtractions.

29 $\sqrt{8} - \sqrt{32} + \sqrt{18} =$

30 $\sqrt{12} + \sqrt{27} - \sqrt{75} =$

31 $\sqrt{20} - \sqrt{45} + \sqrt{245} =$

32 $\sqrt{54} + \sqrt{384} - \sqrt{150} =$

33 $\sqrt{12} - \sqrt{27} - \sqrt{8} + \sqrt{32} =$

34 $\sqrt{50} + \sqrt{180} + \sqrt{98} - \sqrt{20} =$

35 $\sqrt{24} + \sqrt{75} + \sqrt{54} - \sqrt{27} =$

36 $\sqrt{44} - \sqrt{112} - \sqrt{99} + \sqrt{63} =$

37 $\sqrt{\tfrac{1}{3}} + \sqrt{\tfrac{1}{32}} - 5\sqrt{27} + \sqrt{32} =$

38 $\sqrt{\tfrac{2}{3}} + \sqrt{\tfrac{4}{3}} + \sqrt{24} - \sqrt{27} =$

39 $\sqrt{\frac{2}{5}} + \sqrt{\frac{9}{5}} - \sqrt{40} - \frac{4}{5}\sqrt{20} =$

40 $\sqrt{\frac{1}{2}} + \sqrt{\frac{1}{3}} + \sqrt{72} - \sqrt{48} =$

41 $\sqrt[3]{16} + \sqrt{16} - \sqrt[3]{8} + \sqrt{8} =$

42 $\sqrt{147} + \sqrt[3]{192} + \sqrt{192} + \sqrt[3]{375} =$

43 $\sqrt[3]{128} - \sqrt{128} + \sqrt[3]{432} + \sqrt{98} =$

44 $\sqrt{12} + \sqrt[3]{81} + \sqrt{18} - \sqrt[3]{24} =$

Reduce the order of the radicals.

45 $\sqrt[4]{4a^2} =$ **46** $\sqrt[6]{8a^3} =$

47 $\sqrt[9]{27a^3b^6} =$ **48** $\sqrt[8]{16a^6b^4} =$

49 Change $\sqrt{2ab}$ to a radical of order 4.

50 Change $\sqrt[3]{4a^2b}$ to a radical of order 9.

51 Change $\sqrt[3]{5ab^2}$ to a radical of order 6.

52 Change $\sqrt[5]{2a^2b^3c^4}$ to a radical of order 10.

Rationalize the denominator.

53 $\dfrac{\sqrt{a} + 2\sqrt{b}}{2\sqrt{a} - \sqrt{b}} =$

54 $\dfrac{3\sqrt{a} - \sqrt{b}}{\sqrt{a} - 2\sqrt{b}} =$

55 $\dfrac{\sqrt{a} + 2\sqrt{b}}{3\sqrt{a} + 4\sqrt{b}} =$

56 $\dfrac{2\sqrt{a} + 3\sqrt{b}}{3\sqrt{a} - 2\sqrt{b}} =$

Perform the indicated additions and subtractions and simplify each sum.

57 $\dfrac{\sqrt{a} + \sqrt{b}}{\sqrt{a} + 2} + \dfrac{2\sqrt{a} - \sqrt{b}}{\sqrt{a} - 2} =$

58 $\dfrac{\sqrt{a} - 2\sqrt{b}}{3 - \sqrt{a}} + \dfrac{3\sqrt{a} + \sqrt{b}}{3 + \sqrt{a}} =$

59 $\dfrac{2\sqrt{a} + \sqrt{b}}{2 - \sqrt{a}} + \dfrac{2\sqrt{a} + \sqrt{b}}{2 + \sqrt{a}} =$

60 $\dfrac{3\sqrt{a} - \sqrt{b}}{1 - \sqrt{a}} - \dfrac{3\sqrt{a} - \sqrt{b}}{1 + \sqrt{a}} =$

61 $\dfrac{\sqrt{a} + \sqrt{b}}{\sqrt{a} - \sqrt{b}} + \dfrac{\sqrt{a} - \sqrt{b}}{\sqrt{a} + \sqrt{b}} =$

62 $\dfrac{2\sqrt{a} - \sqrt{b}}{\sqrt{a} + 2\sqrt{b}} - \dfrac{\sqrt{a} - 2\sqrt{b}}{2\sqrt{a} + \sqrt{b}} =$

63 $\dfrac{\sqrt{a} + \sqrt{b}}{2\sqrt{a} + 3\sqrt{b}} - \dfrac{3\sqrt{a} + 2\sqrt{b}}{\sqrt{a} - \sqrt{b}} =$

64 $\dfrac{3\sqrt{a} - 2\sqrt{b}}{2\sqrt{a} - 3\sqrt{b}} + \dfrac{2\sqrt{a} + 3\sqrt{b}}{3\sqrt{a} + 2\sqrt{b}} =$

6.11 COMPLEX NUMBERS

We have stated earlier that there is no real square root of a negative number. However, we shall see that numbers of the type $a + b\sqrt{-1}$, where a and b are real, frequently occur in the formal solution of quadratic equations, and in this section we shall explain briefly how the number system is extended to include such numbers.

If we apply the method of Sec. 6.5 to $\sqrt{-49}$, we get $\sqrt{-49} = \sqrt{7^2(-1)} = 7\sqrt{-1}$. Similarly, $\sqrt{-100} = 10\sqrt{-1}$, $\sqrt{-x^2} = x\sqrt{-1}$ if $x > 0$, and $\sqrt{-7} = \sqrt{7}\sqrt{-1}$. We shall now let $i = \sqrt{-1}$. Then $\sqrt{-49} = 7i$, $\sqrt{-100} = 10i$, $\sqrt{-x^2} = xi$ if $x > 0$, and $\sqrt{-7} = \sqrt{7}i$. Such numbers are called **pure imaginary numbers.**

If a pure imaginary number is formally added to a real number, we have a complex number which we define as follows:

A number of the type $a + bi$, where a and b are real and $i = \sqrt{-1}$, is called a **complex number.**

Since $i = \sqrt{-1}$, we have $i^2 = -1$, $i^3 = i \cdot i^2 = -i$, $i^4 = i^3 \cdot i = -i^2 = 1$, $i^5 = i$, $i^6 = -1$, and so on. Because of this cyclical property of powers of i, it is possible to express the sum, product, or quotient of two or more complex numbers in the form $a + bi$. Merely treat i as if it were a variable, and use $i^2 = -1$.

EXAMPLE 1

$2 + 4i + 3 - 2i = 2 + 3 + i(4 - 2) = 5 + 2i$

EXAMPLE 2

$5 + 7i - 3 - 9i = 5 - 3 + i(7 - 9) = 2 - 2i$

EXAMPLE 3

$(3 + 5i)(2 - 4i) = 6 - 12i + 10i - 20i^2$
$= 6 - 2i + 20 = 26 - 2i$

EXAMPLE 4

$(4 + 7i)(-3 + 2i) = -12 + 8i - 21i + 14i^2$
$= -12 - 13i - 14 = -26 - 13i$

EXAMPLE 5

$\dfrac{1 + i}{3 + 2i} = \dfrac{(1 + i)(3 - 2i)}{(3 + 2i)(3 - 2i)}$ multiplying numerator and denominator by $3 - 2i$

$= \dfrac{3 + i - 2i^2}{9 - 4i^2} = \dfrac{3 + i + 2}{9 + 4}$

$= \dfrac{5 + i}{13} = \dfrac{5}{13} + \dfrac{1}{13}i$

EXAMPLE 6

$\dfrac{2 + i}{1 - 2i} + \dfrac{3 + 4i}{2 + 3i}$

$= \dfrac{(2 + i)(2 + 3i) + (3 + 4i)(1 - 2i)}{(1 - 2i)(2 + 3i)}$

$= \dfrac{4 + 6i + 2i - 3 + 3 - 6i + 4i + 8}{2 + 3i - 4i + 6}$

$= \dfrac{12 + 6i}{8 - i}$

$= \dfrac{12 + 6i}{8 - i} \dfrac{8 + i}{8 + i}$

$= \dfrac{96 + 12i + 48i - 6}{64 + 1}$

$= \dfrac{90 + 60i}{65}$

$= \dfrac{18 + 12i}{13}$

STUDENT'S NOTES

EXERCISE 6.6

Perform the indicated operations.

1 $(3 + 2i) + (5 + 4i) =$

2 $(7 - 8i) + (3 + 4i) =$

3 $(5 + 3i) + (1 + 3i) =$

4 $(2 + 5i) + (7 - 6i) =$

5 $(8 + 9i) - (3 + 4i) =$

6 $(7 - 2i) - (-2 + 3i) =$

7 $(4 + 5i) - (2 - 5i) =$

8 $(3 + 7i) - (4 + 9i) =$

9 $(2 + i)(i + 2) =$

10 $(5 + 3i)(4 + 7i) =$

11 $(7 - 2i)(3 + 5i) =$

12 $(4 - i)(5 + 4i) =$

13 $(3 - 7i)(2 - 5i) =$

14 $(6 - 5i)(5 - 6i) =$

15 $(-2 + 3i)(7 - 5i) =$

16 $(-3 - 7i)(5 - 2i) =$

17 $\dfrac{2 + 5i}{3 + 4i} =$

18 $\dfrac{5 + 6i}{6 + 5i} =$

19 $\dfrac{3 - 7i}{7 + 3i} =$

20 $\dfrac{4 + i}{5 - 2i} =$

21. $\dfrac{-2 + 7i}{3 + i} =$

22. $\dfrac{-4 + 3i}{3 - 4i} =$

23. $\dfrac{7 - 6i}{-3 + 2i} =$

24. $\dfrac{-2 - 5i}{5 - i} =$

25. $\dfrac{2 + 3i}{2 - 3i} + \dfrac{4 - 3i}{4 + 3i} =$

26. $\dfrac{3 + 2i}{2 - 3i} + \dfrac{5 + 6i}{6 - 5i} =$

27. $\dfrac{4 + i}{4 - 3i} + \dfrac{4 - 3i}{4 + i} =$

28. $\dfrac{5 - 2i}{2 - 5i} + \dfrac{2 - 5i}{5 - 2i} =$

29. $\dfrac{4 + 7i}{3 - 5i} - \dfrac{3 + 5i}{4 - 7i} =$

30. $\dfrac{5 - 8i}{8 + 5i} - \dfrac{8 - 5i}{5 + 8i} =$

31. $\dfrac{2 - 3i}{3 - 4i} - \dfrac{4 - 5i}{5 - 6i} =$

32. $\dfrac{4 + 3i}{3 - 4i} - \dfrac{4 - 3i}{3 + 4i} =$

6.12
CHAPTER SUMMARY

We began the chapter by recalling the definition and laws of positive integral exponents and followed that with a discussion of negative exponents, roots, and fractional exponents. We then gave work on simplification of radicals, rationalizing monomial denominators, changing the order of a radical, products of polynomials whose terms contain radicals, rationalizing binomial denominators, and addition of radicals. We concluded the chapter with a treatment of roots of negative numbers, then complex numbers.

The definitions and formulas given and developed in the chapter are

$$a \cdot a \cdot a \cdots a \text{ to } n \text{ } a\text{'s} = a^n \tag{6.1}$$

$$a^0 = 1 \qquad a \neq 0 \tag{6.2}$$

$$a^m a^n = a^{m+n} \tag{6.3}$$

$$\frac{a^m}{a^n} = a^{m-n} \qquad a \neq 0 \tag{6.4}$$

$$(ab)^n = a^n b^n \tag{6.5}$$

$$\left(\frac{a}{b}\right)^n = \frac{a^n}{b^n} \qquad b \neq 0 \tag{6.6}$$

$$(a^m)^n = a^{mn} \tag{6.7}$$

$$a^{-n} = \frac{1}{a^n} \qquad a \neq 0 \tag{6.8}$$

$$a^{1/n} = \sqrt[n]{a} \tag{6.9}$$

$$a^{q/p} = (\sqrt[p]{a})^q = \sqrt[p]{a^q} \qquad \text{if } \sqrt[p]{a} \text{ is real} \tag{6.10}$$

$$\sqrt[n]{ab} = \sqrt[n]{a} \sqrt[n]{b} \tag{6.11}$$

$$\sqrt[n]{\frac{a}{b}} = \frac{\sqrt[n]{a}}{\sqrt[n]{b}} \tag{6.12}$$

$$\sqrt[qn]{a^{pn}} = \sqrt[q]{a^p} \tag{6.13}$$

STUDENT'S NOTES

EXERCISE 6.7 REVIEW

Simplify the expression.

1. $2^3(2^5) =$

2. $3^4(3^2) =$

3. $\dfrac{5^8}{5^3} =$

4. $\dfrac{7^5}{7} =$

5. $[3^3(2^2)]^2 =$

6. $[5^0(3^2)]^4 =$

7. $\left(\dfrac{2^3}{3^2}\right)^2 =$

8. $\left(\dfrac{6^4}{6}\right)^2 =$

9. $(2a^3b^2c^0)(3a^2bc) =$

10. $\dfrac{36a^4b^7}{16a^6b^3} =$

11. $\dfrac{(25a^2b^3)^2}{(5ab^2)^3} =$

12. $\left(\dfrac{2a^2b}{c^3}\right)^3 \div \left(\dfrac{4a^3b^2}{c^4}\right)^2 =$

13. $3^{-2} =$

14. $5^{-3} =$

Express each of the following without negative exponents and then simplify.

15 $\dfrac{3^{-1}x^{-2}yw^{-1}}{2^{-2}x^{-4}y^{0}w^{2}} =$

16 $\left(\dfrac{a^{-2}b^{2}}{a^{-1}b^{-3}}\right)^{-2} =$

17 $\dfrac{9x^{-2} - 16y^{-2}}{3x^{-1} - 4y^{-1}} =$

Evaluate the number.

18 $64^{1/3} =$

19 $64^{1/6} =$

20 $0.04^{1/2} =$

21 $0.027^{1/3} =$

22 $32^{3/5} =$

23 $125^{2/3} =$

Write the expression without radicals or fractional exponents.

24 $\sqrt{49x^{4}y^{2}} =$

25 $\sqrt[3]{216x^{3}y^{9}} =$

26 $\sqrt[4]{\dfrac{16x^8}{81y^{40}}} =$

27 $\sqrt[3]{\dfrac{343x^6y^{12}}{64w^3t^0}} =$

Simplify the expression.

28 $\left(\dfrac{64x^3}{27y^{3/2}}\right)^{1/3} =$

29 $\left(\dfrac{81^{1/4}a^{2/5}y^{2/5}}{36^{1/2}a^{-1/5}y^{-2/5}}\right)^{-5} =$

30 $\sqrt{2xy^3}\,\sqrt{8x^{-1}y} =$

31 $\dfrac{\sqrt{72a^5b^{-1}}}{\sqrt{2ab^{-3}}} =$

32 Change $\sqrt[3]{3x^2y^0}$ to a radical of order 9.

33 $\sqrt{27} - \sqrt{48} + \sqrt{12} =$

34 $\sqrt{20} + \sqrt{50} - \sqrt{18} + \sqrt{45} =$

Rationalize the denominator.

35 $\dfrac{2\sqrt{3} - 3\sqrt{2}}{\sqrt{2} + 5\sqrt{3}} =$

36 $\dfrac{\sqrt{7} - \sqrt{5}}{3\sqrt{5} - 2\sqrt{7}} =$

37 $\dfrac{a + 2\sqrt{b}}{a - 3\sqrt{b}} + \dfrac{a - 2\sqrt{b}}{a + 3\sqrt{b}} =$

38 $(3 + 5i)(4 - 7i) =$

39 $\dfrac{6 + 5i}{2 + 3i} =$

40 $\dfrac{3 + 4i}{4 - 3i} - \dfrac{4 + 3i}{3 - 4i} =$

41 Calculate $\dfrac{1}{\sqrt{5}}\left[\left(\dfrac{1 + \sqrt{5}}{2}\right)^3 - \left(\dfrac{1 - \sqrt{5}}{2}\right)^3\right].$

NAME _____ DATE _____ SCORE _____

EXERCISE 6.8   CHAPTER TEST

Simplify the expression.

1 $3^6(3^2) =$

2 $\dfrac{3^6}{3^2} =$

3 $(3^6)^2 =$

4 $[3^6(3^2)]^{1/2} =$

5 $\dfrac{51a^7b^5}{17a^2b} =$

6 $(3a^5b^4)^3 =$

7 $\left(\dfrac{2a^3b^{-1}}{c^2}\right)^3 \div \left(\dfrac{4a^2b^0}{c}\right)^2 =$

8 $\left(\dfrac{a^{-1}b^2}{a^3b^{-2}}\right)^{-3} =$

9 $27^{2/3} =$

10 $81^{-3/4} =$

11 Reduce the order of $\sqrt[6]{729a^6b^9}$.

12 $\sqrt{50} + \sqrt{45} - \sqrt{245} =$

13 $\dfrac{3\sqrt{3} - 2\sqrt{5}}{2\sqrt{5} - \sqrt{3}} =$

14 $\dfrac{3 - 5i}{2 + 5i} =$

15 $\dfrac{1 - 2i}{3 + 4i} + \dfrac{3 - 4i}{1 + 2i} =$

CHAPTER 7

Linear Equations and Inequalities

7.1
OPEN SENTENCES

In the preceding chapters, we discussed problems like the following. Perform the operations indicated in

$(a^2 - ab + b^2)(a + b)$
$(3a + 2b) - (2a - 5b)$
$(a^2 - 3ab + 2b^2) \div (a - 2b)$

These problems can also be formulated as follows: in each of the following statements, find the replacement for x for which the statement is true:

$$(a^2 - ab + b^2)(a + b) = x \quad (1)$$
$$(3a + 2b) - (2a - 5b) = x \quad (2)$$
$$(a^2 - 3ab + 2b^2) \div (a - 2b) = x \quad (3)$$

We can verify that (1) is true if x is replaced by $a^3 + b^3$; (2) is true if x is replaced by $a + 7b$; and (3) is true if x is replaced by $a - b$.

We now consider assertions of the type

$$3 + x = 8 \quad (4)$$

If x is multiplied by 2 and the product is increased by 5, the result is 11. $\quad (5)$

A color in the flag of the United States is x. $\quad (6)$

None of these statements is either true or false as it stands, but (4) is a true statement if x is replaced by 5; (5) is true if x is replaced by 3; and (6) is true if x is replaced by the words red, white, or blue.

In (4) and (5) x stands for a number, and in (6) x stands for a word. Under such circumstances we call the letter x a variable and define this word below.

A **variable** is a symbol, usually a letter, that stands

for one or more elements of a specified set. The specified set is called the **replacement set** for the variable.

Assertions (4) to (6) are called open sentences and illustrate the following definition:

An **open sentence** is an assertion containing a variable that is neither true nor false but becomes a true or a false statement if the variable is replaced by an element chosen from an appropriate set.

If an open sentence becomes a true statement when the variable is replaced by each element of a set T and no others, then T is called the **truth set** for the open sentence.

According to this definition, the truth sets for the open sentences (4), (5), and (6) are, respectively, {5}, {3}, and {red, white, blue}. Furthermore, {2,4,6} is the truth set for the open sentence

x is an even positive number less than 8.

Other examples of open sentences and their truth sets follow:

Open sentence	Truth set
x is an American state larger than California	{Alaska, Texas}
x is a color in a rainbow	{violet, indigo, blue, green, yellow, orange, red}
x is a positive odd integer less than 10	{1,3,5,7,9}
x is an element of $\{a,b,c,d\} \cap \{a,c,d,e,f\}$	{a,c,d}
$x - 8 = 2$	{10}

7.2 EQUATIONS

In the remainder of this chapter, we shall consider some of the procedures for solving equations. Equations are important tools in mathematics, and a great deal of time in introductory algebra courses is devoted to methods for solving them.

Open sentences of the type

$$3x - 5 = 4 + 2x \qquad (1)$$

$$\frac{2x - 3}{x + 2} = \frac{x - 1}{4} \qquad (2)$$

$$x + 1 = x + 5 \qquad (3)$$

$$\frac{x - 1}{3} + \frac{x + 1}{2} = \frac{5x + 1}{6} \qquad (4)$$

are called equations and illustrate the following definition:

An **equation** is an open sentence which states that two expressions are equal. Each of the two expressions is called a **member** of the equation.

If an equation is a true statement after x is replaced by a specific number, that number is called a **root** or **solution** of the equation and is said to satisfy it.

For example, 9 is a root of Eq. (1) since $3(9) - 5 = 4 + 2(9)$. Furthermore, 2 and 5 are roots of Eq. (2) since

$$\frac{2(2) - 3}{2 + 2} = \frac{2 - 1}{4}$$

because each fraction is equal to $\frac{1}{4}$ and

$$\frac{2(5) - 3}{5 + 2} = \frac{5 - 1}{4}$$

because each fraction is equal to 1.

The set of all roots of an equation is called the **solution set** of the equation.

The procedure for finding the solution set of an equation is called **solving** the equation.

As verified above, the solution sets of Eqs. (1) and (2) are {9} and {2,5}, respectively. Equation (3), however, is a false statement for every replacement for x, since the sum of a number and 1 is not equal to the sum of the same number and 5. Hence the solution set of (3) is the empty set $\varnothing$.

In contrast with (1), (2), and (3), Eq. (4) is a true statement for every replacement for x, since if we add the fractions on the left of the equality sign, we get the fraction on the right. Therefore, the solution set of (4) is the set of all numbers. Equations (1) to (3) are called *conditional equations*, and Eq. (4) is an identity. These illustrate the following definitions:

A **conditional equation** is an equation whose solution set contains a finite number of elements or an equation whose solution set is $\varnothing$.

An **identity** is an equation whose solution set is the set of all permissible* numbers.

* Here a permissible number is a number for which a value for each member exists. For example, if in

$$\frac{x + 1}{x - 2} + 3x = \frac{3x^2 - 5x + 1}{x - 2}$$

x is replaced by 2, we get $\frac{3}{0} + 6 = \frac{3}{0}$. Since $\frac{3}{0}$ is not a number, neither member exists if x is replaced by 2. Hence, 2 is not a permissible number.

Since the elements in the solution set of an equation are the replacements for x such that the equation is a true statement, the solution set of (1) is

$\{x | 3x - 5 = 4 + 2x\} = \{9\}$

Similarly for Eq. (2),

$\left\{ x \left| \dfrac{2x - 3}{x + 2} = \dfrac{x - 1}{4} \right.\right\} = \{2, 5\}$

7.3
EQUIVALENT EQUATIONS

The object in solving an equation is to find the roots. The simpler the equation is in form, the easier it is to solve. For example, we shall consider the equations

$$7x - 45 = 5x - 43 \qquad (1)$$

and

$$2x = 2 \qquad (2)$$

At this stage the only way that we can find the root of (1) is to guess at a number and then substitute it for x and see if it satisfies the equation. In (2), however, it is obvious that the root is 1; furthermore, if we substitute 1 for x in (1) and simplify, we get $-38 = -38$. Therefore, 1 is also a root of Eq. (1). Now $7x - 45 = 5x - 43$ and $2x = 2$ are different statements. The first states that the result obtained by multiplying a number x by 7 and then subtracting 45 is the same as the result obtained by multiplying x by 5 and then subtracting 43. The second statement says that twice x is 2. Each statement, however, is true only if x is replaced by 1. Two equations of this type are said to be equivalent and illustrate the definition below.

Two equations are **equivalent** if their solution sets are equal.

Since 1 is a root of Eq. (1), it follows by the additivity and multiplicativity axioms for equalities, that 1 is also a root of

$$7x - 45 + C = 5x - 43 + C \qquad (3)$$

$$N(7x - 45) = N(5x - 43) \qquad N \neq 0 \qquad (4)$$

and

$$\dfrac{7x - 45}{N} = \dfrac{5x - 43}{N} \qquad N \neq 0 \qquad (5)$$

where C is any chosen number and N is a number not equal to 0. Hence (1), (3), (4), and (5) are equivalent equations. This discussion illustrates the following conclusions. We can obtain an equation that is equivalent to a given equation by either or both of these operations: adding the same number to each member of the equation or multiplying or dividing each member of the given equation by the same nonzero number.

In set notation we can state that Eqs. (1), (3), (4), and (5) are equivalent as follows:

$\{x | 7x - 45 = 5x - 43\}$
$\quad = \{x | 7x - 45 + C = 5x - 43 + C\}$
$\quad = \{x | N(7x - 45) = N(5x - 43)\}$
$\qquad\qquad\qquad\qquad\qquad\qquad N \neq 0$
$\quad = \left\{ x \left| \dfrac{7x - 45}{N} = \dfrac{5x - 43}{N} \right.\right\} \quad N \neq 0$

MULTIPLYING BY THE LCM If both members of an equation that contains fractions are multiplied by the lcm of the denominators, the resulting equation will contain no fractions. Furthermore, if the variable is not involved in a denominator in the given equation, the equation obtained by the above multiplication will be equivalent to the given equation. For example, if we multiply each member of

$$\dfrac{3x}{4} - \dfrac{1}{3} = \dfrac{5x}{6} + \dfrac{1}{12}$$

by 12 we get

$$9x - 4 = 10x + 1$$

The essential step in solving an equation in one variable is to obtain an equivalent equation in which the terms that involve the variable are in one member and the constant terms, i.e., the terms that do not involve the variable, are in the other member. This is accomplished by adding the appropriate expression to each member of the equation. For example,

$2x - 5 = x + 2 \qquad$ and $\qquad 2x - x = 2 + 5$

are equivalent expressions, as are

$6x + 5 = -4x - 4 \qquad$ and $\qquad 6x + 4x = -4 - 5$

STUDENT'S NOTES

EXERCISE 7.1

Find the truth set for the open sentence.

1. x is an integer between 2 and 5.

2. x is a positive integer whose spelled-out name contains exactly four letters.

3. x is a state admitted to the United States in 1912.

4. x is the easternmost of the contiguous 48 states.

5. x is a United States president who has served 8 years since the end of World War II.

6. x was the first person to walk on the moon.

7. x sox is the nickname of a major league baseball club.

8. A swimming x and a x table are enjoyable.

9. $x \in \{1,2,5,9\} \cap \{1,3,5,7\}$

10. $x \in \emptyset \cup \{1,5\}$

11. $x^2 = 25$

12. $2x = 25$

Is the equation an identity or a conditional equation?

13. $\dfrac{x}{4} + \dfrac{x}{4} = \dfrac{x}{2}$

14. $\dfrac{x}{1} + \dfrac{x}{3} = \dfrac{x}{4}$

15. $\dfrac{x^2 + 2}{x + 2} + \dfrac{x - 4}{x + 2} = x - 1$

16. $\dfrac{x^2 + 1}{x + 1} = x - \dfrac{x - 1}{x + 1}$

17. $(x + 3)^2 = x^2 + 9$

18. $(2x - 3)^2 = 4x^2 - 12x - 9$

19. $(x + 3)^3 = x^3 + 9x^2 + 27x + 27$

20. $\dfrac{2x^2 + 5x - 1}{x^2 + 2x + 3} = 2$

Chapter 7: Linear Equations and Inequalities

Show that the number or numbers are roots of the equation that follows.

21 $3,\ 2x + 1 = 3x - 2$

22 $5,\ 4x - 3 = 2x + 7$

23 $6,\ \dfrac{x}{2} + \dfrac{x}{3} = 5$

24 $10,\ \dfrac{2x}{5} + \dfrac{x}{2} = x - 1$

25 $-1,\ 2,\ x^2 - x = 2$

26 $1,\ 1.5,\ 2x^2 + 3 = 5x$

27 $-3,\ 2,\ x - 1 = \dfrac{4}{x + 2}$

28 $\dfrac{-2}{3},\ \dfrac{1}{2},\ 2x + 1 = \dfrac{1}{3x - 1}$

State whether the equations in each pair are equivalent.

29 $6x - 3 = 4x + 1$
  $2x = 4$

30 $8x + 5 = 5x + 8$
  $-3 = -3x$

31 $5x = 7$
  $x = 7 - 5$

32 $4x = 2$
  $x = 4 - 2$

33 $\dfrac{2x}{5} - 1 = \dfrac{x}{3}$
  $\dfrac{x}{2} = 1$

34 $\dfrac{3x}{5} - 2 = \dfrac{x}{2}$
  $\dfrac{2x}{3} = 2$

35 $\frac{x}{3} + 2 = \frac{2x}{3}$

$ -\frac{x}{3} = -2$

36 $\frac{x}{3} - \frac{x}{4} = 2$

$ 4x - 3x = 2$

37 $\frac{x}{x-3} = \frac{3}{x-3}$

$ x = 3$

38 $\frac{x}{3} - \frac{x-2}{5} = \frac{1}{15}$

$ 5x - 3x - 6 = 1$

39 $\frac{x}{2} - \frac{3}{4} = \frac{1}{3} - \frac{2x}{3}$

$ 6x - 4 = 9 - 8x$

40 $\frac{x}{2} + 1 = \frac{5x}{6} - 1$

$ 2x = 12$

7.4
LINEAR EQUATIONS IN ONE VARIABLE

A linear equation in one variable is an equation $ax + b = cx + d$, where a, b, c, and d are real numbers. The following equations are linear:

$$3x - 4 = x + 5 \qquad x + 1 - \tfrac{1}{2}(x + 2) = \frac{x - 2}{3}$$

We shall illustrate the procedure for solving a linear equation in one variable by obtaining the solution set of

$$\frac{x - 2}{3} + 3 = \frac{3x}{4} + \frac{2}{3} \qquad (1)$$

1. To get an equation that contains no fractions and that is equivalent to (1), we multiply each member of (1) by 12, the lcm of the denominators, and obtain

$$12\,\frac{x - 2}{3} + 12(3) = 12\,\frac{3x}{4} + 12\,\frac{2}{3} \qquad (2)$$

   Performing the multiplications in (2) gives

   $$4x - 8 + 36 = 9x + 8$$

   and combining similar terms leads to

   $$4x + 28 = 9x + 8 \qquad (3)$$

2. The next step is to obtain an equation that is equivalent to (3) in which all terms containing the variable are in the left member and the constant terms are in the right. This is done by adding $-28 - 9x$ to each member of (3), obtaining

   $$4x + 28 - 28 - 9x = 9x + 8 - 28 - 9x \qquad (4)$$

   or

   $$-5x = -20 \qquad (5)$$

3. Divide each member of Eq. (5) by -5 and get $x = 4$. Hence, the solution set of (1) is $\{4\}$. We can also say the solution, or root, of (1) is 4.
4. Check by substituting 4 for x in each member of (1) and get $(4 - 2)/3 + 3 = (2 + 9)/3 = \tfrac{11}{3}$ for the left member and $\tfrac{12}{4} + \tfrac{2}{3} = 3 + \tfrac{2}{3} = \tfrac{11}{3}$ for the right. Therefore, (1) is a true statement if $x = 4$.

STEPS IN SOLVING A LINEAR EQUATION As the example above illustrates, the process of solving a linear equation in one variable consists of the following steps:

1. If fractions appear in the equation, obtain an equivalent equation by multiplying each member by the lcm of the denominators.
2. By adding the appropriate expression to each member of the equation obtained in step 1, get an equivalent equation in which the terms involving the variable appear in the left member and the constant terms appear in the right member.
3. Combine similar terms in the equation obtained in step 2, thus obtaining an equation of the type $ax = b$.
4. Divide each member of the equation obtained in step 3 by the coefficient of the variable, and thus get the root. Check the root by substituting it for the variable in the original equation.

Steps 1 to 3 prove that every linear equation in one variable is equivalent to an equation of the type $ax = b$. Furthermore, $ax = b$ is a true statement if and only if x is replaced by b/a if $a \neq 0$. Therefore, the solution set of a linear equation in one variable contains only one number, and that number is the root obtained in step 4.

We can accomplish step 2 quickly by **transposing**. This means writing a term on the other side of the equation after changing its sign. Thus in Eq. (3), $9x$ on the right becomes $-9x$ on the left

$$4x + 28 - 9x = 8$$

and 28 on the left becomes -28 on the right

$$4x - 9x = 8 - 28$$

With a little practice we can transpose two or more terms at once. For example the equation

$$4x - 7 = x + 2$$

is equivalent to the equation

$$4x - x = 2 + 7$$

since we have transposed both x and -7.

EXAMPLE 1 The above steps are illustrated in the process of solving

$$5x - 4 = 3x + 8$$

Solution Step 1 is unnecessary, since the equation involves no fractions, so we proceed to step 2, transpose -4 and $3x$, and get

$5x - 3x = 8 + 4$

$\qquad 2x = 12 \qquad$ combining similar terms
$\qquad\ \ x = 6 \qquad$ dividing each member by 2

Therefore, the solution set is $\{6\}$.

Check If we substitute 6 for x in the original equation, we get

$5(6) - 4 = 30 - 4 = 26 \qquad$ for left member
$3(6) + 8 = 18 + 8 = 26 \qquad$ for right member

Consequently, the two members are equal when x is replaced by 6.

EXAMPLE 2 Solve the equation

$$\frac{3x - 1}{3} - \frac{x + 12}{4} = \frac{2x - 5}{6}$$

Solution The lcm of the denominators is 12. Hence, we multiply both members by 12 and complete the solution as follows:

$$12\,\frac{3x - 1}{3} - 12\,\frac{x + 12}{4} = 12\,\frac{2x - 5}{6}$$

$$4(3x - 1) - 3(x + 12) = 2(2x - 5)$$

Performing the indicated multiplication gives

$12x - 4 - 3x - 36 = 4x - 10$

and combining similar terms gives

$9x - 40 = 4x - 10$

When we transpose $4x$ and 40, we get

$9x - 4x = 40 - 10$

Then combining similar terms leads to

$5x = 30$

and dividing each member by 5 gives

$x = 6$

Hence the solution set is $\{6\}$.

Check If x is replaced by 6 in the original equation, we have

$$\frac{18 - 1}{3} - \frac{6 + 12}{4} = \frac{17}{3} - \frac{18}{4} = \frac{68 - 54}{12}$$

$$= \frac{14}{12} = \frac{7}{6} \qquad \text{for left member}$$

and

$$\frac{12 - 5}{6} = \frac{7}{6} \qquad \text{for right member}$$

Hence, the two members of the equation are equal when x is replaced by 6.

EXERCISE 7.2

Circle the correct value.

1. If $3x = 12$, then $\quad x = 9 \quad x = \frac{1}{4} \quad x = 4$

2. If $yx = w$, then $\quad x = w - y \quad x = w/y \quad x = y/w$

3. If $(x - 2)(2x + 1) = 2x + 1$, then $\quad x = 3 \quad x = \frac{1}{2} \quad x = 2$

4. If $(x + 1)(2x + 3) = x + 1$, then $\quad x = -1.5 \quad x = -1 \quad x = 1$

Solve the equation.

5. $3x = 6$

6. $2x = -8$

7. $-5x = 15$

8. $-7x = -28$

9. $2x - 5 = 0$

10. $3x - 9 = 0$

11. $5x + 10 = 0$

12. $4x + 16 = 0$

13. $3x - 1 = 5$

14. $2x + 5 = 11$

15. $5x + 2 = -3$

16. $4x + 26 = 6$

17 $5x - 7 = 2x - 4$

18 $3x + 4 = 6x + 10$

19 $3x + 17 = -2x - 8$

20 $7x + 5 = 2x - 5$

21 $7(2x + 3) = -4(3x - 5)$

22 $5(4x + 3) = 6(3x + 1)$

23 $2(x - 5) = 3(2x + 1)$

24 $8(2x + 3) = 5(3x - 2)$

25 $3(2x - 5) = 2(4x - 3) - 21$

26 $2(5x + 8) = 5(3x + 7) + 1$

27 $3(4 + 3x) = 2(4x + 1) + 14$

28 $5(3x - 2) = 4(2x - 3) + 23$

29 $(x - 3)(x - 7) = (x - 5)(x - 5)$

30 $(x + 4)(x + 9) = (x + 8)(x + 3)$

31 $(x-2)(x-4) = (x-6)(x+4)$

32 $(x+3)(x-3) = (x-5)(x+1)$

33 $\frac{1}{3}(5x-2) = x+2$

34 $\frac{2}{5}(7x-1) = 3x-1$

35 $\frac{3}{4}(3x-2) = 3x$

36 $\frac{3}{7}(4x-7) = 2x-5$

37 $\frac{6x+2}{5} + \frac{4x-9}{3} = 5$

38 $\frac{4x+13}{7} - \frac{3x-5}{10} = 1$

39 $\frac{6x-5}{11} + \frac{4x-5}{3} = -4$

40 $\frac{7x-10}{4} - \frac{3x+4}{5} = -1$

41 $\dfrac{4x-3}{9} + \dfrac{3x+7}{6} - \dfrac{2x+4}{3} = \dfrac{1}{3}$

42 $\dfrac{3x-2}{5} + \dfrac{x+4}{4} - \dfrac{2x+2}{5} = 2$

43 $\dfrac{2x+5}{3} - \dfrac{3x-1}{5} + \dfrac{5x+2}{6} = 4$

44 $\dfrac{3x+4}{5} + \dfrac{2x-1}{3} - \dfrac{x+4}{6} = 2$

45 $\dfrac{x}{a} - b = \dfrac{x - b^2}{a + b}$

46 $\dfrac{x}{2b} + b^2 = a + \dfrac{bx}{2a}$

47 $ax + \dfrac{x}{b^2} = \dfrac{1}{a} + b^2$

48 $\dfrac{x}{b} + ab = \dfrac{ax}{2b} + 2b$

7.5 LINEAR INEQUALITIES

An equation states that two expressions are equal and *an inequality states that two expressions are not necessarily equal.* It does this by saying that one expression is greater than or equal to another or by saying that one expression is less than or equal to another. The symbol $\geq$ is used to indicate greater than or equal to and $\leq$ indicates less than or equal to.

If $f(x)$ and $g(x)$ are two expressions, $f(x) \geq g(x)$ and $f(x) \leq g(x)$ are called **inequalities**. At times we write $f(x) > g(x)$ or $f(x) < g(x)$, read "$f(x)$ is greater than $g(x)$" or "$f(x)$ is less than $g(x)$," and still call them inequalities. If $f(x)$ and $g(x)$ are linear expressions with different coefficients of x, we have a **linear inequality**. Consequently, $2x - 1 > 3x + 5$ and $3x + 4 \leq 8x - 3$ are linear inequalities.

The set of replacements for x for which an inequality is a true statement is called the **solution set** of the inequality. Thus $\{x \mid x > 2\}$ is the solution set of $2x + 1 > 5$. It is also said that the solution of $2x + 1 > 5$ is $x > 2$.

The procedures followed in finding the solution of an inequality are similar to those used in solving an equation. The concept of equivalent inequalities is used and we shall now consider that topic.

Two **inequalities** are **equivalent** if they have the same solution set.

We shall now give without proof several theorems concerning equivalent inequalities.

If $f(x)$, $g(x)$, and $h(x)$ are real-valued expressions, then each of the following is equivalent to $f(x) \geq g(x)$:

$$f(x) + h(x) \geq g(x) + h(x) \quad (7.1)$$

$$f(x)h(x) \geq g(x)h(x) \text{ if } h(x) > 0 \quad (7.2)$$

$$f(x)h(x) \leq g(x)h(x) \text{ if } h(x) < 0 \quad (7.3)$$

Furthermore, if $h(x)$ is a constant, (7.2) and (7.3) reduce to the following:

If k is a positive constant, then $f(x) \geq g(x)$ and $k[f(x)] \geq k[g(x)]$ are equivalent. (7.2a)
If c is a negative constant, then $f(x) \geq g(x)$ and $c[f(x)] \leq c[g(x)]$ are equivalent. (7.3a)

Similar statements are true for $f(x) \leq g(x)$.

Inequality (7.1) can be put in words by saying that the same quantity may be added to each member of an inequality without affecting the sense or direction of the inequality. Inequalities (7.2), (7.3), (7.2a), and (7.3a) can be combined into this statement:

The sense of an inequality is unchanged by multiplying each member by the same positive quantity and is reversed by multiplying by a negative quantity.

EXAMPLE 1 If $5x - 2 > 3x + 4$, it follows from (7.1) that $2x > 6$ since the latter was obtained from the former by addition, specifically, by transposing -2 and $3x$. If we divide each member of $2x > 6$ by the positive number 2, we find that $x > 3$ is the solution of $5x - 2 > 3x + 4$.

EXAMPLE 2 Find the solution set of $2x - 9 < 7x + 6$.

Solution If we transpose -9 and $7x$, we obtain an equivalent inequality by (7.1)

$2x - 7x < 6 + 9$ transposing
$-5x < 15$ collecting similar terms

Now, dividing by the negative coefficient of x and reversing the direction of the inequality as required by (7.3) and (7.3a), we see that $\{x \mid x > -3\}$ is the solution set of the given inequality.

EXAMPLE 3 Solve $\frac{1}{3}x + 1 > \frac{5}{6}x - 2$.

Solution We get x and its coefficient on the left and the constant on the right by transposing 1 and $\frac{5}{6}x$. Thus

$\frac{1}{3}x - \frac{5}{6}x > -2 - 1$
$-\frac{3}{6}x > -3$ collecting similar terms

Now, multiplying by $-\frac{6}{3} = -2$ and reversing the direction of the inequality, we see that $x < 6$ is the solution of the given inequality.

STUDENT'S NOTES

EXERCISE 7.3

Find the solution of the inequality.

1. $3x + 3 > 9$

2. $3x + 2 > 11$

3. $5x + 8 \geq 3$

4. $4x + 27 > 19$

5. $3x + 5 < 14$

6. $2x + 13 \leq 5$

7. $9x - 4 < 5$

8. $5x - 2 \leq 8$

9. $4x + 3 \geq 2x + 7$

10. $9x - 7 > 6x + 5$

11. $7x - 2 > 3x + 10$

12. $6x - 2 \geq 4x + 6$

13. $6x + 8 < 3x - 1$

14. $8x + 5 < 3x - 10$

15. $8x - 3 \leq 5x + 9$

16. $7x + 4 < 4x + 13$

17. $4x - 3 > 7x + 6$

18. $2x + 3 \geq 9x - 4$

19. $3x + 8 > 8x + 3$

20. $5x - 1 > 8x + 5$

21. $2x - 5 \leq 5x - 2$

22. $3x + 6 < 7x - 6$

23. $5x + 8 < 9x - 4$

24. $4x - 3 \leq 6x + 7$

25. $\frac{4}{5}x - 3 > \frac{2}{5}x + 1$

26. $\frac{3}{7}x - 1 < \frac{1}{7}x + 5$

27. $\frac{7}{8}x - 5 \leq \frac{3}{8}x - 2$

28. $\frac{7}{9}x - 6 > \frac{2}{9}x + 4$

29. $\frac{2}{3}x + 7 < \frac{7}{6}x + 4$

30. $\frac{8}{9}x - 2 > \frac{1}{3}x + 3$

31. $\frac{3}{4}x + 1 > \frac{5}{6}x - 2$

32. $\frac{4}{5}x - 4 \leq \frac{2}{3}x - 2$

7.6 SOLUTION OF STATED PROBLEMS

A stated problem is a description of a situation that involves both known and unknown quantities and also certain relations between these quantities. If the problem is solvable by means of an equation, it must be possible to find two combinations of the quantities in the problem that are equal. Furthermore, at least one of the combinations must involve the unknown.

STEPS IN SOLVING A STATED PROBLEM
The method of attacking such a problem has four steps:

1. Read and study the problem until you thoroughly understand the situation.
2. Let x represent one of the unknowns in the problem and then express the other unknowns in terms of x.
3. Study the relations between the quantities in the problem and find two combinations of the quantities that are equal.
4. Set the two quantities equal to each other and solve the equation thus obtained.

We shall illustrate the method by solving several problems.

EXAMPLE 1 Two brothers earned $545 during their summer vacation. The older boy worked 55 days and the younger worked 60 days. If the older earned 50 cents more per day than the younger, find the daily wage of each.

Solution We shall tabulate the quantities involved in the problem and enter the value of each that is known.

Boy	Time worked, days	Daily wage	Amount earned
Younger	60	$	$
Older	55		
Total			$545

The combinations that are equal are found as follows: The sum of the earnings of the two boys is $545. We know that the younger boy earned 50 cents less per day than the older, so we let x = the daily wage in dollars of the older, and then $x - 0.50$ = the daily wage in dollars of the younger. The total amount earned by the younger is $60(x - 0.50)$, and the amount earned by the older is $55x$. Therefore, $60(x - 0.50) + 55x = 545$.

If we enter these amounts in the above table, it will appear as below.

Boy	Time worked, days	Daily wage	Amount earned
Younger	60	$x - 0.50$	$60(x - 0.50)$
Older	55	x	$55x$
Total			$545

Combinations that are equal:

$$60(x - 0.50) + 55x = 545$$

Solution of equation:

$$60x - 30 + 55x = 545$$
$$60x + 55x = 545 + 30$$
$$115x = 575$$
$$x = 5$$
$$x - 0.50 = 4.50$$

Hence, the older boy earned $5 per day and the younger $4.50.

EXAMPLE 2 The dramatic club at a certain college sold 500 tickets to their annual play and received a total of $850. If the prices for the balcony tickets and main-floor tickets were $1.50 and $2, respectively, find the number of each type that was sold.

Solution We shall tabulate the quantities in the problem and then explain how those involving x were found.

	Number of tickets	Price	Money received
Balcony	x	$1.50	$1.50x$
Main floor	$500 - x$	2.00	$2(500 - x)$
Total	500		$850

Combinations that are equal: amount from balcony + amount from main floor = $850, or

$$1.50x + 2(500 - x) = 850$$

Since we know that 500 tickets were sold, we start by letting x = the number of balcony tickets sold; then $500 - x$ = the number of main-floor tickets sold. Hence, $1.50x$ = the amount in dollars received for balcony tickets, and $2(500 - x)$ = the amount in dollars received for the main-floor tickets; furthermore, the sum of these two amounts is $850.

We shall now rewrite the equation and solve it.

$$1.50x + 2(500 - x) = 850$$
$$1.50x + 1000 - 2x = 850$$
$$1.50x - 2x = 850 - 1000$$
$$-0.50x = -150$$
$$x = 300 \quad \text{balcony}$$
$$500 - x = 200 \quad \text{main floor}$$

EXAMPLE 3 A rancher made a trip of 690 mi by traveling a part of the way in a car that averaged 40 mi/h and the remainder of the way on a train that averaged 60 mi/h. If the entire trip required 12 h and there were no delays, find the distances traveled in the car and on the train.

	Distance, mi	Rate, mi/h	Time, h
In car	x	40	$\dfrac{x}{40}$
On train	$690 - x$	60	$\left(\dfrac{690 - x}{60}\right)$
Total	690		12

Solution Combinations that are equal: time in car + time on train = 12 h, or

$$\frac{x}{40} + \frac{690 - x}{60} = 12$$

We know the total distance, the two speeds, and the total time. Hence, if we let x = the distance in miles traveled in the car, then $690 - x$ = distance in miles traveled on the train. Since distance ÷ speed = time, we have $x/40$ = time spent in the car, and $(690 - x)/60$ = time spent on the train. Finally, since the total traveling time was 12 h, we have

$$\frac{x}{40} + \frac{690 - x}{60} = 12$$

which we solve as follows:

$3x + 2(690 - x) = 1440 \quad$ multiplying each member by 120
$3x + 1380 - 2x = 1440$
$3x - 2x = 1440 - 1380$
$x = 60 \quad$ miles in car
$690 - x = 630 \quad$ miles on train

EXAMPLE 4 A boy who had the job of taking care of three lawns found that he could mow them in 8 h by using a hand mower. After buying a power mower, he found that he could do the same work in 4 h. On one occasion he started the work with the power mower, and after it broke down, he finished with the hand mower. If at this time it took him 6 h to mow the lawns, how long did he use the power mower?

Mower	Time required to mow three lawns, h	Part mowed in 1 h	Time each mower was used, h	Part mowed with each mower
Power	4	$\dfrac{1}{4}$	x	$\dfrac{x}{4}$
Hand	8	$\dfrac{1}{8}$	$6 - x$	$\dfrac{6 - x}{8}$

Solution Combinations that are equal: part mowed by power mower + part mowed by hand mower = whole job, or

$$\frac{x}{4} + \frac{6 - x}{8} = 1$$

The unknown quantities in this problem are the periods of time each mower was used. However, since we know that the sum of the two periods is 6 h, we let x = the number of hours the power mower was used; then $6 - x$ = the number of hours the hand mower was used. Since one-fourth of the job was done by the power mower in 1 h and one-eighth of the job was done by the hand mower in 1 h, we have $x/4$ = part of the job done by the power mower in x hours and $(6 - x)/8$ = part of the job done by the hand mower in $6 - x$ hours. Furthermore, since the sum of these two parts is the entire job, we have

$$\frac{x}{4} + \frac{6 - x}{8} = 1$$

then

$2x + 6 - x = 8$
$2x - x = 8 - 6$
$x = 2 \quad$ time power mower was used

EXAMPLE 5 A car radiator that holds 6 gal is full of a solution 20% of which is alcohol. How much of the solution must be drained out and replaced by pure alcohol in order to bring the alcohol strength up to 30%?

	Radiator contents before draining	Solution drained out	Alcohol added	Radiator contents after refilling
Solution, gal	6	x		6
Alcohol, %	20	20	100	30
Alcohol, gal	1.2	$0.2x$	x	1.8

Solution Combinations that are equal: alcohol in radiator − alcohol drained out + alcohol added = 30% of 6 gal, or $1.2 - 0.2x + x = 1.8$.

The unknown in this problem is the number of gallons of the solution drained out, and this is of course equal to the amount of alcohol added. The alcohol in the solution before draining was 20% of 6 gal, or $0.20 \times 6 = 1.2$ gal. The alcohol drained out was 20% of x gal, or $0.2x$; the number of gallons added was x gallons; and the number of gallons of alcohol after refilling was 30% of 6 gal, or 1.8 gal. Hence, we have

$$1.2 - 0.2x + x = 1.8$$
$$-0.2x + x = 1.8 - 1.2$$
$$0.8x = 0.6$$
$$x = 0.75 \quad \text{gallon drained out}$$

STUDENT'S NOTES

EXERCISE 7.4

1. Divide 67 into two parts such that twice the larger part exceeds 3 times the smaller by 4.

2. Three-fourths of 3 times a number exceeds twice the number by 3. Find the number.

3. Find two numbers such that their sum is 40 and their quotient is $\frac{2}{3}$.

4. Mr. Johnson needed 278 ft of fencing to fence his yard. He used block wall for the portion facing the street and grape stake fencing for the rest. The block wall cost $4.25 a linear foot, and the grape stake fencing cost $2 a linear foot. If the total cost was $837.25, how many feet of block wall did he need and how many feet of grape stake? Complete the tabulation of data, find the equation, and then solve.

	Fence length	Price	Cost
Block	x	$4.25	
Grape stake	$278 - x$	2.00	
Equation			

5 Sally had twice as many dimes as nickels in her piggy bank. After she took out two dimes and put in one nickel, her number of dimes equaled her number of nickels. How much money was in her piggy bank when she finished?

6 Stewart is twice as old as his son George. In 11 years George's age will be three-fifths that of his father. How old are Stewart and George now?

7 Fred spent one-third of his summer vacation traveling. He then took care of a neighbor's yard for 13 days and at the end of that time half his vacation was gone. How long was his vacation?

8 Bruce and Rose earned the same amount of money each month from their paper routes. One month Bruce paid Rose $5 for some coins for his collection and Rose spent $20 for a small tape recorder. If Rose then had three-fourths as much money left from her monthly earnings as Bruce had, what were the monthly earnings of each?

9 At the beginning of the term in a certain high school there were two algebra classes and 74 students taking algebra. Eight students dropped out of the first class and three students transferred from the second class to the first. If each class then contained the same number of students, how many students were in each class at the beginning of the term?

10 Airfields A and C are due south and due east, respectively, of airfield B. A pilot flew a plane from A to B in $1\frac{2}{3}$ h. The next day he flew the same plane from B to C in $1\frac{3}{4}$ h. On the first day the wind blew from the south at 15 mi/h and on the second it blew from the east at 10 mi/h. If the total distance flown was 725 mi, find the speed of the plane in still air. Complete the tabulation of the data and then solve the problem.

	Speed of plane in still air	Actual speed of plane	Time flown, h	Distance flown, mi
First day	x	$x + 15$	$\frac{5}{3}$	
Second day	x	$x - 10$	$\frac{7}{4}$	
Total distance flown				
Equation				

11 Mrs. Billings flew her own plane from her home town to a city due east where she attended a business meeting and then flew back home. She spent 6 h in the city, her plane in still air had a speed of 156 mi/h, and the wind blew from the west at 12 mi/h all day. If $10\frac{1}{3}$ h elapsed from the time she left her home airport until she returned, how far was the city from her town?

12 Two girls on a boating trip could row 9 mi/h in still water. One day they rowed down a stream $3\frac{1}{3}$ h before it ran into a river. They then rowed $2\frac{1}{2}$ h down the river before making camp. They traveled 15 mi more in the stream than in the river. If the current in the stream was 2 mi/h faster than that of the river, find the speed of the current in both the stream and the river.

13 A vacationing family drove 200 mi on a highway to a national park and then 20 mi in the park to a campground. Their speed in the national park was one-half that on the highway. Find their highway speed if the entire trip took 4 h.

14 Two families who lived 250 mi apart decided to meet for a picnic at a park located on the highway between them. They left home at the same time; both drove for 2 h and arrived at the park at the same time. What was the average speed of each if one traveled 7 mi/h faster than the other?

15 John started on a bicycle trip riding at a speed of 12 mi/h. When he had been gone 45 min, Jennifer started after him on a minibike at a speed of 30 mi/h. How long had John traveled when Jennifer caught up with him?

16 Mr. Alexius used a 0.5% solution of insecticide to spray his trees. When he finished, he had $1\frac{1}{4}$ gal of solution left in his 5-gal spray tank. With what percent insecticide solution should he fill his spray tank so that he would have a 0.2% insecticide solution to spray his roses? Complete the tabulation of the data and then solve the problem.

	Quantity of solution, gal	Insecticide, %
After spraying trees	$1\frac{1}{4}$	0.5
To refill tank	$3\frac{3}{4}$	x
Refilled tank	5	0.2
Equation		

17 Ms. Murrill wanted a 10% acid solution but found she had made 500 mL of a 5% acid solution instead. How much 60% acid solution must she add to obtain the desired solution?

$$500 \text{ ml} \quad\quad x \text{ ml} \quad\quad x+500$$
$$[\ 5\%\ ] + [\ \ ]\ 60\% = [\ \ ]\ 10\%$$

$$.05(500) + .60(x) = .10(x+500)$$
$$25 + .6x = .1x + 50$$
$$.5x = 25$$
$$x = 50 \text{ ml}$$

18 How many pounds of chocolates worth $5 per pound must be mixed with 14 lb of a quality worth $6.50 per pound in order to obtain a mixture worth $6 per pound?

Chapter 7: Linear Equations and Inequalities

19 Mr. Boone paid $91 for 14 shrubs of two varieties. If one variety cost $5 apiece and the other variety cost $8.50 apiece, how many of each type did he buy?

20 Two blocks of alloy containing 20 and 40% copper are melted together to obtain 100 lb of an alloy that is 33% copper. Find the weight of each block.

21 Mrs. Windell mixed 8 doz of one type cookie with 12 doz of another type to make a mixture worth $1.18 per doz. If the price of the first type cookie was 20 cents per dozen more than the price of the second, find the price of each type.

22. A maintenance man can wash the windows in a certain building in 9 h. His helper requires 12 h to wash the same windows. One day they worked together for a while, and then the maintenance man left for another job. The helper finished in 5 h. How long did they work together? Complete the tabulation of the data and then solve the problem.

	Time to wash windows alone, h	Part done in 1 h	Time worked, h	Part done by each	Time to finish job, h
Maintenance man	9	$\frac{1}{9}$	x		
Helper	12		$x + 5$		
Equation					

23. A cow can eat the grass of a certain field in 48 days, but a goat requires 80 days to eat the same amount of grass. If both animals are put in the same field, how long will it support them? Assume that no new grass grows during this time.

24. A swimming pool is fed by two intake pipes that can fill it in 6 and 9 h individually. The smaller pipe is opened and allowed to run for 1 h, and then the larger pipe is opened. How much longer will it take to fill the pool?

25 Miss Johnson prepared a group of form letters for mailing in 6 h while Miss Ellis required 8 h to prepare a similar group. Miss Ellis worked alone on a third group for 1 h before Miss Johnson started working with her. How long did it take them to finish the work?

26 Janice and JoBeth decided to make a group of posters for a campus election. Working alone, Janice could have made the posters in 10 h, JoBeth in 12 h. Janice worked alone for 1 h before JoBeth came, and then worked alone for 2 h while Janice attended a seminar. How long after Janice came back did it take them to finish the posters?

27 An irrigation reservoir can be filled in 10 h and drained in 8 h. At the start of an irrigation job the reservoir was full when the owner opened the outlet pipe. After irrigating for a while he opened the intake pipe and continued irrigating until the reservoir was empty. If the intake pipe was open for exactly one-fourth the time required for the whole job, how long was it open?

28 In 1970 Mr. Holmes bought some stock for $3600. By 1980 the stock had tripled in value. He sold part of it for $5400 and had 75 shares left. What was his purchase price per share?

7.7
CHAPTER SUMMARY

The first three sections of the chapter were devoted to the concept of an equation and to conditions under which two equations are equivalent. We then solved linear equations in one variable and followed that treatment by a discussion of linear inequalities. The chapter ended with a discussion of stated or word problems. The following statements concerning inequalities were given without proof and then used in solving inequalities.

If $f(x)$, $g(x)$, and $h(x)$ are real-valued expressions, then (7.1) to (7.3) are equivalent to $f(x) \geq g(x)$.

$$f(x) + h(x) \geq g(x) + h(x) \tag{7.1}$$

$$f(x)h(x) \geq g(x)h(x) \quad \text{if } h(x) > 0 \tag{7.2}$$

$$f(x)h(x) \leq g(x)h(x) \quad \text{if } h(x) < 0 \tag{7.3}$$

If k is a positive constant, then $f(x) \geq g(x)$ and $k[f(x)] \geq k[g(x)]$ are equivalent. (7.2a)

If c is a negative constant, then $f(x) \geq g(x)$ and $c[f(x)] \leq c[g(x)]$ are equivalent. (7.3a)

Similar statements are true for $f(x) \leq g(x)$.

STUDENT'S NOTES

EXERCISE 7.5 REVIEW

1. Find the truth set of the statement that x is all presidents of the United States who served more than 4 and less than 8 years between 1947 and 1976, inclusive.

2. For what values of x is the following statement true: x is a two-digit number that is divisible by 2, 3, and 5.

3. Is $\dfrac{x}{3} + \dfrac{x}{5} = \dfrac{8}{15}x$ a conditional equation or an identity?

4. Is $\dfrac{3x^2 + 5x - 9}{x^2 + 2x - 3} = 3$ a conditional equation or an identity?

5. Show that 4 is a root of $3x - 5 = x + 3$.

6. Show that 10 is not a root of $\dfrac{x}{2} - \dfrac{x}{5} = 2$.

7. Are $7x + 4 = 4x + 7$ and $-3x = -3$ equivalent equations?

8. Are $3x = 7$ and $x = 7 - 3$ equivalent equations?

9. Are $\dfrac{x}{3} + 1 = \dfrac{5x}{6} - 1$ and $3x = 18$ equivalent equations?

Solve the equations.

10. $5(3x - 4) = 2(2x + 1)$

11. $(2x - 1)(x - 1) = (x + 2)(2x - 4)$

12 $\dfrac{x}{a} + \dfrac{x}{b} = a + b$

13 $\dfrac{3x - 4}{8} = \dfrac{2x + 3}{9} + 1$

Solve the inequalities.

14 $2x + 5 > 11$

15 $3 - 5x \leq -17$

16 $3x + 7 < 5x + 13$

17 $\frac{2}{5}x + 7 \geq \frac{2}{3}x + 3$

18 Find two numbers whose sum is 49 and whose quotient is $\frac{5}{2}$.

19 Mrs. Seely earned $18,350 in 2 years. How much did she earn each year if her beginning salary was $850 less than her second-year salary?

NAME _____ DATE _____ SCORE _____

EXERCISE 7.6 CHAPTER TEST

1. Find the truth set of the statement that x is a man who served as Vice President while Richard Nixon was president.

2. If x is a two-digit number less than 20 that is not divisible by 2, 3, or 17, find x.

3. Is $4x = 3$ equivalent to $\dfrac{2x}{3} + 1 = \dfrac{1}{2}$?

4. Are $5x = 9$ and $x = \dfrac{5}{9}$ equivalent equations?

5. Divide 36 into two parts whose quotient is $\dfrac{5}{7}$.

Solve each equation and inequality.

6 $3(4x + 5) = 2(3x + 7) + 1$

7 $\dfrac{2x + 1}{3} - \dfrac{5x - 2}{4} + \dfrac{3}{2} = 0$

8 $\dfrac{x}{ab} + \dfrac{x}{b} = a(a + 1)$

9 $3x - 2 > x + 8$

10 $(3x + 4)(x - 1) \leq (x + 3)(3x - 4)$

11 $\dfrac{3x - 5}{8} - \dfrac{2x + 3}{11} \geq 2$

CHAPTER 8

Fractional Equations and Inequalities

8.1 FRACTIONAL EQUATIONS

A *fractional equation* is an equation that contains fractions in which the variable appears in one or more denominators. For example,

$$\frac{1}{x-1} - \frac{3}{x} = \frac{5-x}{x(x-1)} \qquad (1)$$

is a fractional equation.

Since neither member of (1) is a number if x is replaced by 0 or 1, these replacements for x must be excluded. In any fractional equation, we exclude all replacements for x for which a denominator is 0.

The first step in obtaining the solution set of (1) is to derive an equation that contains no fractions and whose solution set is the solution set of (1) or has the solution set of (1) as a subset. By use of the multiplicativity axiom of equalities, we have the following conclusion: if a is a root of Eq. (1), then a is also a root of

$$x(x-1)\left(\frac{1}{x-1} - \frac{3}{x}\right) = x(x-1)\frac{5-x}{x(x-1)} \qquad (2)$$

obtained by multiplying each member of (1) by the lcm of the denominators $x(x-1)$. Hence the solution set of (1) is either equal to, or is a subset of, the solution set of (2). In fact, the only numbers which could be solutions of (2) but not (1) are the ones which make $x(x-1) = 0$, that is, $x = 0$ or $x = 1$.

Now we perform the indicated multiplication in (2) and get

$$x - 3(x-1) = 5 - x$$

This is a linear equation, and we now complete the solution:

$$x - 3x + 3 = 5 - x \quad \text{removing parentheses}$$
$$x - 3x + x = 5 - 3 \quad \text{transposing 3 and } -x$$
$$-x = 2 \quad \text{combining similar terms}$$
$$x = -2 \quad \text{dividing each member by } -1$$

Hence, the solution set of (2) is $\{-2\}$.

Now we replace x in each member of (1) by -2 to determine whether or not -2 is a root of the equation. Thus, from the left member, we get

$$\frac{1}{-2-1} - \frac{3}{-2} = -\frac{1}{3} + \frac{3}{2} = \frac{-2+9}{6} = \frac{7}{6}$$

and from the right member

$$\frac{5-(-2)}{-2(-2-1)} = \frac{7}{6}$$

Therefore, since the members of (1) are equal when x is replaced by -2, the solution set of (1) is $\{-2\}$.

SOLVING A FRACTIONAL EQUATION As illustrated by this example, the steps in solving a fractional equation are as follows:

1. Multiply each member of the given equation by the lcm of the denominators and equate the products. The resulting equation will be clear of fractions, and we shall call it the **cleared equation.**
2. Solve the cleared equation to get the solution set.
3. Check the solution set obtained in step 2 by substituting the number (or numbers) in it in the given equation.

Step 3 is an essential step in solving a fractional equation. Since the lcm of the denominators in a fractional equation contains the variable, *the cleared equation may have roots that are not roots of the given equation.* Hence, after the solution set of the cleared equation is found, we must substitute the number (or numbers) in it into the given equation in order to determine whether or not this set is also the solution set of the given equation.

EXAMPLE 1 Find the solution set of the equation

$$\frac{1}{x-1} - \frac{1}{x} = \frac{2}{x(x+2)} \quad (1)$$

Solution We multiply each member of (1) by the lcm of the denominators, $x(x-1)(x+2)$ and get

$$x(x-1)(x+2)\left(\frac{1}{x-1} - \frac{1}{x}\right)$$
$$= x(x-1)(x+2)\frac{2}{x(x+2)}$$

Then

$$x(x+2) - (x-1)(x+2) = 2(x-1)$$

Multiplying as indicated gives

$$x^2 + 2x - x^2 - x + 2 = 2x - 2$$

and combining similar terms leads to

$$x + 2 = 2x - 2$$

Then transposing x and -2 gives

$$4 = x$$

Check Since each member of (1) is equal to $\frac{1}{12}$ if x is replaced by 4, the solution set of (1) is $\{4\}$.

EXAMPLE 2 Find the set

$$\left\{x \mid \frac{2}{x-1} - \frac{1}{x} = \frac{2}{x(x-1)}\right\} \quad (1)$$

Solution The required set is the solution set of the equation

$$\frac{2}{x-1} - \frac{1}{x} = \frac{2}{x(x-1)} \quad (2)$$

and we shall now solve this equation. Multiplying each member of (2) by the lcm of the denominators, we get

$$x(x-1)\left(\frac{2}{x-1} - \frac{1}{x}\right) = x(x-1)\frac{2}{x(x-1)} \quad (3)$$

Performing the multiplication indicated in (3) gives

$$2x - (x-1) = 2 \quad (4)$$

Then removing parentheses, combining terms, and transposing 1, we get

$$2x - x + 1 = 2 \quad x = 1$$

Hence, the solution set of (4) is $\{1\}$.

Now if we attempt to check this solution set by replacing x by 1 in (2), we get

$$\frac{2}{0} - \frac{1}{1} = \frac{2}{1(0)}$$

However, since neither $\frac{2}{0}$ nor $2/[1(0)]$ is a number, $x = 1$ is not a root of (2). Furthermore, if Eq. (2) has a root, this root must be included in the solution set of (4). Consequently, since the only element of this set is not a root of (2), we conclude that the solution set of (2) is the empty set $\varnothing$.

EXERCISE 8.1

Find the solutions of the equations.

1. $\dfrac{1}{x} = \dfrac{1}{4}$

2. $\dfrac{1}{x} - \dfrac{5}{4x} = \dfrac{1}{12}$

3. $\dfrac{3}{x} - \dfrac{2}{3x} - \dfrac{3}{2x} = \dfrac{1}{6}$

4. $\dfrac{3}{2x} - \dfrac{1}{x} - \dfrac{2}{x} = \dfrac{3}{8}$

5. $\dfrac{2}{3x+1} = \dfrac{1}{x}$

6. $\dfrac{2}{x} = \dfrac{11}{6x-5}$

7. $\dfrac{1}{6x} + \dfrac{1}{x+7} = 0$

8. $\dfrac{1}{x} = \dfrac{1}{3x+4}$

9. $\dfrac{2}{3x+1} = \dfrac{5}{8x+1}$

10. $\dfrac{2}{x+1} = \dfrac{11}{6x+1}$

11. $\dfrac{8}{5x-4} = \dfrac{5}{3x-1}$

12. $\dfrac{2}{3x+5} = \dfrac{1}{x+1}$

13. $\dfrac{x-2}{x+1} = \dfrac{x-3}{x-1}$

14. $\dfrac{2x-3}{4x-3} = \dfrac{4x+1}{8x+5}$

15. $\dfrac{2x-5}{4x-1} = \dfrac{3x-4}{6x+9}$

16. $\dfrac{6x-8}{9x+8} = \dfrac{4x-5}{6x+6}$

17. $\dfrac{x-5}{x+1} = \dfrac{x-4}{x+4}$

18. $\dfrac{x-1}{x+5} = \dfrac{x-2}{x+2}$

19. $\dfrac{x-3}{x+4} = \dfrac{x-4}{x+2}$

20. $\dfrac{6x-1}{4x+1} = \dfrac{3x+5}{2x+5}$

21. $\dfrac{1}{2x+3} - \dfrac{3}{x-3} = \dfrac{3}{(2x+3)(x-3)}$

22. $\dfrac{2}{x+2} + \dfrac{1}{2x-1} = \dfrac{5}{(x+2)(2x-1)}$

23. $\dfrac{5}{3x-1} - \dfrac{1}{5x-7} = \dfrac{11x-1}{(3x-1)(5x-7)}$

24. $\dfrac{1}{2x+9} + \dfrac{1}{x+5} = \dfrac{2}{(x+5)(2x+9)}$

25. $\dfrac{2}{x+1} - \dfrac{3}{2x+3} = \dfrac{2}{4x+1}$

26. $\dfrac{1}{2x+3} + \dfrac{1}{x+2} = \dfrac{15}{10x+18}$

27. $\dfrac{2}{2x-1} - \dfrac{1}{3x-9} = \dfrac{2}{3x+6}$

28. $\dfrac{1}{x+4} + \dfrac{3}{3x-8} = \dfrac{2}{x-1}$

Show that the equation has no solution.

29. $\dfrac{6x+1}{5-2x} = -3$

30. $\dfrac{5}{x+3} - \dfrac{4}{x-2} = \dfrac{x-12}{(x+3)(x-2)}$

31. $\dfrac{2}{(x-1)(x+2)} = \dfrac{1}{x-1} - \dfrac{1}{x+2}$

32. $\dfrac{6}{x+3} - \dfrac{5}{3x-1} = \dfrac{13x-20}{(3x-1)(x+3)}$

8.2
FRACTIONAL INEQUALITIES

We shall consider only those fractional inequalities which are the quotient of two linear functions. In solving fractional inequalities, we must use the fact that *a quotient of two numbers is positive if both are of the same sign and negative if they have opposite signs*. Furthermore, the quotient is 0 if and only if the numerator is 0 and does not exist if the denominator is 0.

EXAMPLE 1 Solve $(3x - 4)/(x + 2) > 0$.

Solution Since the quotient is to be positive, the two linear expressions must both be positive or must both be negative. Consequently, we must solve

$$3x - 4 > 0 \quad \text{and} \quad x + 2 > 0 \qquad (1)$$

simultaneously *and also* must solve

$$3x - 4 < 0 \quad \text{and} \quad x + 2 < 0 \qquad (2)$$

simultaneously. Now solving the pair of inequalities in (1), we have $x > \frac{4}{3}$ and $x > -2$ simultaneously. Both these are satisfied if $x > \frac{4}{3}$. Both inequalities in (2) are satisfied if $x < \frac{4}{3}$ and $x < -2$; hence if $x < -2$. Consequently the given inequality is a true statement if $x > \frac{4}{3}$ or if $x < -2$. In set notation, the solution is $\{x|x > \frac{4}{3}\} \cup \{x|x < -2\}$.

EXAMPLE 2 Solve $(2x - 5)/(x + 3) \le 0$.

Solution Since the quotient of two linear factors is to be negative, they must be of opposite signs. Thus, we must solve

$$2x - 5 \le 0 \quad \text{and} \quad x + 3 > 0 \qquad (1)$$

simultaneously and also must solve

$$2x - 5 \ge 0 \quad \text{and} \quad x + 3 < 0 \qquad (2)$$

simultaneously. Both linear inequalities in (1) are satisfied if $x \le \frac{5}{2}$ and $x > -3$ simultaneously. Thus, $\{x| -3 < x \le \frac{5}{2}\}$ is the simultaneous solution of the pair of inequalities in (1). The two inequalities in (2) are satisfied if $x \ge \frac{5}{2}$ and $x < -3$ simultaneously. These two requirements are contradictory; hence, the solution is the null set $\varnothing$. Therefore, the solution of the given inequality is $\{x| -3 < x \le \frac{5}{2}\} \cup \{\varnothing\}$ or $-3 < x \le \frac{5}{2}$.

By use of the method of Sec. 5.6, we can sketch the graph of y equal to the quotient of two linear functions of x and can then see the range of values of x for which y is positive or negative.

EXAMPLE 3 Sketch the graph of $y = (2x - 3)/(3x + 1)$ and find the values of x for which $y < 0$ and for which $y > 0$.

Solution We shall begin by assigning values to x and finding each corresponding value of y. We find the zero, $x = -\frac{1}{3}$, of the denominator and then assigning several values larger than $-\frac{1}{3}$ and several smaller. Thus,

x	large	5	4	3	2	1	0	$-\frac{1}{6}$	$-\frac{1}{3}$
y	near $\frac{2}{3}$	$\frac{7}{16}$	$\frac{5}{13}$	$\frac{3}{10}$	$\frac{1}{7}$	$-\frac{1}{4}$	-3	$-\frac{20}{3}$	no val.

	$-\frac{1}{2}$	-1	-2	-3	-4	-5	large, negative
	8	$2\frac{1}{2}$	$1\frac{2}{5}$	$\frac{9}{8}$	1	$\frac{13}{14}$	near $\frac{2}{3}$

Before locating the points determined by the above number pairs, we shall notice that as x gets very large, the value of y gets close to $\frac{2}{3}$. Now drawing the dotted lines $y = \frac{2}{3}$ and $x = -\frac{1}{3}$, locating the points determined by the table, and drawing a smooth curve for those with $x > -\frac{1}{3}$ and another for those with $x < -\frac{1}{3}$, we have the curve shown in Fig. 8.1. The only place that the curve crosses the X axis is for $x = \frac{3}{2}$. Now from the figure we can see that $y < 0$ for $-\frac{1}{3} < x < \frac{3}{2}$ and that $y > 0$ for $\{x|x > \frac{3}{2}\} \cup \{x|x < -\frac{1}{3}\}$.

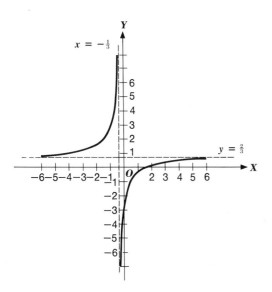

Fig. 8.1

STUDENT'S NOTES

EXERCISE 8.2

Solve the inequality either algebraically or graphically.

1. $\dfrac{3x+2}{x-2} > 0$

2. $\dfrac{5x-4}{x+3} > 0$

3. $\dfrac{2x+5}{x-3} > 0$

4. $\dfrac{4x-7}{x+2} > 0$

5. $\dfrac{6x-10}{x+4} \geq 0$

6. $\dfrac{3x+5}{x-1} \geq 0$

7. $\dfrac{4x-9}{x+1} \geq 0$

8. $\dfrac{7x+13}{x-3} \geq 0$

9. $\dfrac{2x-11}{3x-1} \geq 0$

10. $\dfrac{2x+7}{7x+2} \geq 0$

11. $\dfrac{5x+4}{4x+9} \geq 0$

12. $\dfrac{6x-11}{2x-1} \geq 0$

13 $\dfrac{x+5}{2x-3} < 0$

14 $\dfrac{x+4}{3x-8} < 0$

15 $\dfrac{x-6}{3x+10} < 0$

16 $\dfrac{x-3}{4x+7} < 0$

17 $\dfrac{2x+7}{7x-2} \le 0$

18 $\dfrac{5x+9}{9x-5} \le 0$

19 $\dfrac{3x+8}{2x-7} \le 0$

20 $\dfrac{4x+9}{2x-5} \le 0$

21 $\dfrac{4x+11}{2x+1} \le 0$

22 $\dfrac{7x+4}{2x+7} \le 0$

23 $\dfrac{5x-9}{2x-1} \le 0$

24 $\dfrac{2x-8}{3x-2} \le 0$

8.3 CHAPTER SUMMARY

Only two topics have been considered in this chapter; fractional equations were treated in Sec. 8.1 and fractional inequalities in Sec. 8.2.

STUDENT'S NOTES

EXERCISE 8.3 REVIEW

Solve the equation.

1. $\dfrac{1}{x} + \dfrac{2}{3x} = \dfrac{5}{6}$

2. $\dfrac{1}{x} + \dfrac{1}{2x} + \dfrac{1}{3x} = \dfrac{11}{72}$

3. $\dfrac{1}{x+1} = \dfrac{1}{3x-9}$

4. $\dfrac{1}{x} = \dfrac{2}{3x-4}$

5. $\dfrac{3x+4}{4x+4} = \dfrac{6x+4}{8x+3}$

6. $\dfrac{5x-3}{10x+8} = \dfrac{2x-3}{4x-1}$

7. $\dfrac{4x-4}{2x+1} - \dfrac{2x-2}{x+2} = \dfrac{3x}{(2x+1)(x+2)}$

8. $\dfrac{x+1}{x+3} - \dfrac{5}{(x+3)(x-2)} = \dfrac{x-3}{x-2}$

9. $\dfrac{2}{x+2} + \dfrac{3}{x+6} = \dfrac{5}{x+4}$

10. $\dfrac{1}{x+5} + \dfrac{3}{3x-1} = \dfrac{2}{x+1}$

Solve the inequality.

11. $\dfrac{x-3}{x+1} > 0$

12. $\dfrac{x+2}{x-1} < 0$

13. $\dfrac{2x+11}{x-1} < 0$

14. $\dfrac{3x-10}{x+2} > 0$

15. $\dfrac{3x+1}{2x-5} \leq 0$

16. $\dfrac{4x-13}{2x+1} \geq 0$

17. $\dfrac{5x-1}{2x-8} \geq 0$

18. $\dfrac{6x-12}{5x+6} \leq 0$

NAME _____ DATE _____ SCORE _____

EXERCISE 8.4   CHAPTER TEST

Solve the fractional equation.

1 $\dfrac{1}{x} - \dfrac{2}{3x} = \dfrac{1}{24}$

2 $\dfrac{3}{x+4} = \dfrac{1}{x-2}$

3 $\dfrac{3x+2}{9x-4} = \dfrac{4x+5}{12x+1}$

4 $\dfrac{2}{3x+1} = \dfrac{10}{(3x+1)(4x-12)} + \dfrac{1}{4x-12}$

5 $\dfrac{2}{x+5} + \dfrac{6}{3x+5} = \dfrac{4}{x+3}$

Solve the fractional inequality.

6 $\dfrac{2x-9}{3x+2} < 0$

7 $\dfrac{5x-17}{2x+1} > 0$

8 $\dfrac{4x+7}{3x+1} \geq 0$

9 $\dfrac{6x-19}{2x-3} \leq 0$

CHAPTER

Ratio, Proportion, and Variation

9.1 RATIOS

The **ratio** of the number a to the number b is the quotient obtained by dividing a by b. It is expressed as $a \div b$, $\frac{a}{b}$, a/b, or $a:b$. The expression $a:b$ is read "a is to b."

The ratio $a:b$ is the number that represents the portion of a that corresponds to one unit of b. For example, if Fred has $100 in his savings account and Mary has $50, then Fred's account: Mary's account $= 100:50 = 100/50 = 2$. This means that for every dollar in Mary's account, there are two dollars in Fred's. This ratio can also be considered as a comparison; i.e., Fred's account is twice Mary's.

Many ratios are of sufficient importance in physical and social sciences to be given names. For example, the *pitch* of a roof is defined as the ratio of the height to the span, and the *specific gravity* of a body is defined, as the ratio of the weight of the body to the weight of an equal body of water. In each of these ratios the two numbers involved represent quantities of the same kind. Frequently, however, we see ratios expressed between different kinds of numbers. As examples, we have

$$\text{Velocity} = \frac{\text{distance}}{\text{time}}$$
$$= \text{number of linear units traveled in one unit of time}$$

$$\text{Density} = \frac{\text{mass of body}}{\text{volume of body}}$$
$$= \text{mass of unit volume of body}$$

$$\text{Pressure} = \frac{\text{normal force acting on area}}{\text{area}}$$
$$= \text{normal force acting on unit area}$$

Gasoline mileage
$$= \frac{\text{number of miles traveled}}{\text{gallons of gasoline consumed}}$$
$$= \text{miles per gallon}$$

If we know the ratio of two numbers and the value of one of them, we can find the other.

EXAMPLE 1 If $a:b = \frac{2}{3}$ and $b = 27$, then since $a/27 = \frac{2}{3}$, $3a = 54$, and $a = 18$.

EXAMPLE 2 If the pitch of a roof is $\frac{1}{2}$ and the span is 30 ft, find the height.

Solution

$$\text{Pitch} = \frac{\text{height}}{\text{span}} = \frac{1}{2}$$

Therefore,

$$\frac{\text{Height}}{30 \text{ ft}} = \frac{1}{2}$$

and

$$\text{Height} = 30(\tfrac{1}{2}) = 15 \text{ ft}$$

EXAMPLE 3 The specific gravity of aluminum is 2.6, and water weighs 62.5 lb/ft³. Find the weight of a cubical block of aluminum with each edge 2 ft in length.

Solution The volume of the aluminum is $2^3 = 8$ ft³. The weight of 8 ft³ of water is $8(62.5) = 500$ lb, and the specific gravity of aluminum is 2.6. Therefore, since the specific gravity of a body is the ratio of the weight of the body to the weight of an equal volume of water, we have

$$2.6 = \frac{\text{weight of alluminum}}{500 \text{ lb}}$$

Hence, weight of aluminum = 2.6(500 lb) = 1300 *lb*.

EXAMPLE 4 Divide 88 into two parts which are in the ratio 4:7.

Solution If we let the two parts be a and b, then $a + b = 88$ and $a:b = 4:7$. Thus

$$\frac{a}{b} = \frac{4}{7} \quad \text{and so} \quad a = \frac{4b}{7}$$

Hence

$$\frac{4b}{7} + b = 88 \quad \frac{11b}{7} = 88 \quad \frac{b}{7} = 8$$
$$\text{and} \quad b = 56$$

Therefore

$$a = \frac{4(56)}{7} = 32$$

Another way to solve the problem is to let $a = 4x$ and $b = 7x$. Hence

$$4x + 7x = 88$$
$$11x = 88$$
$$x = 8$$

and it follows that $a = 4x = 32$ and $b = 7x = 56$.

Although the ratio $a:b$ is the fraction a/b, we also use the term "ratio" for three or more things. If the top speeds of a runner going uphill, level, and downhill are 12, 15, and 18 mi/h, then these speeds are in the ratio 12:15:18, which can be simplified by division by 3 to 4:5:6.

EXAMPLE 5 An investment portfolio is composed of blue-chip stocks, speculative stocks, and municipal bonds in the ratio 9:4:7. What percentage of each one is there?

Solution The amount of each can be represented as $9x$, $4x$, and $7x$, respectively, and since the total is 100%, we have

$$9x + 4x + 7x = 100$$
$$20x = 100$$
$$x = 5$$

Hence the percentage of blue chips is $9(5) = 45$, of speculative stocks is $4(5) = 20$, and of municipal bonds is $7(5) = 35$.

EXERCISE 9.1

Express the ratio as a fraction and simplify.

1. $240:840$

2. $252:378$

3. $\dfrac{2}{3}:\dfrac{16}{21}$

4. $\dfrac{48}{9}:\dfrac{32}{27}$

5. 6 weeks to 4 months

6. 6 quarters to 5 dimes

7. 4 h to 80 min

8. 1 m to 1 cm

9. $x^4y^2:x^3y^3$

10. $a^2 - b^2 : a + b$

11. $x^3 + y^3 : x + y$

12. $a^2b^3 : a^3b^2$

13 The radii of two spheres are 12 in. [1 ft] and 3 in. Find the ratio of the volume of the larger sphere to the volume of the smaller.

14 The radii of two circles are 12 and 8 units. Find the ratio of the area of the larger circle to the area of the smaller.

15 Find two numbers whose ratio is 3:4 and whose difference is 4.

16 Divide 180 into two parts that have the ratio 2:3.

17 The sum of two numbers is twice their difference. Find the ratio of the numbers.

18 The ratio of two numbers is $3:5$. If 6 is added to each number, the ratio of the sums is $3:4$. Find the numbers.

19 Divide 65 into three parts which are in the ratio $4:3:6$.

20 The public bet \$13,630 on Sinko in the fifth race. If the win, place, and show bets were in the ratio $14:9:6$, how much was bet on her to show?

21 The specific heat of a substance is the number of calories required to raise the temperature of 1 g of the substance 1°C. If 3.20 cal is required to raise the temperature of 12 g of brass 3°C, find the specific heat of brass.

22 The freezing and boiling points on the Celsius scale are 0 and 100°, respectively. On the Fahrenheit scale the freezing point is 32° and the boiling point is 212°. Find the ratio of 1°F to 1°C.

23 The specific gravity of a body is the ratio of the weight of the body to the weight of an equal volume of water. If 1 ft^3 of water weighs 62.5 lb and 1 ft^3 of iron weighs 490 lb, find the specific gravity of iron.

24 If the specific gravity of mercury is 13.6 and the specific gravity of lead is 11.35, find the difference between the weights of 1 ft^3 of mercury and 1 ft^3 of lead. (See Prob. 23.)

9.2 PROPORTION

A statement of the type

$$a:b = c:d \quad \text{or} \quad \frac{a}{b} = \frac{c}{d} \qquad (9.1)$$

is called a **proportion.** In other words, a proportion is a statement that two ratios are equal. The proportion is read "a is to b as c is to d," or "a divided by b is equal to c divided by d."

In any proportion, the first and fourth terms are called the **extremes,** and the second and third are called the **means.**

If three of the four numbers in a proportion are known, the other can be found by use of the property stated and proved below.

In any proportion, the product of the extremes is equal to the product of the means. That is

$$\text{if } a:b = c:d \quad \text{then} \quad ad = bc \qquad (9.2)$$

PROOF We first write the proportion in the form $a/b = c/d$ and then multiply each fraction by bd and have $abd/b = cbd/d$, or $ad = bc$.

EXAMPLE 1 Find x if $4:3 = 8:x$.

Solution Since the product of the means is equal to the product of the extremes, we have

$$4x = 24 \qquad x = 6$$

EXAMPLE 2 Find x if $5:x = 2:3$.

Solution

$$2x = 15 \qquad x = 7\tfrac{1}{2}$$

Proportions can be used to solve a variety of problems in everyday living, and they also have numerous applications in geometry, physics, and the other sciences. The following examples illustrate some of their uses.

EXAMPLE 3 If 15 gal of gasoline is required to operate a family car for 12 days, how many gallons will be required to operate the car for 6 weeks?

Solution We shall let x represent the required number of gallons. Now since the ratio of the two periods of time the car is in operation is equal to the ratio of the two corresponding amounts of gasoline consumed, we have, since 6 weeks = 42 days,

12 days : 42 days = 15 gal : x gal

Therefore

$$12x = 630 \quad \text{and} \quad x = 52\tfrac{1}{2} \text{ gal}$$

In Example 3 we set up the proportion

$$\frac{12 \text{ days}}{42 \text{ days}} = \frac{15 \text{ gal}}{x \text{ gal}}$$

We could also have written days : gal or gal : days as long as the same unit is in corresponding places in the proportion. For example

$$\frac{15 \text{ gal}}{12 \text{ days}} = \frac{x \text{ gal}}{42 \text{ days}}$$

which is solved by setting

$$\frac{15}{12} = \frac{x}{42} \quad \text{hence } x = 42\left(\frac{15}{12}\right) = \frac{42(5)}{4} = 52\tfrac{1}{2}$$

EXAMPLE 4 On August 1, Jones sold a piece of property to Brown. If the taxes on the property amount to $500 per year, find the amount that should be paid by each man.

Solution Since Jones owned the property 7 months of the year, the taxes should be divided in the ratio 7:5. We shall let

x = amount paid by Brown

Then

$500 - x$ = amount paid by Jones

We now have the proportion

$$7:5 = (500 - x):x$$

Hence,

$$7x = 2500 - 5x$$

and we solve this by adding $5x$ to each member and getting

$$12x = 2500$$
$$x = 208.33$$
$$500 - x = 291.67$$

Thus, Jones' part is $291.67 and Brown's is $208.33.

Blueprints for construction projects and maps are drawn according to a scale. For example, if a line 1 in long in a plan for a house represents 10 ft, the scale may be indicated by stating that 1 in =

10 ft, or the scale may be indicated as a ratio. In the latter case, the two distances must be expressed in terms of the same unit; the scale 1 in represents 10 ft would be expressed as the ratio 1:120.

EXAMPLE 5 If the floor plan for a house is drawn to the scale 1:144, find the dimensions of a room represented in the plan by a rectangle $1\frac{1}{2}$ by $2\frac{1}{4}$ in.

Solution We shall let w and l represent the width and length of the room, respectively. We find w by use of the proportion

$1:144 = 1\frac{1}{2}:w$

Hence,

$w = 1\frac{1}{2} \times 144 = 216$ in $= 18$ ft

Likewise,

$1:144 = 2\frac{1}{4}:l$

and

$l = 2\frac{1}{4} \times 144 = 324$ in $= 27$ ft

Two triangles or two polygons that have the same shape are said to be **similar.** The ratios of the corresponding sides of two similar figures are equal. For example, in the similar triangles in Fig. 9.1

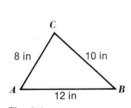

Fig. 9.1

$$\frac{AB}{A'B'} = \frac{BC}{B'C'} = \frac{CA}{C'A'}$$

EXAMPLE 6 Two boys in a geometry class were given the problem of finding the height of a flagpole. One of the boys was 4 ft 8 in tall, and they found that his shadow was 3 ft long. They immediately measured the shadow of the flagpole and found that it was 18 ft long. How could they calculate the height of the pole from this information?

Solution By use of similar triangles, they would know by Fig. 9.2 that

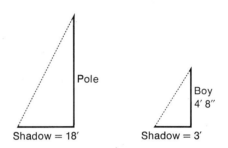

Fig. 9.2

$$\frac{\text{Height of boy}}{\text{Height of pole}} = \frac{\text{length of shadow of boy}}{\text{length of shadow of pole}}$$

They could then substitute the known quantities and, since 4 ft 8 in $= 4\frac{2}{3}$ ft, get

$$\frac{4\frac{2}{3}}{\text{Height of pole}} = \frac{3}{18}$$

3 (height of pole) $= (4\frac{2}{3})(18) = 84$
Height of pole $= 28$ ft

EXERCISE 9.2

Find the value of x.

1. $12:18 = 8:x$

2. $15:20 = x:16$

3. $\dfrac{x}{17} = \dfrac{9}{51}$

4. $\dfrac{9}{24} = \dfrac{x}{16}$

5. $5\tfrac{1}{3}:3\tfrac{1}{5} = 25:x$

6. $4\tfrac{4}{5}:24 = \tfrac{1}{15}:x$

7. $25:x = x:9$

8. $16:x = x:4$

9. $x:2\tfrac{1}{2} = 10:x$

10. $15:x = x:2\tfrac{2}{5}$

11. $9:6 = (10 - x):4$

12. $12:8 = 9:(x + 2)$

13 If $x + y = 14$ and $x:y = 3:4$, find x and y.

14 If $x - y = 1$ and $x:y = 24:36$, find x and y.

15 If $x + 2y = 15$ and $x:y = 2\frac{1}{2}:5$, find x and y.

16 If $x + y = 10$ and $(x + y):(x - y) = 5:3$, find x and y.

17 If $x - y = 4$ and $(x + y):(x - y) = 5:1$, find x and y.

18 If $x + 2y = 16$ and $(x - 2y):6 = 2:3$, find x and y.

19 Find x if $(x-1):5 = 3:(x+1)$.

20 Find x if $(x-2):(x+1) = (x-3):(x+3)$.

21 The areas of two circles are proportional to the squares of the radii. If the radius of a larger circle is twice the radius of the smaller and the area of the larger circle is 616 in², find the area of the smaller.

22 The volumes of two spheres are proportional to the cubes of their radii. If the volume of one sphere is 8 times the volume of another and the radius of the smaller is 2, find the radius of the larger.

23 A map is drawn to the scale $2\frac{1}{4}$ in to 8 mi. Find the distance between two towns if their locations on the map are 9 in apart.

Chapter 9: Ratio, Proportion, and Variation

24 A boy $4\frac{2}{3}$ ft tall standing near a telephone pole casts a shadow $3\frac{1}{2}$ ft long. If the shadow of the pole is 30 ft long, how tall is the pole?

25 The amounts of water discharged by two circular orifices under the same pressure are proportional to the squares of the radii of the orifices. If an orifice with a radius of 8 cm discharges water at 20.9 L/s, how many liters will be discharged in 1 s by an orifice of radius 3 cm?

[Handwritten work:]
$r = 8$ $\quad w = Kr^2$
$w = 20.9$ $\quad 20.9 = K \cdot 64$
$\frac{20.9}{64} \cdot r^2 = w$
$\frac{20.9}{64} \cdot 9 = w$
$2.93 \text{ L/s} = w$

26 The annual tax on Smith's residence is $800. On June 1 he sold the property to Brown. How much of the tax should each pay?

27 Two weights balanced on a lever are inversely proportional to their respective distances from the fulcrum. If 40 kg 4 m from the fulcrum balances 70 kg, find the distance of the latter weight from the fulcrum.

28 The weights of two blocks of metal with the same volume are proportional to their specific gravities. If a cubic inch of iron weighs 0.238 lb, find the weight of a cubic inch of mercury if the specific gravities of iron and mercury are 7.84 and 13.6, respectively.

9.3 VARIATION

If two nonzero quantities change in such a way that their quotient never changes, then one of the quantities is said to **vary as** the other. For example, consider the following situation: If the valve of the intake pipe of an empty water tank is opened and 2 gal/min of water flows into the tank, the amount A of water in the tank at the end of t min is equal to $2t$ gal. As t changes, A changes but at all times $A/t = 2$. Hence, A varies as t.

If y *varies as* x, then $y/x = k$, where k is a constant, and $y = kx$. The constant k is called the **constant of variation.**

Many physical laws state that one quantity varies as another or as certain combinations of others, and these statements must be translated into equations before the methods of algebra can be applied to them. As an example, Charles' law states that if the pressure is constant, the volume of any mass of gas varies as the absolute temperature. Hence, if we let V equal the volume of a mass of gas and T equal the absolute temperature, then, since V varies as T, we have $V = kT$, where k is a constant.

The four types of variation are direct, inverse, joint, and combined variation. We shall next explain the meaning of each of these types and the method for obtaining the equation of variation for each.

Direct variation: If y varies *directly* as x, then $y = kx$. The word "directly" is usually omitted when referring to direct variation.

Inverse variation: The statement "y varies *inversely* as x" means that y varies directly as the reciprocal of x. Therefore $y = k(1/x) = k/x$.

Joint variation: If a quantity varies *jointly* as two or more others, then it varies directly as their product. For example, if y varies jointly as x and w, then $y = kxw$.

Combined variation: In this type of variation we have a combination of direct, joint, and inverse variation. Newton's law of gravitation for the force of attraction between two bodies is an example of this type, since F varies jointly as the product of the masses and inversely as the square of the distance. Thus $F = kmm'/d^2$. If y varies as the product of x and w and inversely as z, then $y = k(xw/z)$.

A problem that can be solved by the method of variation consists of the following three parts:

1. A statement of the law that is operating in the problem
2. A set of data consisting of a value for each variable in the problem
3. A second set of data that contains a value for all but one of the variables

STEPS IN SOLVING A VARIATION PROBLEM

The process of solving such a problem consists of the following three steps:

1. Represent each of the quantities in the problem by a letter and then use the statement of the law in the problem to formulate the equation of variation.
2. Substitute the values in the first set of data for the letters in the equation obtained in step 1 and then solve for k.
3. Substitute the values in the second set of data together with the value of k in the equation and solve for the remaining variable.

EXAMPLE 1 The horsepower (hp) required to propel a ship varies as the cube of the speed. If the horsepower required for a speed of 12 knots is 5184, what is the horsepower required for a speed of 15 knots?

Solution

1. Let P = required horsepower
   s = speed

   Then, since P varies as s^3, we have

   $$P = ks^3 \tag{1}$$

2. If $s = 12$ knots, then $P = 5184$. On substituting these values in (1), we get

   $$5184 = k(12^3)$$

   Therefore,

   $$k = \frac{5184}{12^3} = \frac{5184}{1728} = 3$$

   and the equation of variation is $P = 3s^3$.

3. For $s = 15$ and $k = 3$, we have

   $$P = 3(15^3) = 3(3375) = 10{,}125 \text{ hp}$$

EXAMPLE 2 The safe load on a beam supported at both ends varies directly as the product of the width and the square of the depth of the beam and inversely as the length of the beam between supports. If the safe load on a beam 2 in wide, 6 in deep, and 10 ft long between supports is 1440 lb, find the safe load for a beam 3 in wide, 4 in deep, and 12 ft long between supports.

Solution

1. Let w = width
   d = depth
   l = length between supports
   L = safe load

   Now, by using the statement of the law in the problem, we have

   $$L = k\frac{wd^2}{l} \qquad (1)$$

2. If $w = 2$, $d = 6$, and $l = 10$, then $L = 1440$. By substituting these values in (1), we get

   $$1440 = \frac{k(2)(6^2)}{10}$$

   Performing indicated operations and multiplying each member by 10 gives

   $$14{,}400 = 72k$$

   Therefore $k = 200$ and the equation of variation is

   $$L = 200\frac{wd^2}{l} \qquad (2)$$

3. In order to find L when $w = 3$, $d = 4$, and $l = 12$, we substitute these values in (2) and obtain

$$L = 200\frac{3(4^2)}{12} = 800 \text{ lb}$$

We can also solve Example 2 without finding k, for Eq. (2) can be written

$$\frac{Ll}{wd^2} = k$$

This means that the left member of the equation is constant, so that we can equate the left members for two sets of data. Using the first datum on the right side and the second on the left gives

$$\frac{L(12)}{3(4^2)} = \frac{(1440)(10)}{2(6^2)}$$

$$L = \frac{1440}{12}\frac{10}{2}\frac{3(4)(4)}{6(6)} = \frac{120(5)(2^2)}{3} = 800$$

EXAMPLE 3 If y varies inversely with x, fill in the missing values in the table.

x	3	6		60
y	10		2	

Solution Since $y = k/x$, then $10 = k/3$ gives $k = 3(10) = 30$. Thus $y = 30/x$. Using this with $x = 6$ gives $y = 5$, with $y = 2$ gives $x = 15$, and with $x = 60$ gives $y = \frac{1}{2}$.

EXERCISE 9.3

Express the statement as an equation.

1. a varies directly as b.

2. a varies directly as the square of b.

3. z varies jointly as x and y.
   $z = kxy$

4. a varies jointly as b and the square of c.

5. a varies inversely as b.
   $a = k\frac{1}{b} \quad a = \frac{k}{b}$

6. y varies inversely as the square of x.
   $y = \frac{k}{x^2}$

7. z varies inversely as the square root of w.
   $z = \frac{k}{\sqrt{w}} \quad z = k\frac{1}{\sqrt{w}}$

8. y varies jointly as x and v and inversely as w.
   $y = \frac{kxv}{w}$

9. a varies jointly as b and c and inversely as the product of d and e.
   $a = \frac{kbc}{d \times e}$

10. y varies directly as the square of z and inversely as the sum of v and w.
    $y = \frac{kz^2}{v-w}$

11. The bend b of a bar under a given force F varies inversely as the width w of the bar.

12. The power p available in a jet of water varies jointly as the cube of the water's velocity V and the cross-sectional area a of the jet.

Set up the equation that involves the variables, evaluate k using the given data, and complete the problem as directed.

13. If y varies directly as x and $y = 10$ when $x = 4$, find y if $x = 12$.

14. If y varies jointly as x and z and $y = 12$ when $x = 6$ and $z = 3$, find y if $x = 4$ and $z = 9$.

15. If y varies jointly as x and z and inversely as w, and if $y = 12$ when $x = 4$, $z = 9$, and $w = 6$, find y if $x = 3$, $z = 12$, and $w = 18$.

16 If y varies as the sum of x and z and inversely as w, and if $y = 2$ when $x = 7$, $z = 9$, and $w = 4$, find y if $x = 5$, $z = 13$, and $w = 3$.

17 If y varies directly as the sum of a and b, fill in the missing values in the table.

a	1	3	4	
b	2	5		8
y	8		4	2

18 Complete the table if y varies inversely as x^2.

x	2	6	
y	9		4

19 Fill in the blanks in the table if p varies jointly as q and the square root of r.

q	3	2		4
r	4	1	5	
p	6		5	16

20 Complete the table if d varies jointly as e^2 and f and inversely as $2 + g$.

d	4	1	
e	4	3	2
f	2		5
g	46	52	8

21 The heat produced in a heater coil varies as the square of the current. If a current of 12 A produces 68.0 cal of heat, find the heat produced by a current of 8 A.

22 The centrifugal force on a plane making a turn with a given radius varies as the square of the velocity. If the centrifugal force at 200 mi/h is 12,800 lb, find the force at 170 mi/h.

23 If the resistance due to air is neglected and the initial velocity is zero, the distance that a compact body will fall varies as the square of the number of seconds the body falls. If a body falls 1030.4 ft in 8 s, how far will another body fall in 3 s?

24 The exposure time necessary to produce a good photographic negative varies directly as the square of the f number of the camera lens. If $\frac{1}{200}$ s produces a good negative at $f/8$, find the exposure time for $f/10$.

25 The distance in miles that can be seen from a plane at an altitude of h feet varies directly as the square root of h. If a pilot can see 24.5 mi from a plane flying at an altitude of 400 ft, how far can he see when the plane is at an altitude of 8100 ft?

26 The increase in the length of an iron rod due to heating varies directly as the difference in temperatures. If an iron rail is 32 ft long at 0°C and 32.015 ft long at 40°C, find its length at 30°C.

27 The weight of an object at a distance of d mi from the surface of the earth varies inversely as the square of the distance from the center of the earth. If an astronaut weighs 200 lb on the surface, how much would he weigh in a capsule 16,000 mi from the surface of the earth? Assume the radius of the earth to be 4000 mi.

28 The gravitational attraction between two bodies varies inversely as the square of the distance between them. If the attraction is 3.330×10^{-2} dyn when the distance is 100 cm, find the attraction when the distance is 10 cm.

9.4 CHAPTER SUMMARY

The ratio of a to b is written $a:b$, and its value is the fraction a/b. We can also speak of x, y, and z being in the ratio $x:y:z$.

A proportion is the statement

$$\frac{a}{b} = \frac{c}{d}$$

that two ratios are equal. The numbers a and d are called the extremes of the proportion, and b and c are called the means. In any proportion, we have

$$ad = bc$$

which means that the product of the extremes equals the product of the means. A correct proportion must have the same units in corresponding places in the proportion.

If y varies directly as x, then

$$y = kx$$

If y varies inversely as x, then

$$y = \frac{k}{x}$$

If y varies jointly as x and w, then

$$y = kxw$$

STUDENT'S NOTES

EXERCISE 9.4 REVIEW

Express the following ratios as fractions and simplify.

1. $504 : 546$

2. $\frac{15}{98} : \frac{25}{14}$

3. $14 \text{ cm} : 0.5 \text{ m}$

4. $4000 \text{ s} : 1 \text{ h}$

5. $a^2 - b^2 : a^2 + ab - 2b^2$

6. If the radii of two circles are in the ratio of $2:5$, what is the ratio of their areas?

7. Find two numbers whose sum is 60 if they are in the ratio of $2:3$.

8. Find three numbers whose sum is 60 if they are in the ratio of $3:4:5$.

9 If a rich uncle left $7,650,000 to be divided among three people in the ratio of $2:3:4$, how much does each receive after 15% is deducted for various fees?

Find the value of x.

10 $x:2.5 = 10:25$

11 $\frac{5}{2} : \frac{13}{4} = \frac{15}{2} : x$

12 $(x - 3):8 = 25:40$

13 $x:y = 5:7$ and $x + y = 36$

14 The volume of a right circular cylinder is proportional to the product of its altitude and the square of its radius. If a cylinder of radius r and altitude h has volume V, what is the volume of a cylinder with radius $2r$ and altitude $h/2$?

15 A body immersed in water is buoyed up by a force equal to the weight of the water displaced. If a block of lead weighs 2.0525 kg in air and 1.8717 kg in water, find the specific gravity of lead. See Prob. 23 in Exercise 9.1.

16 If the distance is constant, the gravitational attraction between two bodies varies jointly as their masses. If the attraction between a 100-kg sphere and a 50-kg sphere is $\frac{10}{3}$ dyn, find the attraction between a 75-kg sphere and a 25-kg sphere.

17 If a compact body is attached to a string and whirled in a circle, the pull on the string in poundals varies jointly as the mass, the radius, and the square of the number of revolutions per second. If a 2-lb ball that is whirled on a 4-ft string at 4 r/s exerts a pull of 1040 poundals, find the pull of a 1-lb ball that is whirled on a 2-ft string at 3 r/s.

18 The electrical resistance of a wire varies directly as its length and inversely as the square of its diameter. If the resistance of a wire 50 ft long with a diameter of $\frac{1}{16}$ in is $\frac{1}{75}$ ohm, find the resistance of a wire 40 ft long with a diameter of $\frac{1}{32}$ in.

19 The safe load for a horizontal beam supported at both ends varies jointly as the width and the square of the depth and inversely as the distance between the supports. If a beam 4 in wide, 6 in deep, and 49 ft long supports 450 lb, find the safe load of a beam 6 in wide, 4 in deep, and 15 ft long.

20 The force of wind on a flat surface perpendicular to the direction of the wind varies as the area of the surface and the square of the velocity of the wind. When the wind is blowing 16 mi/h, the force on a 3 by 4 ft window is 15 lb. Find the force on a 4 by 6 ft sign in a wind blowing 8 mi/h.

21 The gravitational attraction between two bodies varies directly as the product of their masses and inversely as the square of the distance between them. Two spheres whose masses are 5×10^5 g and 2×10^5 g are 10^3 cm apart, and the gravitational attraction between them is 6.66×10^{-3} dyn. Find the force of attraction between two spheres of masses 4×10^4 and 3×10^3 g that are 10^2 cm apart.

22 If 2 chickens lay 6 eggs in 3 days, how many eggs do 5 chickens lay in 6 days?

NAME _____ DATE _____ SCORE _____

EXERCISE 9.5 CHAPTER TEST

1 Express the ratio 5 lb to 12 oz as a fraction and simplify.

2 Find the quotient of 4 doz cookies divided by 16 boys and interpret the result.

3 If 85 g of silver nitrate will react with 37 g of potassium bromide, how much potassium bromide will react with 255 g of silver nitrate?

4 If $6:15 = x:50$, find the value of x.

5 If $x - y = 7$ and $x:y = 12:8$, find x and y.

6 Griggs used three sacks of cement to mix concrete for a sidewalk 3 ft wide, 4 in thick, and 12 ft long. Find how much cement he will need to mix concrete for a similar sidewalk 16 ft long.

7   The centrifugal force at any point of a revolving body varies as the radius of the circle in which the body is revolving. If the centrifugal force of a revolving body is 450 lb when the radius of the circle is 12 in, at what radius is the force 600 lb?

8   If the lever arm is constant, the mechanical advantage of a jackscrew varies inversely as the pitch of the screw. If the mechanical advantage of a jackscrew with a pitch of 0.12 is 125, what is the mechanical advantage of a jackscrew with the same lever arm that has a pitch of 0.15?

9   The kinetic energy of a body varies jointly as the weight and the square of the velocity of the body. If a 2000-lb car has a kinetic energy of 242,000 ft·lb when traveling 60 mi/h (88 ft/s), what is its kinetic energy when it is traveling 45 mi/h (66 ft/s)?

10   The current passing through a system varies directly as the electromotive force and inversely as the resistance of the system. If in a certain system a current of 32 A flows through a resistance of 5 ohms, with an electromotive force of 160 V, find the current that 120 V will send through the system.

CHAPTER 10

Systems of Linear Equations and Determinants

10.1 LINEAR EQUATIONS IN TWO VARIABLES

We shall discuss linear equations in two and three variables in this chapter.

An equation of the type $ax + by = c$ is a **linear equation in two variables** or unknowns.

Examples are $3x + 5y = -4$ and $y = -4x + 8$.

A **solution** or **solution pair** of $ax + by = c$ is an ordered pair of numbers such that if x is replaced by the first number in the pair and y is replaced by the second, a true statement is obtained.

The **solution set** of $ax + by = c$ is the set of all solutions of the equation.

Two linear equations in two variables are **equivalent** if their solution sets are equal.

To illustrate the above definitions, we consider the equation

$$2x + 3y = 9$$

This equation has $a = 2$, $b = 3$, and $c = 9$. We can obtain a solution pair by assigning a number to x and then solving the resulting equation for y or by replacing y by a number and solving the resulting equation for x. For example, if we assign 3 to x, we have $6 + 3y = 9$; then $y = 1$, so $(3,1)$ is a solution pair. Similarly, if we assign 5 to y and solve for x, we get $2x + 15 = 9$; then $x = -3$, so $(-3,5)$ is a solution pair. By use of this method we can obtain the six solution pairs tabulated as follows:

x	y	Solution pair
−6	7	(−6,7)
−3	5	(−3,5)
0	3	(0,3)
3	1	(3,1)
$4\frac{1}{2}$	0	$(4\frac{1}{2},0)$
6	−1	(6,−1)

Since we can assign any real number to x and solve for y or any real number to y and solve for x, the solution set of (2) contains an infinitude of solution pairs.

10.2 GRAPHS OF LINEAR EQUATIONS IN TWO VARIABLES

Since a solution pair of an equation in two variables is an ordered pair of numbers, each pair determines a unique point in a cartesian plane. We therefore have the following definition, as in Chap. 5:

The **graph** of an equation in two variables is the set of all points such that the coordinates of each is a solution pair of the equation.

If we plot the points determined by the solution pairs in the table of Sec. 10.1, we obtain the points in Fig. 10.1. They appear to lie on a straight line,

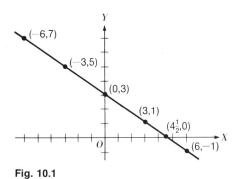

Fig. 10.1

and, in fact, they do, since it is proved in analytic geometry that the graph of a linear equation in two variables is a straight line. Now we draw a straight-line segment through these points, extend it upward and downward in the space allowed, and

thus obtain the graph of $2x + 3y = 9$. The graph is a line of unlimited length, and only a portion of it can be drawn.

Since a straight line is uniquely determined by two points on it, it is necessary to obtain only two solution pairs of a linear equation in order to construct the graph. It is advisable, however, to determine a third solution pair as a check. Usually the first step in obtaining the graph of an equation is to determine the coordinates of the points of intersection of the graph with the coordinate axes. The abscissa of the intersection with the X axis is called the X **intercept,** and the ordinate of the intersection with the Y axis is the Y **intercept.** Since the ordinate of any point on the X axis is 0, we get the X intercept by replacing y by 0 in the equation and then solving for x. Similarly, to get the Y intercept, we replace x by 0 and solve for y. The procedure is illustrated in Example 1. Note that an equation for the X axis is $y = 0$.

EXAMPLE 1 Construct the graph of $3x + 4y = 12$.

Solution We get the X and Y intercepts as follows. Assign 0 to y, solve for x, and get

$$3x + 4(0) = 12$$
$$x = 4$$

Assign 0 to x, solve for y, and get

$$3(0) + 4y = 12$$
$$y = 3$$

Consequently the graph intersects the X and Y axes at the points (4,0) and (0,3), respectively. To get a third point, we assign any number other than 4 or 0 to x and solve for y. We arbitrarily choose 8, and get

$$3(8) + 4y = 12$$
$$4y = 12 − 24 = −12$$
$$y = −3$$

Thus, we have the three solution pairs arranged in tabular form, as follows:

x	0	4	8
y	3	0	−3

Now we plot the points determined by the ordered pairs in the above table, draw a straight-line segment through them, and obtain the graph in Fig. 10.2.

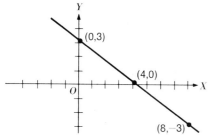

Fig. 10.2

10.3 SYSTEMS OF TWO LINEAR EQUATIONS: GRAPHICAL METHOD

We stated in Sec. 10.1 that a linear equation in two variables has an unlimited number of solution pairs. In this section we shall consider a system of two linear equations and explain the graphical method for finding the pair of numbers that satisfy both equations. The set of such pairs is called the **simultaneous solution set** of the system.

To illustrate the method, we consider the equations

$$3x - 5y = 15 \qquad (1)$$

$$3x + y = 6 \qquad (2)$$

The coordinates of each point on the graph of each of these equations is a solution pair of the equation. Hence if the equations have the same solution pair, it is the coordinates of a point on both graphs and is therefore the coordinates of their point of intersection. Therefore to obtain the simultaneous solution set, we construct the graphs of (1) and (2) and estimate the coordinates of their point of intersection. We use the solution pairs in the following tables to construct the graphs:

x	-5	0	5
y	-6	-3	0

for (1)

and

x	0	2	4
y	6	0	-6

for (2)

The graphs are shown in Fig. 10.3. They intersect at the point P whose coordinates appear to be $(2\frac{1}{2}, -1\frac{1}{2})$. Hence, we estimate that the simultaneous solution set of the two equations is $\{(2\frac{1}{2}, -1\frac{1}{2})\}$. We can check the accuracy of this

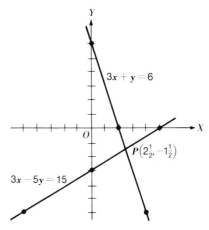

Fig. 10.3

estimation by replacing x by $2\frac{1}{2}$ and y by $-1\frac{1}{2}$ in each of the equations. Thus $3(2\frac{1}{2}) - 5(-1\frac{1}{2}) = \frac{15}{2} + \frac{15}{2} = 15$ from (1) and $3(2\frac{1}{2}) + (-1\frac{1}{2}) = \frac{15}{2} - \frac{3}{2} = \frac{12}{2} = 6$ from (2). Consequently, since the right members of (1) and (2) are 15 and 6, respectively, $\{(2\frac{1}{2}, -1\frac{1}{2})\}$ is the exact solution set.

EXAMPLE 1 Solve simultaneously

$$2x - 3y = 6 \qquad (3)$$

and

$$x + 2y = 4 \qquad (4)$$

Solution We use the following tables of solution pairs to construct the graphs:

x	3	0	-3
y	0	-2	-4

for (3)

and

x	-4	0	4
y	4	2	0

for (4)

Now by referring to Fig. 10.4, we see that the abscissa of the point of intersection of the graphs ap-

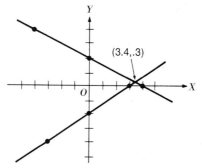

Fig. 10.4

pears to be slightly less than $3\frac{1}{2}$, and the ordinate is about $\frac{1}{3}$. Hence, to the nearest tenth, we estimate the simultaneous solution set to be $\{(3.4, .3)\}$. Now if we replace x by 3.4 and y by .3 in (3) and (4), we get

$$2(3.4) - 3(0.3) = 6.8 - 0.9 = 5.9 \quad \text{from (3)}$$

and

$$3.4 + 2(0.3) = 3.4 + 0.6 = 4 \quad \text{from (4)}$$

Therefore, since the right members of (3) and (4) are 6 and 4, respectively, we are fairly safe in stating that our estimate is correct.

EXAMPLE 2 Solve graphically the system

$$x + 3y = 7 \tag{5}$$
$$2x - 5y = 3 \tag{6}$$

Solution We use the following tables to construct the graphs:

x	7	1	-5
y	0	2	4

for (5)

and

x	-6	-1	4
y	-3	-1	1

for (6)

Now by Fig. 10.5, the point of intersection appears to be (4,1). This is the exact solution, as can be checked in each equation.

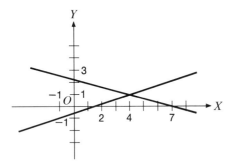

Fig. 10.5

If an exact solution is not obtained graphically, the solution can be estimated from the graph. The degree of precision depends on the accuracy of the graph and the choice of scale on each axis.

In Secs. 10.5 and 10.6 two methods are presented which give exact solutions.

Problems 1 to 20 in Exercise 10.1 may be done now.

10.4 INDEPENDENT, INCONSISTENT, AND DEPENDENT EQUATIONS

Obviously, if the graphs of two linear equations are *parallel* lines but not the same, the equations have *no* simultaneous solution pair. Furthermore, if the two graphs *coincide*, every solution pair of one equation is a solution pair of the other. If the graphs are *neither parallel nor coincident*, they intersect in exactly one point, and therefore the simultaneous solution set consists of only *one* ordered pair of numbers. This illustrates the following definition:

Two linear equations in two variables are **independent** if their simultaneous solution set consists of one ordered pair of numbers, they are **inconsistent** if their simultaneous solution set is the empty set, and they are **dependent** if every solution pair of one equation is a solution pair of the other.

The following theorem enables us to decide whether two linear equations are independent, inconsistent, or dependent.

Two linear equations $ax + by = c$ and $Ax + By = C$ are

1. Independent if and only if $A/a \neq B/b$
2. Inconsistent if and only if $A/a = B/b \neq C/c$
3. Dependent if and only if $A/a = B/b = C/c$

EXAMPLE If we apply this theorem to each of the following systems of equations (see Fig. 10.6), we see that

$$2x - 3y = 4$$
$$5x + 2y = 8$$

are independent, since $\frac{2}{5} \neq -\frac{3}{2}$;

$$3x - 9y = 1$$
$$2x - 6y = 2$$

are inconsistent, since $\frac{3}{2} = -9/(-6) \neq \frac{1}{2}$; and

$$2x - 4y = 12$$
$$3x - 6y = 18$$

are dependent, since $\frac{2}{3} = -4/(-6) = \frac{12}{18}$.

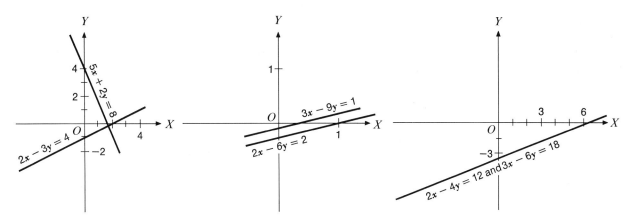

Fig. 10.6

STUDENT'S NOTES

EXERCISE 10.1

Find the exact solution pairs graphically; graph paper is suggested.

1. $2x - 3y = 2$
   $x + y = 6$

2. $2x + y = 4$
   $x - 3y = 9$

3. $2x + 5y = -1$
   $3x - y = -10$

4. $6x - y = -1$
   $3x + 2y = -13$

5. $x + y = 8$
   $x - y = -2$

6. $x + 2y = -8$
   $2x - y = 9$

7. $7x + y = 4$
   $5x - 3y = 14$

8. $4x + 3y = 7$
   $-2x + y = 9$

9. $2x - y = 9$
   $x - 2y = 0$

10. $-3x + y = 13$
    $2x - y = -11$

11. $3x - 4y = -17$
    $-2x + y = -2$

12. $4x + 3y = -7$
    $3x + 2y = -3$

Find each coordinate of each solution pair graphically to the nearest integer; graph paper is suggested.

13. $5x - 17y = 3$
    $2x + 7y = 14$

14. $8x - 7y = -19$
    $5x + 2y = 19$

15. $9x + 8y = 13$
    $-6x + 5y = -61$

16. $8x + 7y = 3$
    $7x - 6y = -57$

17. $16x + 3y = 55$
    $4x - 5y = -32$

18. $6x + 11y = -48$
    $8x - 5y = 75$

19. $3x + 8y = 60$
    $5x - 7y = 0$

20. $12x + 7y = 25$
    $10x - 3y = 4$

Show that the equations are independent.

21. $2x + 9y = 4$
    $3x - y = -5$

22. $4x - 7y = 6$
    $-5x + 8y = 6$

23. $6x - 5y = 4$
    $5x - 4y = 3$

24. $-2x + 7y = 1$
    $-3x + 12y = 1$

Show that the equations are dependent.

25. $2x - 5y = 7$
    $4x - 10y = 14$

26. $3x + 2y = 7$
    $9x + 6y = 21$

27. $2x + 4y = 10$
    $3x + 6y = 15$

28. $6x - 9y = 6$
    $10x - 15y = 10$

Show that the equations are inconsistent.

29. $2x + 7y = 5$
    $2x + 7y = 4$

30. $3x - 4y = 6$
    $6x - 8y = 11$

31. $6x - 4y = 12$
    $15x - 10y = 20$

32. $15x + 12y = 6$
    $20x + 16y = -8$

State whether the equations are independent, dependent, or inconsistent.

33. $2x + 6y = 7$
    $4x + 12y = 12$

34. $2x + 6y = 5$
    $4x + 12y = 10$

35. $5x - 3y = 6$
    $10x - 7y = 4$

36. $8x + 12y = 2$
    $12x + 18y = 3$

37. $4x + 7y = 6$
    $6x + 12y = 9$

38. $4x + 8y = 6$
    $6x + 12y = 9$

39. $4x + 8y = 7$
    $6x + 12y = 9$

40. $4x + 9y = 6$
    $6x + 12y = 9$

10.5 ELIMINATION BY ADDITION OR SUBTRACTION

The procedure for finding the simultaneous solution of a system of equations is called *solving the system*. In the remainder of this chapter we shall discuss algebraic methods for accomplishing this.

EXAMPLE 1 Solve the system of equations

$$2x + 3y = 13 \tag{1}$$
$$x - 3y = -7 \tag{2}$$

Solution Since $\frac{2}{1} \neq 3/(-3)$, the equations are independent, so one and only one solution pair exists. Now by adding corresponding members of Eqs. (1) and (2), it follows that (x,y) is a solution pair of

$$(2x + 3y) + (x - 3y) = 13 + (-7) \tag{3}$$

or, combining terms,

$$2x + x + 3y - 3y = 13 - 7 \tag{4}$$

Therefore, $3x = 6$, and $x = 2$. Consequently, 2 is the first number in the solution pair. We obtain the second number by replacing x by 2 in either (1) or (2) and solving the resulting equation for y. We use (2) and get

$$2 - 3y = -7$$
$$-3y = -7 - 2 = -9$$
$$y = \frac{-9}{-3} = 3$$

Hence, the simultaneous solution of (1) and (2) is (2,3).

Check Replacing x by 2 and y by 3 in the given equations, we have

$$2(2) + 3(3) = 4 + 9 = 13 \quad \text{from (1)}$$
$$2 - 3(3) = 2 - 9 = -7 \quad \text{from (2)}$$

Note that in the above procedure we combined Eqs. (1) and (2) to obtain one equation in one variable whose root was one of the numbers in the solution pair. This procedure is called **eliminating a variable.**

In Eqs. (1) and (2) the coefficients of y are 3 and -3, respectively, so we eliminate y by equating the sums of the corresponding members of the equations. Usually the coefficients of x or the coefficients of y are not equal. In such cases we multiply each equation by a constant as the first step in the process of solving. We illustrate the procedure in Example 2.

EXAMPLE 2 Solve the equations

$$x - 2y = 6 \tag{5}$$
$$3x + 5y = 7 \tag{6}$$

Solution If we simply add (5) and (6), neither x nor y will be eliminated. We can eliminate x by multiplying (5) by -3 and then adding (6). This is equivalent to multiplying (5) by 3 and then subtracting (6). Multiplying (5) by -3 gives

$$-3x + 6y = -18 \tag{7}$$
$$3x + 5y = 7 \tag{8}$$

where (8) is the same as (6). Adding (7) and (8), we have

$$11y = -11 \quad y = -1$$

Substituting $y = -1$ in (5) gives $x - 2(-1) = 6$, $x = 6 - 2 = 4$.

Thus the solution is $(4, -1)$, which should be checked in both original equations (5) and (6).

Of course we can eliminate either variable as the first step in solving a system of equations, and it is usually advisable to choose the variable whose coefficients have the smaller lcm, as is illustrated in Example 3.

EXAMPLE 3 Simultaneously solve the following equations:

$$5x + 14y = 12 \tag{9}$$
$$-3x + 8y = 1 \tag{10}$$

Solution Since the lcm of the coefficients of x is 15, and the lcm of the coefficients of y is 56, the computation will be easier if we eliminate x instead of y. Hence, we proceed as follows:

$$\begin{aligned} 15x + 42y &= 36 \quad \text{multiplying (9) by 3} \tag{11}\\ -15x + 40y &= 5 \quad \text{multiplying (10) by 5} \tag{12}\\ \hline 82y &= 41 \quad \text{adding corresponding} \\ & \qquad \text{members of (11)} \\ & \qquad \text{and (12)} \end{aligned}$$

$$y = \frac{41}{82} = \frac{1}{2}$$

$$5x + 14(\tfrac{1}{2}) = 12 \quad \text{replacing } y \text{ by } \tfrac{1}{2} \text{ in (9)}$$
$$5x = 12 - 7 = 5$$
$$x = 1$$

Hence the solution is $(1, \tfrac{1}{2})$.

Problems 1 to 12 in Exercise 10.2 may be done now.

10.6 ELIMINATION BY SUBSTITUTION

If one of a given pair of linear equations in two variables is easily solved for one variable in terms of the other, it is frequently more efficient to solve the equations simultaneously by the method of substitution. The method is illustrated in Examples 1 and 2.

EXAMPLE 1 Solve the following equations simultaneously by the method of substitution:

$$3x - 7y = 13 \quad (1)$$
$$x + 2y = 0 \quad (2)$$

Solution If we solve (2) for x in terms of y, we get

$$x = -2y \quad (3)$$

Now we substitute $-2y$ for x in (1) and solve the resulting equation for y, as indicated by

$$\begin{aligned} 3(-2y) - 7y &= 13 \\ -6y - 7y &= 13 \\ -13y &= 13 \\ y &= -1 \end{aligned}$$
replacing x by $-2y$ in (1)
multiplying $-2y$ by 3

Finally, we replace y by -1 in (3) and get

$$x = -2(-1) = 2$$

Hence, the solution is $(2, -1)$.

EXAMPLE 2 We can solve Example 2 of Sec. 10.5 by substitution.

Solution To do so, we solve (5) for x, getting $x = 6 + 2y$. Using this in (6) gives

$$\begin{aligned} 3(6 + 2y) + 5y &= 7 \\ 18 + 6y + 5y &= 7 \\ 11y &= -11 \\ y &= -1 \end{aligned}$$

The solution is now completed as before, giving $(4, -1)$ as the solution.

In deciding whether to use addition or substitution, remember that both methods work. To avoid fractions, use substitution if one of the variables in the system has a coefficient of 1. If no coefficient is 1, use addition.

EXERCISE 10.2

Solve the following equations simultaneously by addition or subtraction.

1. $2x + 5y = 1$
   $3x - 5y = 14$

2. $5x - 2y = 20$
   $3x + 2y = -4$

3. $2x + 7y = 13$
   $2x - 5y = -23$

4. $5x + 2y = -18$
   $5x - 3y = 2$

5. $3x - 4y = 7$
   $4x + 2y = 13$

6. $5x + 9y = 19$
   $2x + 3y = 8$

7. $10x - 5y = 1$
   $5x + 2y = 5$

8. $8x + 3y = 15$
   $4x - 2y = -3$

9   $3x + 2y = 2$
     $2x + 3y = 1$

10   $4x + 8y = 3$
      $3x + 7y = 2$

11   $5x - 12y = 2$
     $7x - 8y = 5$

12   $16x - 6y = 3$
      $12x + 5y = 7$

Solve the following equations simultaneously by substitution.

13   $3x + y = 3$
     $5x + 2y = 4$

14   $6x - 5y = 2$
      $2x + y = 6$

15   $x + 5y = 8$
     $4x - 2y = -1$

16   $7x + 5y = 1$
      $x - 3y = 2$

17 $2x - y = 3$
$4x - 3y = -1$

18 $x - 2y = -4$
$5x - 6y = 4$

19 $2x + y = 11$
$6x + 7y = 5$

20 $x - 3y = 28$
$8x + 5y = -8$

21 $4x - 3y = 18$
$3x + 2y = 5$

22 $3x + 8y = 26$
$4x + 7y = 31$

23 $8x + 3y = -4$
$12x - 11y = 25$

24 $8x + 3y = 21$
$12x + 5y = 33$

Solve the following equations simultaneously by either algebraic method.

25 $2x + 5y = 10$
$5x - 3y = -37$

26 $6x - 15y = 21$
$7x + 10y = 19$

27 $9x + 8y = 23$
$6x + 10y = 27$

28 $8x - 15y = -7$
$12x - 10y = -3$

29 $12x - 6y = -19$
$18x + 8y = -3$

30 $6x + 5y = 3$
$4x + 3y = 5$

31 $7x + 12y = 3$
$5x + 9y = 2$

32 $3x - 10y = 2$
$4x - 15y = 7$

33 $10x - 3y = 6$
$16x + 5y = 4$

34 $7x + 6y = 5$
$8x + 9y = 10$

35 $18x - 2y = 3$
$15x + 7y = 9$

36 $3x + 4y = -2$
$6x - 4y = 5$

37 $\dfrac{x}{3} + \dfrac{2x - y}{2} = 2$
$2x + 3y = 18$

38 $\dfrac{x}{5} + \dfrac{4x - 3y}{4} = 3$
$x + 5y = 25$

39 $\dfrac{2x+y}{2} + \dfrac{4x-y}{4} = 5$

$\dfrac{3x-y}{2} + \dfrac{x+2y}{5} = 3$

40 $\dfrac{x+3y}{3} + \dfrac{4x+y}{4} = 1$

$\dfrac{2x-y}{5} + \dfrac{2x+3y}{3} = 0$

Solve the following problems first for $1/x$ and $1/y$ and then find the values of x and y.

41 $\dfrac{3}{x} + \dfrac{4}{y} = 2$

$\dfrac{6}{x} - \dfrac{4}{y} = 1$

42 $\dfrac{4}{x} + \dfrac{6}{y} = 0$

$\dfrac{6}{x} - \dfrac{3}{y} = -4$

43 $\dfrac{5}{x} + \dfrac{6}{y} = 4$

$\dfrac{7}{x} + \dfrac{10}{y} = 6$

44 $\dfrac{14}{x} - \dfrac{8}{y} = -5$

$\dfrac{2}{x} + \dfrac{4}{y} = 1$

10.7 STATED PROBLEMS

Frequently, more than one unknown quantity is involved in a stated problem, and in such cases it may be easier to obtain the equations for solving the problem if more than one letter is introduced. The method for solving such problems in general is the same as that of Sec. 7.6. However, one must obtain *the same number of equations as there are unknowns* introduced. We shall illustrate the method with three examples.

EXAMPLE 1 A man traveled a total of 1000 mi in a car, on a train, and on a plane, and the entire journey required 10 h. The average speed of the car was 50 mi/h; of the train, 60 mi/h; and of the plane, 200 mi/h. Find the time in hours required for each part of the trip if the traveler spent 3 h more on the train than in the car.

Solution We first tabulate the information given in the problem and then explain the method of solution.

	Time, h	Rate, mi/h	Distance, mi
Car		50	
Train		60	
Plane		200	
Total	10		1000

From this tabulation, we can see that the unknown quantities are the number of hours spent on each conveyance and the number of miles traveled by each method. We know the rates of travel, so we can find the distances if we have the number of hours devoted to each part of the trip. Furthermore, we know that the man was on the train 3 h longer than in the car. Hence, we shall let

x = number of hours in car

Then

$x + 3$ = number of hours on train

We have no information concerning the time spent on the plane, so we shall let y = number of hours he flew. Now, since distance = rate $\times$ time, we can complete the tabulation as below.

	Time, h	Rate, mi/h	Distance, mi
Car	x	50	$50x$
Train	$x + 3$	60	$60(x + 3)$
Plane	y	200	$200y$
Total	10		1000

Now we can see that

$$x + (x + 3) + y = 10 \tag{1}$$

$$50x + 60(x + 3) + 200y = 1000 \tag{2}$$

If we simplify the above equations, we get from (1)

$$2x + y = 7 \tag{3}$$

and from (2)

$$110x + 200y = 820 \tag{4}$$

Since the coefficient of y in (3) is 1, the method of substitution is preferable for solving these equations. This method yields $x = 2$, $y = 3$, and therefore $x + 3 = 5$. Hence the man traveled 2 h in the car, 5 h on the train, and 3 h on the plane.

The use of two variables is especially advantageous for solving problems involving rates in opposite directions, such as problems dealing with an airplane flying with or against the wind or a boat moving with or against the current. If a plane can fly x miles per hour in still air, then it can fly $x + y$ miles per hour with a wind that is moving y miles per hour, and $x - y$ miles per hour against the wind.

EXAMPLE 2 A pilot flew from field A to field B 630 mi due east of A in 3 h. After lunch he returned to field A in $3\frac{1}{2}$ h. The wind was blowing from the west during the entire trip, but its velocity during the westward flight was double its velocity during the eastward flight. Find the speed of the plane in still air and the speed of the wind on each flight.

Solution We shall let x = the speed in miles per hour of the plane in still air and y = the speed in miles per hour of the wind during the eastward flight; then we can tabulate the data of the problem as follows:

Flight	Speed of plane in still air, mi/h	Speed of wind, mi/h	Actual speed of plane, mi/h	Time, h	Distance, mi
Eastward	x	y	$x + y$	3	630
Westward	x	$2y$	$x - 2y$	$3\frac{1}{2}$	630

Now, since speed $\times$ time = distance, we have

$$3(x + y) = 630 \tag{5}$$

$$(3\tfrac{1}{2})(x - 2y) = 630 \tag{6}$$

We can simplify these two equations by dividing (5) by 3 and (6) by $3\tfrac{1}{2}$ and thereby obtain

$$x + y = 210 \tag{7}$$
$$x - 2y = 180 \tag{8}$$

The simultaneous solution of (7) and (8) is $x = 200$ and $y = 10$. Hence, the speed of the plane in still air was 200 mi/h, the speed of the wind during the eastward flight was 10 mi/h, and the wind speed during the westward flight was 20 mi/h.

EXAMPLE 3 One term Susan received 22 quality points by earning the grade of A or B in each of the six subjects she studied. If each A was worth four quality points and each B was worth three quality points, in how many subjects did she earn the grade of A and in how many subjects did she earn the grade of B?

Grade	Number of subjects	Quality points for each	Total quality points
A	x	4	
B	y	3	
Total	6		22

Solution From the above table we see that if she made A in x subjects and B in y subjects,

$$x + y = 6 \tag{9}$$

The number of quality points due to A courses is $4x$, while the number due to B courses is $3y$. Since the total quality points is 22,

$$4x + 3y = 22 \tag{10}$$

Putting $y = 6 - x$ from (9) in (10) gives

$$4x + 3(6 - x) = 22$$
$$x = 22 - 18 = 4$$
$$y = 6 - 4 = 2$$

10.7 Stated Problems   333

EXERCISE 10.3

In these problems complete the tabulation of the data or tabulate the data and then find the unknown quantities.

1. A high school club earned a net profit of $91.60 selling candy apples and suckers which cost them 16 cents apiece at a basketball game. If they sold 480 candy apples and 610 suckers, and 50 cents bought both a sucker and a candy apple, what did they charge for each?

	Number of items sold	Price of item	Cost of item	Profit from item	Profit from sale
Candy apples	480	x	$0.16	$x-16$	
Suckers	610	y	0.16	$y-16$	
Both					$91.60

2. Mrs. Stewart spent $5.45 for $2\frac{1}{2}$ qt of cream to make ice cream. If whipping cream cost 60 cents per half pint and half-and-half cost 65 cents per pint, how much of each type cream did she use?

	Cost per half pint	Amount used, qt	Total cost
Whipping cream	$0.60	x	
Half-and-half	0.65	y	
Total		$2\frac{1}{2}$	$5.45

3   A music teacher charged $5 for each ½-h organ lesson and $3.50 for each ½-h piano lesson. If in 4 h of teaching she earned $32.50, how many lessons of each type did she teach?

$3.50x + 5.00y = 32.50$

4   Tickets for a banquet were $8 for a single ticket or $15 for a couple. If 144 people attended the banquet and $1098 was collected from ticket sales, how many couples went and how many people went without dates?

5 An apartment building contained 20 units, consisting of one-bedroom apartments, which rented for $110 a month, and two-bedroom apartments, which rented for $135 a month. If the rental from 17 apartments for one month was $2045 and three apartments were vacant, how many apartments of each type were rented?

6 Frank found that he could drive from the campus to his home in 5 h by averaging 55 mi/h. However, on one trip, after he had averaged 55 mi/h for a while, he encountered bad weather and was forced to reduce his speed to 40 mi/h. If that trip required $5\frac{3}{4}$ h, how many miles did he travel at each speed?

7 A man had 6 gal of paint to cover 2380 ft² of fencing. If 1 gal of paint will cover 470 ft² with one coat or 250 ft² with two coats, and if he used all his paint, how many square feet of the fence would be covered with two coats and how many with one?

8 Two different routes between two cities differ by 20 mi. Two women made the trip between the cities in exactly the same time. One traveled the shorter route at 50 mi/h, and the other traveled the longer route at 55 mi/h. Find the length of each route.

9 On the first day of homecoming weekend a campus organization earned $650 by selling 450 college pennants and 100 corsages. On the second day they sold the 150 pennants they had left but found the remaining 50 corsages had wilted so that they lost as much per corsage as they had earned on the previous day. If on the second day they earned $50, how much did they earn on each pennant and each corsage sold?

10 A biology class of 35 students took a field trip including a hike of 8 mi. Part of the class also investigated a side trail which added 3 mi to their hike. If the class walked a total of 331 student-miles, how many students took the longer hike and how many took the shorter? [If a group of 20 students walk 10 mi, the group has walked a total of 10(20) = 200 student-miles.]

11 Jack and Dick signed a lease to rent an apartment for 9 months. At the end of 6 months Dick got married and moved out. He paid the landlord an amount equal to the difference between double-occupancy rental of the apartment and single-occupancy rental for the remaining 3 months, and Jack paid the single-occupancy rate for the 3 months. If the 9 months' rental cost Jack $870 and Dick $570, what were the single and double monthly rates of the apartment?

12 The sum of the seven digits in a telephone number is 30. Counting from the left, the first three digits and the last digit are the same. The fourth digit is twice the first, the sum of the fifth digit and the first is 7, and the sixth digit is twice the fifth. Find the telephone number.

13 Three volunteer women assembled 741 newsletters for a bulk mailing. The first could assemble 124 per hour, the second 118 per hour, and the third 132 per hour. They worked a total of 6 woman-hours. If the first worker worked 2 h, how long did each of the others work?

14 A class of 32 students was made up of people who were 18, 19, or 20 years of age. The average of their ages was 18.5. How many of each age were in the class if the number of 18-year-olds was 6 more than the number of 19- and 20-year-olds?

15 The average cost per roll for 12 rolls of color film was $4.90. Some rolls cost $4.60 per roll, others $4.80 per roll, and still others $5.80 per roll. How many rolls were bought at each price if two more rolls were bought at $4.80 than at $4.60?

16 Tickets to a service club breakfast cost $3 for adults, $2.25 for 7- to 12-year-olds, and 75 cents for children under 7. A total of $930 was collected from the 400 tickets sold. How many tickets in each age group were sold if there were two-thirds as many children under 7 as over 7?

10.8 DETERMINANTS OF THE SECOND ORDER

A notation, called a **determinant,** that is very useful in solving systems of linear equations was invented or discovered independently by Leibniz in 1693 and by the Japanese mathematician Kiowa in 1683. In 1750 Cramer rediscovered the notation and stated the rule for applying it to systems of linear equations. A determinant is a compact way for expressing certain types of polynomials. We shall define a determinant of the second order and then show how a determinant is used in solving two linear equations in two unknowns.

A square array of numbers of the type $\begin{vmatrix} a_1 & b_1 \\ a_2 & b_2 \end{vmatrix}$ is called a **determinant of the second order,** and it stands for the binomial $a_1 b_2 - a_2 b_1$.

The binomial $a_1 b_2 - a_2 b_1$ is called the **expansion** of the determinant, and it is obtained as indicated below.

$$\begin{vmatrix} a_1 & b_1 \\ a_2 & b_2 \end{vmatrix} = +a_1 b_2 - a_2 b_1$$

We shall illustrate the definition in Example 1.

EXAMPLE 1

$$\begin{vmatrix} 2 & 4 \\ 6 & 5 \end{vmatrix} = +(2)(5) - (4)(6) = 10 - 24 = -14$$

$$\begin{vmatrix} -3 & 2 \\ -7 & 6 \end{vmatrix} = +(-3)(6) - (2)(-7) = -18 + 14 = -4$$

$$\begin{vmatrix} 7 & 5 \\ 8 & -3 \end{vmatrix} = +(7)(-3) - (5)(8) = -21 - 40 = -61$$

The determinants in Example 1 illustrate the fact that when a_1, a_2, b_1, and b_2 are replaced by numbers, the expansion $a_1 b_2 - a_2 b_1$ is equal to a single number. This number is called the **value of the determinant.**

Problems 1 to 16 of Exercise 10.4 may be worked now.

Now we shall solve the two linear equations below, using the method of addition and subtraction and then show how the solution set can be expressed by using determinants.

$$a_1 x + b_1 y = c_1 \qquad (1)$$
$$a_2 x + b_2 y = c_2 \qquad (2)$$

The method for obtaining the solution set of the equations is as follows:

$a_1 b_2 x + b_1 b_2 y = b_2 c_1$ multiplying (1) by b_2 **(3)**

$a_2 b_1 x + b_1 b_2 y = b_1 c_2$ multiplying (2) by b_1 **(4)**

$\overline{a_1 b_2 x - a_2 b_1 x = b_2 c_1 - b_1 c_2}$ subtracting (4) from (3)   **(5)**

$$x = \frac{b_2 c_1 - b_1 c_2}{a_1 b_2 - a_2 b_1} \qquad \text{solving (5) for } x$$

Similarly,

$$y = \frac{a_1 c_2 - a_2 c_1}{a_1 b_2 - a_2 b_1}$$

You can now verify that the numerators of the values of x and of y are, respectively, the expansions of the determinants.

$$\begin{vmatrix} c_1 & b_1 \\ c_2 & b_2 \end{vmatrix} \quad \text{and} \quad \begin{vmatrix} a_1 & c_1 \\ a_2 & c_2 \end{vmatrix}$$

Furthermore, in each case, the denominator is the expansion of

$$\begin{vmatrix} a_1 & b_1 \\ a_2 & b_2 \end{vmatrix}$$

Hence, the solution set of Eqs. (1) and (2) can be expressed in the form

$$x = \frac{\begin{vmatrix} c_1 & b_1 \\ c_2 & b_2 \end{vmatrix}}{\begin{vmatrix} a_1 & b_1 \\ a_2 & b_2 \end{vmatrix}} \qquad y = \frac{\begin{vmatrix} a_1 & c_1 \\ a_2 & c_2 \end{vmatrix}}{\begin{vmatrix} a_1 & b_1 \\ a_2 & b_2 \end{vmatrix}} \qquad (10.1)$$

This is known as **Cramer's rule** for two variables. The procedure consists of the following steps:

1. Arrange the terms in the equations so that the unknowns appear in the same order in the

left member and the constant terms appear in the right.
2. Form the determinant D whose rows consist of the coefficients of the unknowns in the order in which they occur in the equations.
3. Form the determinant D_x by replacing the column of coefficients of x in D by the column of constant terms.
4. Form the determinant D_y by replacing the column of coefficients of y in D by the column of constant terms.
5. Then the solution pair of the given equations is $x = D_x/D$, $y = D_y/D$, provided that $D \neq 0$. If $D = 0$, the equations are not independent.

We shall illustrate the process with examples.

EXAMPLE 2 Solve the following equations by determinants.

$2x + y = 10$
$3x + 2y = 17$

Solution

1. The terms in the equations are arranged in the proper order, so we go immediately to step 2.
2. $D = \begin{vmatrix} 2 & 1 \\ 3 & 2 \end{vmatrix} = 4 - 3 = 1$
3. $D_x = \begin{vmatrix} 10 & 1 \\ 17 & 2 \end{vmatrix} = 20 - 17 = 3$
4. $D_y = \begin{vmatrix} 2 & 10 \\ 3 & 17 \end{vmatrix} = 34 - 30 = 4$
5. $x = D_x/D = \frac{3}{1} = 3$, $y = D_y/D = \frac{4}{1} = 4$

Hence the solution is $(3,4)$. The solution can be checked by replacing x by 3 and y by 4 in the given equations.

EXAMPLE 3 By use of determinants obtain the solution of the equations

$4x - 3y = -4$
$2x + 5y = 11$

Solution

$D = \begin{vmatrix} 4 & -3 \\ 2 & 5 \end{vmatrix} = 20 + 6 = 26$

$D_x = \begin{vmatrix} -4 & -3 \\ 11 & 5 \end{vmatrix} = -20 + 33 = 13$

$D_y = \begin{vmatrix} 4 & -4 \\ 2 & 11 \end{vmatrix} = 44 + 8 = 52$

$x = \dfrac{D_x}{D} = \dfrac{13}{26} = \dfrac{1}{2}$

$y = \dfrac{D_y}{D} = \dfrac{52}{26} = 2$

Hence, the solution is $(\frac{1}{2}, 2)$.

EXERCISE 10.4

Fill in the blanks.

1 $\begin{vmatrix} 2 & 4 \\ 5 & 7 \end{vmatrix} = +(2)() - (4)() = - =$

2 $\begin{vmatrix} 3 & -7 \\ 2 & -4 \end{vmatrix} = +3() - (-7)() = - =$

3 $\begin{vmatrix} 4 & -2 \\ -3 & 6 \end{vmatrix} = +()(6) - ()(-3) = - =$

4 $\begin{vmatrix} 5 & 3 \\ -4 & -9 \end{vmatrix} = +()(-9) - (3)() = - =$

Find the value of the following determinants.

5 $\begin{vmatrix} 3 & 1 \\ 2 & 8 \end{vmatrix} =$

6 $\begin{vmatrix} 4 & -1 \\ 3 & -2 \end{vmatrix} =$

7 $\begin{vmatrix} 5 & 4 \\ -2 & -3 \end{vmatrix} =$

8 $\begin{vmatrix} -1 & 3 \\ -2 & -5 \end{vmatrix} =$

9 $\begin{vmatrix} 7 & -8 \\ -5 & 7 \end{vmatrix} =$

10 $\begin{vmatrix} -9 & 2 \\ 13 & -4 \end{vmatrix} =$

11 $\begin{vmatrix} 7 & -9 \\ 5 & -8 \end{vmatrix} =$

12 $\begin{vmatrix} 7 & 9 \\ -8 & -12 \end{vmatrix} =$

Verify the following equations.

13. $\begin{vmatrix} 4 & -5 \\ 8 & -11 \end{vmatrix} = - \begin{vmatrix} 4 & 5 \\ 8 & 11 \end{vmatrix}$

14. $\begin{vmatrix} 3 & -7 \\ 2 & 1 \end{vmatrix} = - \begin{vmatrix} -3 & 7 \\ 2 & 1 \end{vmatrix}$

15. $\begin{vmatrix} 4 & 3 \\ 8 & 6 \end{vmatrix} = 0$

16. $\begin{vmatrix} 3 & 7 \\ 8 & 10 \end{vmatrix} = \begin{vmatrix} 3 & 8 \\ 7 & 10 \end{vmatrix}$

Find the value of the determinants D, D_x, and D_y, and solve for x and y by using Cramer's rule.

17. $2x + y = 9$
    $5x - y = -2$

18. $3x + 2y = 9$
    $5x + y = 1$

19. $5x + 6y = 9$
    $-3x + 7y = 37$

20. $4x - 3y = 14$
    $3x + 4y = -2$

Solve the following equations using Cramer's rule.

21 $2x + 3y = 4$
$x + 2y = 1$

22 $3x - 2y = 6$
$4x - 3y = 2$

23 $2x + 4y = 5$
$3x + 6y = 2$

24 $5x + 2y = 1$
$8x + 3y = 2$

25 $5x + 3y = 2$
$4x + 2y = 3$

26 $6x - 5y = 3$
$3x - 2y = 1$

27 $6x + 5y = 4$
       $3x + 2y = 2$

28 $4x + 8y = 1$
       $x + 2y = 1$

29 $3x - 9y = 1$
       $2x + 4y = -1$

30 $4x - 2y = 3$
       $2x + 5y = 6$

31 $7x - 14y = -3$
       $3x + 2y = 1$

32 $5x + 2y = 3$
       $7x - 5y = -1$

10.9
SYSTEMS OF THREE LINEAR EQUATIONS

In this section we shall explain two algebraic methods for solving a system of equations of the type

$$a_1x + b_1y + c_1z = d_1 \quad (1)$$
$$a_2x + b_2y + c_2z = d_2 \quad (2)$$
$$a_3x + b_3y + c_3z = d_3 \quad (3)$$

ELIMINATION BY ADDITION AND SUBTRACTION We shall first illustrate the elimination method by addition and subtraction by obtaining the solution set for the system

$$2x - 3y + 4z = 19 \quad (4)$$
$$x + 2y - 2z = -6 \quad (5)$$
$$3x + y + z = 8 \quad (6)$$

The method consists of the following steps:

1. We eliminate z from (4) and (5) by addition and obtain an equation in x and y.
2. We eliminate z from Eqs. (5) and (6) and obtain a second equation in x and y.
3. We solve the two equations obtained in steps 1 and 2 for x and y.
4. Finally, we replace x and y in one of the given equations by the values obtained in step 3 and solve the resulting equation for z.

The details of the computation follow.

$$\begin{array}{ll} 2x - 3y + 4z = 19 & \text{rewriting (4)} \\ \underline{2x + 4y - 4z = -12} & \text{multiplying (5) by 2} \quad (7) \\ 4x + y \phantom{{}-4z} = 7 & \text{adding (4) and (7)} \quad (8) \end{array}$$

$$\begin{array}{ll} x + 2y - 2z = -6 & \text{rewriting (5)} \\ \underline{6x + 2y + 2z = 16} & \text{multiplying (6) by 2} \quad (9) \\ 7x + 4y \phantom{{}+2z} = 10 & \text{adding (5) and (9)} \quad (10) \end{array}$$

We now solve (8) and (10) by determinants and have

$$D = \begin{vmatrix} 4 & 1 \\ 7 & 4 \end{vmatrix} = 16 - 7 = 9$$

$$D_x = \begin{vmatrix} 7 & 1 \\ 10 & 4 \end{vmatrix} = 28 - 10 = 18$$

$$D_y = \begin{vmatrix} 4 & 7 \\ 7 & 10 \end{vmatrix} = 40 - 49 = -9$$

Consequently, $x = \frac{18}{9} = 2$ and $y = -\frac{9}{9} = -1$.

Finally, we replace x by 2 and y by -1 in (4) and solve the resulting equation for z. Thus

$$2(2) - 3(-1) + 4z = 19$$
$$4z = 19 - 4 - 3 = 12$$
$$z = 3$$

Hence, the solution is $(2, -1, 3)$.

In using the addition and subtraction method, any one of the three variables may be eliminated, giving two equations in two unknowns. These two equations are then solved by any method.

Problems 1 to 4 in Exercise 10.5 may be done now.

ELIMINATION BY SUBSTITUTION A second method for solving (4) to (6) is elimination by substitution. This works easiest when one variable in one equation has 1 for a coefficient. Solving (5) for x gives

$$x = 2z - 2y - 6 \quad (11)$$

Substituting this in (4) and (6), we have

$$2(2z - 2y - 6) - 3y + 4z = 19$$
$$3(2z - 2y - 6) + y + z = 8$$

and simplifying gives

$$-7y + 8z = 31 \quad (12)$$
$$-5y + 7z = 26 \quad (13)$$

The solution to (12) and (13) is $y = -1$, $z = 3$. Using these values in (11) gives $x = 2$.

Problems 5 to 8 in Exercise 10.5 may be done now.

10.10
DETERMINANTS OF THE THIRD ORDER

In this section we shall define a determinant of the third order and explain the method for obtaining the expansion.

A determinant that contains three rows and three columns is called a **determinant of the third order.** We shall consider the determinant

$$D = \begin{vmatrix} a_1 & b_1 & c_1 \\ a_2 & b_2 & c_2 \\ a_3 & b_3 & c_3 \end{vmatrix}$$

and define the **expansion** of D to be the polynomial

$$a_1b_2c_3 + a_2b_3c_1 + a_3b_1c_2 \\ - a_3b_2c_1 - a_2b_1c_3 - a_1b_3c_2 \quad (1)$$

Notice that each term in the expansion (1) of D is the product of three elements a_i, b_j, c_k, where i, j, k is an arrangement of the integers 1, 2, and 3.

The symbols $a_i, b_i, c_i, i = 1, 2,$ and 3 are called the **elements** of D, and each stands for a number. If these numbers are known, the value of each product in (1) can be computed and the results combined into a single number. This number is the **value of the determinant** D.

We shall presently show that the expansion (1) can be expressed in several compact forms, each of which involves three determinants of order 2. Each such determinant is called a minor of an element of D and illustrates the following definition.

The **minor** of a given element of D is the determinant that remains after the row and column that contains the given element is deleted from D.

Thus the minor of c_2 is

$$\begin{vmatrix} a_1 & b_1 & \cancel{c_1} \\ \cancel{a_2} & \cancel{b_2} & \cancel{c_2} \\ a_3 & b_3 & \cancel{c_3} \end{vmatrix} = \begin{vmatrix} a_1 & b_1 \\ a_3 & b_3 \end{vmatrix} = m(c_2)$$

We shall denote the minors of a_i, b_i, and $c_i, i = 1, 2,$ and 3, by $m(a_i), m(b_i),$ and $m(c_i)$, respectively.

EXPANSION BY FIRST ROW We next show that the expansion (1) of D can be expressed in terms of the minors of the elements in the first row. For this purpose we rearrange and group the terms so that those containing a_1 are in the first group, those containing b_1 are in the second, and those containing c_1 are in the third. Thus we get

$$D = (a_1 b_2 c_3 - a_1 b_3 c_2) - (a_2 b_1 c_3 - a_3 b_1 c_2) \\ + (a_2 b_3 c_1 - a_3 b_2 c_1)$$

Next we factor each group and obtain

$$D = a_1(b_2 c_3 - b_3 c_2) - b_1(a_2 c_3 - a_3 c_2) \\ + c_1(a_2 b_3 - a_3 b_2)$$

Now, we note that

$$b_2 c_3 - b_3 c_2 = \begin{vmatrix} b_2 & c_2 \\ b_3 & c_3 \end{vmatrix}$$

$$a_2 c_3 - a_3 c_2 = \begin{vmatrix} a_2 & c_2 \\ a_3 & c_3 \end{vmatrix}$$

$$a_2 b_3 - a_3 b_2 = \begin{vmatrix} a_2 & b_2 \\ a_3 & b_3 \end{vmatrix}$$

Hence,

$$D = a_1 \begin{vmatrix} b_2 & c_2 \\ b_3 & c_3 \end{vmatrix} - b_1 \begin{vmatrix} a_2 & c_2 \\ a_3 & a_3 \end{vmatrix} + c_1 \begin{vmatrix} a_2 & b_2 \\ a_3 & b_3 \end{vmatrix} \quad (2)$$

EXPANSION BY SECOND COLUMN Similarly, if we group the terms in (1) so that the factors of the first, second, and third groups are $b_1, b_2,$ and b_3, respectively, we get

$$D = -(a_2 b_1 c_3 - a_3 b_1 c_2) + (a_1 b_2 c_3 - a_3 b_2 c_1) \\ -(a_1 b_3 c_2 - a_2 b_3 c_1)$$

$$= -b_1(a_2 c_3 - a_3 c_2) + b_2(a_1 c_3 - a_3 c_1) \\ -b_3(a_1 c_2 - a_2 c_1) b_3$$

$$= -b_1 \begin{vmatrix} a_2 & c_2 \\ a_3 & c_3 \end{vmatrix} + b_2 \begin{vmatrix} a_1 & c_1 \\ a_3 & c_3 \end{vmatrix} - b_3 \begin{vmatrix} a_1 & c_1 \\ a_2 & c_2 \end{vmatrix} \quad (3)$$

Now if we compare the expansions (2) and (3) with D, we see that each determinant of order 2 is the minor of the element that precedes it. That is, the determinants in (2) are the minors of $a_1, b_1,$ and c_1.

Equation (2) is called the **expansion** of D in terms of the first row, and (3) is the expansion in terms of the elements of the second column.

By similar arguments it can be verified that the expansion (1) can be expressed in terms of any row or column. We suggest that the following be used in obtaining an expansion:

$$\begin{matrix} + & - & + \\ - & + & - \\ + & - & + \end{matrix}$$

This diagram indicates the signs associated with the elements in any row or column. For example, the signs of the elements of the second row are $- + -$, and the signs of the elements of the third column are $+ - +$.

We illustrate the method of expanding a determinant with three examples:

EXAMPLE 1 Expand

$$D_1 = \begin{vmatrix} 3 & 2 & 4 \\ 1 & 5 & 2 \\ 4 & 7 & 6 \end{vmatrix}$$

in terms of the first row.

Solution Referring to the above diagram, we see that the signs of the first row are $+ - +$. The minors of 3, 2, and 4 are

$$\begin{vmatrix} 5 & 2 \\ 7 & 6 \end{vmatrix} \quad \begin{vmatrix} 1 & 2 \\ 4 & 6 \end{vmatrix} \quad \text{and} \quad \begin{vmatrix} 1 & 5 \\ 4 & 7 \end{vmatrix}$$

respectively. Therefore

$$D_1 = +3 \begin{vmatrix} 5 & 2 \\ 7 & 6 \end{vmatrix} - 2 \begin{vmatrix} 1 & 2 \\ 4 & 6 \end{vmatrix} + 4 \begin{vmatrix} 1 & 5 \\ 4 & 7 \end{vmatrix}$$
$$= 3(30 - 14) - 2(6 - 8) + 4(7 - 20)$$
$$= 3(16) - 2(-2) + 4(-13)$$
$$= 48 + 4 - 52 = 0$$

Check We check this result by expanding D_1 in terms of the second column. According to the diagram, the signs are $- + -$. Hence we have

$$D_1 = -2 \begin{vmatrix} 1 & 2 \\ 4 & 6 \end{vmatrix} + 5 \begin{vmatrix} 3 & 4 \\ 4 & 6 \end{vmatrix} - 7 \begin{vmatrix} 3 & 4 \\ 1 & 2 \end{vmatrix}$$
$$= -2(6 - 8) + 5(18 - 16) - 7(6 - 4)$$
$$= -2(-2) + 5(2) - 7(2)$$
$$= 4 + 10 - 14 = 0$$

EXAMPLE 2 Expand

$$D_2 = \begin{vmatrix} -2 & 4 & 3 \\ 1 & -5 & -6 \\ 3 & 1 & -2 \end{vmatrix}$$

in terms of the third column and check the result by using the second row.

Solution Using the third column, we have

$$D_2 = 3 \begin{vmatrix} 1 & -5 \\ 3 & 1 \end{vmatrix} - (-6) \begin{vmatrix} -2 & 4 \\ 3 & 1 \end{vmatrix}$$
$$+ (-2) \begin{vmatrix} -2 & 4 \\ 1 & -5 \end{vmatrix}$$
$$= 3(1 + 15) - (-6)(-2 - 12) + (-2)(10 - 4)$$
$$= 48 - 84 - 12 = -48$$

Check Using the second row, we have

$$D_2 = - \begin{vmatrix} 4 & 3 \\ 1 & -2 \end{vmatrix} + (-5) \begin{vmatrix} -2 & 3 \\ 3 & -2 \end{vmatrix}$$
$$- (-6) \begin{vmatrix} -2 & 4 \\ 3 & 1 \end{vmatrix}$$
$$= -(-8 - 3) + (-5)(4 - 9) - (-6)(-2 - 12)$$
$$= 11 + 25 - 84 = -48$$

If one or more of the elements of a determinant are 0, it is advisable to expand the determinant in terms of the minors of the elements of the row or column that contains the greatest number of 0s.

EXAMPLE 3 Expand

$$D_3 = \begin{vmatrix} 3 & 2 & 4 \\ 0 & 2 & 0 \\ 1 & 3 & 2 \end{vmatrix}$$

Solution Here the second row contains two 0s. Hence, we shall use this row to get the expansion:

$$D_3 = -0 \begin{vmatrix} 2 & 4 \\ 3 & 2 \end{vmatrix} + 2 \begin{vmatrix} 3 & 4 \\ 1 & 2 \end{vmatrix} - 0 \begin{vmatrix} 3 & 2 \\ 1 & 3 \end{vmatrix}$$
$$= -0(4 - 12) + 2(6 - 4) - 0(9 - 2)$$
$$= 0 + 4 - 0 = 4$$

We shall not discuss determinants of order greater than 3 in this book. The method of expanding a determinant in terms of minors, however, can be applied to a determinant of any order. Although determinants have a wide variety of applications in mathematics, we shall consider only their use in solving systems of linear equations.

Problems 9 to 16 in Exercise 10.5 may be done now.

10.11
CRAMER's RULE

In Sec. 10.8 we stated that Cramer devised a method for using determinants to solve a system of linear equations, and we used the method for solving a system of two equations. In this section we shall illustrate the use of Cramer's rule for solving a system of three linear equations in three variables.

The following are the formal steps in the application of **Cramer's rule for three variables:**

1. Arrange the terms in the given equations so that the terms involving x, y, and z are in the same order in the left members of the equations and the constant terms are in the right members.
2. Set up the determinant D whose elements are the coefficients of the variables.
3. Set up D_x, D_y, and D_z by replacing the columns of coefficients of x, y, and z in D by the column of constant terms.
4. Then the solution is $x = D_x/D$, $y = D_y/D$, $z = D_z/D$.

EXAMPLE 1 Solve the system of equations by using determinants.

$$2x + y + 3z = 1 \tag{1}$$
$$x + 4y + 6z = 9 \tag{2}$$
$$4x + 3y + 9z = 5 \tag{3}$$

Solution We shall form the determinants D, D_x,

D_y, and D_z and expand each in terms of the first row. Thus, we get

$$D = \begin{vmatrix} 2 & 1 & 3 \\ 1 & 4 & 6 \\ 4 & 3 & 9 \end{vmatrix} = 2\begin{vmatrix} 4 & 6 \\ 3 & 9 \end{vmatrix} - 1\begin{vmatrix} 1 & 6 \\ 4 & 9 \end{vmatrix} + 3\begin{vmatrix} 1 & 4 \\ 4 & 3 \end{vmatrix}$$
$$= 2(36 - 18) - 1(9 - 24) + 3(3 - 16)$$
$$= 36 + 15 - 39 = 12$$

$$D_x = \begin{vmatrix} 1 & 1 & 3 \\ 9 & 4 & 6 \\ 5 & 3 & 9 \end{vmatrix} = 1\begin{vmatrix} 4 & 6 \\ 3 & 9 \end{vmatrix} - 1\begin{vmatrix} 9 & 6 \\ 5 & 9 \end{vmatrix} + 3\begin{vmatrix} 9 & 4 \\ 5 & 3 \end{vmatrix}$$
$$= (36 - 18) - (81 - 30) + 3(27 - 20)$$
$$= 18 - 51 + 21 = -12$$

$$D_y = \begin{vmatrix} 2 & 1 & 3 \\ 1 & 9 & 6 \\ 4 & 5 & 9 \end{vmatrix} = 2\begin{vmatrix} 9 & 6 \\ 5 & 9 \end{vmatrix} - 1\begin{vmatrix} 1 & 6 \\ 4 & 9 \end{vmatrix} + 3\begin{vmatrix} 1 & 9 \\ 4 & 5 \end{vmatrix}$$
$$= 2(81 - 30) - (9 - 24) + 3(5 - 36)$$
$$= 102 + 15 - 93 = 24$$

$$D_z = \begin{vmatrix} 2 & 1 & 1 \\ 1 & 4 & 9 \\ 4 & 3 & 5 \end{vmatrix} = 2\begin{vmatrix} 4 & 9 \\ 3 & 5 \end{vmatrix} - 1\begin{vmatrix} 1 & 9 \\ 4 & 5 \end{vmatrix} + 1\begin{vmatrix} 1 & 4 \\ 4 & 3 \end{vmatrix}$$
$$= 2(20 - 27) - (5 - 36) + (3 - 16)$$
$$= -14 + 31 - 13 = 4$$

Now we have $x = D_x/D = (-12)/12 = -1$, $y = D_y/D = \frac{24}{12} = 2$, and $z = D_z/D = \frac{4}{12} = \frac{1}{3}$. Hence, the solution is $(-1, 2, \frac{1}{3})$. The solution can be checked by replacing x, y, and z, respectively, by -1, 2, and $\frac{1}{3}$ in Eqs. (1) to (3).

EXAMPLE 2 Solve the equations by use of determinants.

$$2x + y = 4$$
$$3x + 2z = 17$$
$$ 2y + z = 0$$

Solution We first rewrite the above equations in the form

$$2x + y + 0z = 4$$
$$3x + 0y + 2z = 17$$
$$0x + 2y + z = 0$$

Now we have

$$D = \begin{vmatrix} 2 & 1 & 0 \\ 3 & 0 & 2 \\ 0 & 2 & 1 \end{vmatrix} = 2\begin{vmatrix} 0 & 2 \\ 2 & 1 \end{vmatrix} - \begin{vmatrix} 3 & 2 \\ 0 & 1 \end{vmatrix} + 0\begin{vmatrix} 3 & 0 \\ 0 & 2 \end{vmatrix}$$
$$= 2(-4) - 3 + 0(6)$$
$$= -8 - 3 = -11$$

$$D_x = \begin{vmatrix} 4 & 1 & 0 \\ 17 & 0 & 2 \\ 0 & 2 & 1 \end{vmatrix} = 4\begin{vmatrix} 0 & 2 \\ 2 & 1 \end{vmatrix} - \begin{vmatrix} 17 & 2 \\ 0 & 1 \end{vmatrix}$$
$$+ 0\begin{vmatrix} 17 & 0 \\ 0 & 2 \end{vmatrix}$$
$$= 4(-4) - (17) + 0(34)$$
$$= -16 - 17 = -33$$

$$D_y = \begin{vmatrix} 2 & 4 & 0 \\ 3 & 17 & 2 \\ 0 & 0 & 1 \end{vmatrix}$$

Since two elements of the third row are 0, we shall expand D_y in terms of this row and get

$$D_y = 0\begin{vmatrix} 4 & 0 \\ 17 & 2 \end{vmatrix} - 0\begin{vmatrix} 2 & 0 \\ 3 & 2 \end{vmatrix} + \begin{vmatrix} 2 & 4 \\ 3 & 17 \end{vmatrix}$$
$$= 0 - 0 + (34 - 12) = 22$$

Similarly

$$D_z = \begin{vmatrix} 2 & 1 & 4 \\ 3 & 0 & 17 \\ 0 & 2 & 0 \end{vmatrix} = 0\begin{vmatrix} 1 & 4 \\ 0 & 17 \end{vmatrix} - 2\begin{vmatrix} 2 & 4 \\ 3 & 17 \end{vmatrix} + 0\begin{vmatrix} 2 & 1 \\ 3 & 0 \end{vmatrix}$$
$$= 0 - 2(34 - 12) + 0 = -44$$

Consequently, $x = D_x/D = -33/(-11) = 3$, $y = D_y/D = 22/(-11) = -2$, and $z = D_z/D = -44/(-11) = 4$. Hence, the solution is $(3, -2, 4)$.

EXERCISE 10.5

Solve by addition and subtraction.

1. $2x + y - z = -1$
   $x - 3y + 2z = 8$
   $4x - 2y - 3z = 0$

2. $x + y + z = 3$
   $x + 2y - z = -1$
   $2x + 3y - z = 1$

3. $x + y - z = -1$
   $2x + y + 3z = 4$
   $x + 2y + z = 7$

4 $x + 2y - z = -4$
$2x + 3y + z = 3$
$x + 4y - 2z = -6$

Solve by substitution.

5 $2x - y + 3z = 2$
$4x + 2y - 5z = 1$
$6x - 3y + 4z = 1$

6 $2x + 3y + z = 2$
$4x - 6y + 5z = 5$
$x + 9y + 2z = -3$

7 $3x + 4y + 2z = -2$
$2x - y + 8z = 5$
$x + 5y - 4z = -4$

8 $2x + 3y + z = 3$
$6x - y - 2z = -1$
$2x + 5y + 3z = 6$

Find the value of each determinant in two ways: expand in terms of one row and then in terms of one column.

9 $\begin{vmatrix} 2 & 1 & 4 \\ 1 & 3 & 5 \\ 3 & 1 & 2 \end{vmatrix} =$

10 $\begin{vmatrix} 4 & 5 & 6 \\ 2 & 3 & 2 \\ 1 & 2 & 3 \end{vmatrix} =$

11 $\begin{vmatrix} 3 & -5 & 7 \\ 2 & 6 & -4 \\ -4 & 1 & 5 \end{vmatrix} =$

12 $\begin{vmatrix} -2 & 3 & -4 \\ -1 & 5 & -2 \\ 6 & 2 & -2 \end{vmatrix} =$

Verify the equation by calculation.

13. $\begin{vmatrix} a & b & c \\ 1 & 3 & 1 \\ 4 & 2 & 5 \end{vmatrix} = \begin{vmatrix} a & b & c \\ 1+2a & 3+2b & 1+2c \\ 4 & 2 & 5 \end{vmatrix}$

14. $\begin{vmatrix} 1 & -1 & 3 \\ 2 & -5 & -1 \\ 7 & 4 & 3 \end{vmatrix} = \begin{vmatrix} 1 & 2 & 7 \\ -1 & -5 & 4 \\ 3 & -1 & 3 \end{vmatrix}$

15 $\begin{vmatrix} 6 & 1 & 2 \\ -1 & 3 & 4 \\ 2 & 2 & -3 \end{vmatrix} = - \begin{vmatrix} -1 & 3 & 4 \\ 6 & 1 & 2 \\ 2 & 2 & -3 \end{vmatrix}$

16 $\begin{vmatrix} a & b & c \\ 1 & 2 & 3 \\ a+1 & b+2 & c+3 \end{vmatrix} = 0$

Solve by Cramer's rule.

17 $5x + 2y + z = 1$
$x + 4y - 2z = 3$
$4x - 2y - z = 2$

18 $x + 2y + z = 4$
$x - 6y + 2z = 1$
$2x + 4y - 3z = -2$

19 $2x + y = 3$
$4y + z = -1$
$3x - 4z = -6$

20 $x + y = 1$
$3y - 2z = -2$
$5x + 6z = 9$

10.12 CHAPTER SUMMARY

We found four ways to solve two equations

$$a_1 x + b_1 y = c_1 \tag{1}$$

$$a_2 x + b_2 y = c_2 \tag{2}$$

in two unknowns. Since the graph of $ax + by = c$ is a straight line, we can solve graphically by estimating the point of intersection of the two lines, if there is one. Two equations (1) and (2) are called independent if the lines are not parallel and intersect in exactly one point. They are called dependent if the lines are parallel and intersect in infinitely many points (they are the same lines). They are are called inconsistent if they are parallel and do not intersect.

A second way to solve (1) and (2) is by addition and subtraction. A third way is by substitution. Both these methods eliminate one of the variables. Either method may be used but if in doubt, use addition and subtraction so as to avoid fractions. The fourth way is Cramer's rule.

In solving stated problems, the problem must be translated into algebraic equations. Always choose as an unknown one of the things requested in the problem.

In order to solve three equations of the form $ax + by + cz = d$ in three unknowns, all the above methods can be used except graphing.

A second-order determinant is an expression

$$\begin{vmatrix} a & b \\ c & d \end{vmatrix} = ad - bc$$

A third-order determinant is defined in terms of second-order determinants as

$$\begin{vmatrix} a_1 & a_2 & a_3 \\ b_1 & b_2 & b_3 \\ c_1 & c_2 & c_3 \end{vmatrix} = a_1[m(a_1)] - a_2[m(a_2)] + a_3[m(a_3)]$$

where

$$m(a_1) = \begin{vmatrix} b_2 & b_3 \\ c_2 & c_3 \end{vmatrix}$$

is the determinant obtained by striking out the row and column containing a_1. A third-order determinant can be expanded in terms of the elements of any row or column.

Cramer's rule expresses each component of the solution of a system of equations as a quotient $x = D_x/D$, etc., where D is the determinant of coefficients. The determinant D_x is obtained from D by replacing the column of coefficients of x by the column of constants, and D_y and D_z are obtained similarly.

STUDENT'S NOTES

EXERCISE 10.6 REVIEW

Solve graphically, finding each coordinate to the nearest integer.

1. $3x + y = 14$
   $4x - 3y = -3$

2. $5x + 4y = 16$
   $5x - 8y = -5$

3. $7x + 6y = 9$
   $-7x + 18y = -5$

4. $3x + 2y = 18$
   $6x - 3y = 1$

Graph the equations and state whether they are independent, inconsistent, or dependent.

5. $4x - 3y = 6$
   $-12x + 9y = -18$

6. $9x + 6y = 18$
   $12x + 8y = 22$

7   $-12x + 10y = -16$
    $30x - 25y = 30$

8   $10x + 15y = 35$
    $14x + 21y = 49$

Solve by addition or subtraction.

9   $4x + 3y = 7$
    $3x + 4y = 0$

10 $3x + 5y = 3$
    $6x - 5y = 0$

11 $6x + 6y = 1$
    $4x - 9y = -8$

Solve by substitution.

12 $2x + y = 8$
$3x - 5y = -14$

13 $4x + 3y = 2$
$6x - 6y = -11$

14 $-8x - y = 4$
$12x + 2y = -9$

Verify the following equations by expanding the determinants.

15 $\begin{vmatrix} a & b \\ -c & -d \end{vmatrix} = -\begin{vmatrix} a & c \\ b & d \end{vmatrix}$

16 $\begin{vmatrix} a & b \\ c & d \end{vmatrix} = \begin{vmatrix} a & b \\ c + 3a & d + 3b \end{vmatrix}$

17 $\begin{vmatrix} 2 & 3 & 5 \\ 4 & 1 & 3 \\ 6 & 4 & 8 \end{vmatrix} = 0$

18 $\begin{vmatrix} a & 2a & x \\ b & 2b & y \\ c & 2c & z \end{vmatrix} = 0$

Solve the following equations by Cramer's rule.

19    $6x - 7y = -22$
       $3x + y = 7$

20    $5x + 4y = 6$
       $7x - 8y = 5$

21
$$3x + y + 2z = 4$$
$$6x - 3y + 2z = -18$$
$$3x + 2y - 8z = 4$$

Solve the following equations by addition or substitution.

22
$$4x + 3y + z = 2$$
$$5x - 9y + 3z = -7$$
$$7x + 6y + 2z = 3$$

23
$$2x + 6y - 5z = 2$$
$$-x - 6y + 10z = 3$$
$$-3x - 12y + 10z = 0$$

24 Show that if t is any real number, then $x = 10t + 1$, $y = 9t + 4$, $z = 7t + 2$ is a solution of the equations

$$x + 2y - 4z = 1$$
$$2x - 3y + z = -8$$
$$4x + y - 7z = -6$$

25 Show that the graph of

$$\begin{vmatrix} x & -y \\ y & x \end{vmatrix} = 1$$

is a circle.

26 Show that the graph of

$$\begin{vmatrix} x & -b \\ y & a \end{vmatrix} = c$$

is a straight line.

NAME _____ DATE _____ SCORE _____

EXERCISE 10.7 CHAPTER TEST

1. Show that $8x - 12y = 4$ and $-6x + 9y = -2$ are inconsistent. Graph them, and notice that the lines are parallel.

2. For which value of b are the following equations dependent?
$10x + 12y = -8$
$15x + by = -12$

Solve the following equations.

3. $4x + 5y = 31$
$5x + 4y = 32$

4. $-3x + y = 11$
$x + 3y = 13$

5. $3x + 5y = -1$
$6x - 5y = 4$

6 $2x + 3y = 5$
$2x - 3y = -4$

7 $4x + 2y - 5z = 1$
$7x - 8y + 2z = 1$
$5x \quad\quad - 4z = 1$

8 $5x + 3y + 2z = 3$
$7x - 6y + 4z = -2$
$3x - 3y + 6z = 1$

CHAPTER 11

Quadratic Equations

11.1 INTRODUCTION

Many problems lead to equations in which the variable appears to a power greater than the first. Equations of the form

$$ax^2 + bx + c = 0 \quad a \neq 0 \quad (11.1)$$

are called **quadratic equations,** where a, b, and c represent real numbers. Every quadratic equation has two roots, counting repeated roots and complex numbers.

A quadratic equation of the form

$$ax^2 + bx = 0$$

can readily be solved by factoring.

EXAMPLE 1 Solve $x^2 + 3x = 0$.

Solution We can factor as in Chap. 3.

$$\begin{aligned} x^2 + 3x &= 0 && \text{given} \\ x(x + 3) &= 0 && \text{factoring} \end{aligned}$$

Recall from (2.25) that if the product is 0, at least one of the factors must be 0. Thus either

$$\begin{aligned} x = 0 \quad &\text{or} \quad x + 3 = 0 \\ &\phantom{\text{or}} \quad x = -3 \end{aligned}$$

The solutions are 0 and -3, which should be checked in the given equation.

EXAMPLE 2 Solve $4x^2 - 20x = 0$.

Solution

$$\begin{aligned} 4x^2 - 20x &= 0 && \text{given} \\ 4x(x - 5) &= 0 && \text{factoring} \\ 4x = 0 \quad &\text{or} \quad x - 5 = 0 \\ &&& \text{each factor} = 0 \\ x = 0 \quad &\text{or} \quad x = 5 && \text{solutions} \end{aligned}$$

A quadratic equation of the form

$ax^2 + c = 0$

can also be solved readily. The nature of the roots is different for $c > 0$ and $c < 0$.

EXAMPLE 3 Solve $x^2 - 25 = 0$.

Solution

$(x - 5)(x + 5) = 0$ factoring
$x - 5 = 0$ or $x + 5 = 0$
                                each factor = 0
$x = 5$ or $x = -5$
                                  solutions

Notice that we could also have solved $x^2 - 25 = 0$ by writing

$x^2 = 25$
$x = \pm\sqrt{25} = \pm 5$

In doing so, we must be sure to use both plus and minus signs.

EXAMPLE 4 Solve $x^2 + 36 = 0$.

Solution

$x^2 = -36$ transposing 36
$x = \pm\sqrt{-36}$ taking square roots

In Sec. 6.11 we saw that $\sqrt{-36} = \sqrt{36(-1)} = 6i$. Thus the solutions are $x = \pm 6i$, which are complex numbers.

EXAMPLE 5 Solve $2x^2 + 7 = 0$.

Solution

$2x^2 = -7$ transposing 7
$x^2 = -\frac{7}{2}$ dividing by 2
$x = \pm\sqrt{-\frac{7}{2}}$ taking square roots
$x = \pm i\sqrt{\frac{7}{2}}$ definition of i

We could go farther and write

$\sqrt{\frac{7}{2}} = \sqrt{\frac{7(2)}{2(2)}} = \frac{1}{2}\sqrt{14}$

and thus have

$x = \pm\frac{i}{2}\sqrt{14}$

11.2
SOLUTION BY FACTORING

In the previous section we saw how to solve the special case of $ax^2 + bx + c = 0$, where $c = 0$ or $b = 0$. If none of the coefficients a, b, and c is zero, we can sometimes factor and solve. The method depends on the following fact [see Eq. (2.25)]:

The product of two factors is 0 if and only if at least one of the factors is 0.   **(11.2)**

We shall see later in this chapter that if a, b, and c are integers, then $ax^2 + bx + c$ has *factors* $px + q$ and $rx + s$ with p, q, r, and s integers *if and only if $b^2 - 4ac$ is a perfect square* (that is, 0, 1, 4, 9, 16, 25, 36, . . .). This only says that there are factors; it does not tell how to find them.

EXAMPLE 1 Solve $x^2 - 3x + 2 = 0$.

Solution

$(x - 2)(x - 1) = 0$ factoring
$x - 2 = 0$ or $x - 1 = 0$ by (11.2)
$x = 2$ or $x = 1$ solutions

EXAMPLE 2 Solve $3x^2 - x + 2 = x^2 + 3$.

Solution

$2x^2 - x - 1 = 0$ transposing
$(2x + 1)(x - 1) = 0$ factoring
$2x + 1 = 0$ or $x - 1 = 0$ by (11.2)
$x = -\frac{1}{2}$ or $x = 1$ solutions

EXAMPLE 3 Solve $6x^2 + 7x = 20$.

Solution

$6x^2 + 7x - 20 = 0$ transposing
$(3x - 4)(2x + 5) = 0$ factoring
$3x - 4 = 0$ or $2x + 5 = 0$
                                 by (11.2)
$x = \frac{4}{3}$ or $x = -\frac{5}{2}$
                                 solutions

In Example 3, it is also important to know what *not* to do. We could have written

$x(6x + 7) = 20$ factoring
$x = 20$ or $6x + 7 = 20$
                                 not by (11.2)

The last step is wrong! We use (11.2) only if one member is 0. It is also *wrong* to say, for example, that from

$x(6x + 7) = (4)(5)$ factoring

it follows that $x = 4$ or $6x + 7 = 5$. Equating *nonzero* factors like this is wrong.

EXAMPLE 4 Solve $2x^2 - 5x - 6 = 0$.

Solution Here $a = 2$, $b = -5$, and $c = -6$, so $b^2 - 4ac = (-5)^2 - 4(2)(-6) = 25 + 48 = 73$. Since this is not a perfect square, we cannot factor $2x^2 - 5x - 6$. We shall develop a method of solution in the next section.

SOLVING QUADRATIC EQUATIONS BY FACTORING The general process for the use of factoring in solving a quadratic equation is as follows:

1. By adding the same number to each member of the given equation, obtain an equivalent equation in which the right member is 0.
2. Factor the left member if possible.
3. Set each factor equal to 0 and solve each of the resulting equations.
4. Check the roots obtained in step 3 by replacing the variable in the given equation by each root.

STUDENT'S NOTES

EXERCISE 11.1

Solve the following quadratic equations.

1. $x^2 + 5x = 0$

2. $x^2 - 7x = 0$

3. $x^2 + 21x = 0$

4. $x^2 - \pi x = 0$

5. $4x^2 - 7x = 0$

6. $3x^2 - 100x = 0$

7. $7x^2 + 8x = 0$

8. $12x^2 + 5x = 0$

9. $3x^2 = 2x$

10. $2x^2 + x = -4x$

11. $6x - x^2 = 6x^2$

12. $8x^2 + 7x = 6x$

13 $x^2 - 4 = 0$

14 $x^2 - 49 = 0$

15 $4x^2 - 81 = 0$

16 $9x^2 - 1 = 0$

17 $25x^2 - 27 = 0$

18 $9x^2 - 32 = 0$

19 $16x^2 - 5 = 0$

20 $49x^2 - 6 = 0$

21 $x^2 + 4 = 0$

22 $x^2 + 144 = 0$

23 $9x^2 + 64 = 0$

24 $81x^2 + 25 = 0$

25 $x^2 + 3 = 0$

26 $4x^2 + 11 = 0$

27 $81x^2 = -5$

28 $100x^2 - 23 = -50$

29 $x^2 - 3x + 2 = 0$

30 $x^2 - 5x + 6 = 0$

31 $x^2 + x - 6 = 0$

32 $x^2 + x - 2 = 0$

33 $2x^2 - 7x + 3 = 0$

34 $3x^2 - 7x + 2 = 0$

35 $2x^2 - 3x - 2 = 0$

36 $3x^2 - 8x - 3 = 0$

37 $5x^2 - 24x - 5 = 0$

38 $3x^2 - 14x + 8 = 0$

39 $5x^2 - 17x + 6 = 0$

40 $5x^2 - 32x + 12 = 0$

41 $7x^2 - 33x = 10$

42 $4x^2 - 9x = 9$

43 $3x^2 - 4 = 4x$

44 $5x^2 - 6 = -13x$

45 $5x^2 + 9x - 5 = 2x + 1$

46 $3x^2 + 5x - 23 = 4x - 21$

47 $11x^2 - 13x + 5 = 5x^2 - 1$

48 $7x^2 + 5x - 9 = x^2 - 3$

11.3 SOLUTION BY COMPLETING THE SQUARE

It can be proved that any quadratic equation is equivalent to an equation of the form

$$(x + d)^2 = k \qquad (1)$$

Furthermore, $x = -d + \sqrt{k}$ and $x = -d - \sqrt{k}$ are the solutions of (1), as can be seen by substituting these values for x in it. The two values of x just given can be obtained from (1) by equating the square roots of its members and then solving for x. Consequently, we can solve any quadratic equation if we can put it in the form (1). From our work on special products and factoring, we know that a quadratic trinomial with 1 as leading coefficient is a **perfect square** provided the constant term is the square of half the coefficient of x. We shall now give some examples.

EXAMPLE 1 $x^2 - 6x + 9$ is the square of a binomial since $9 = [\frac{1}{2}(-6)]^2$, and the principal square root is $x - 3$.

EXAMPLE 2 $x^2 + \frac{2}{3}x + \frac{1}{9}$ is the square of $x + \frac{1}{3}$, since $\frac{1}{9} = [\frac{1}{2}(\frac{2}{3})]^2$.

EXAMPLE 3 $x^2 - \frac{1}{4}x + \frac{1}{64}$ is the square of $x - \frac{1}{8}$, since $\frac{1}{64} = [\frac{1}{2}(-\frac{1}{4})]^2$.

In Chap. 2 we stated that if the same number is added to each member of a given equation or if each member is multiplied or divided by the same nonzero constant, the resulting equation is equivalent to the given equation.

The process of adding the square of one-half the coefficient of x to each member of an equation in the form (11.1) with $a = 1$ is called **completing the square**.

We shall conclude this section by using the method of completing the square to solve four quadratic equations.

EXAMPLE 4 Solve the equation $x^2 = 8x + 20$.

Solution

$$\begin{aligned} x^2 &= 8x + 20 & \text{given equation} \\ x^2 - 8x &= 20 & \text{adding } -8x \text{ to each member} \end{aligned}$$

This equation is in proper form so we complete the square and proceed as follows. Adding $[\frac{1}{2}(-8)]^2 = 16$ to each member gives

$$x^2 - 8x + 16 = 20 + 16$$

Since $x^2 - 8x + 16 = (x - 4)^2$ and $20 + 16 = 36$, we get

$$\begin{aligned} (x - 4)^2 &= 36 \\ x - 4 &= \pm 6 & \text{by (1)} \\ x &= 4 \pm 6 & \text{adding 4 to each member} \end{aligned}$$

Consequently,

$$x = 4 + 6 = 10 \quad \text{and} \quad x = 4 - 6 = -2$$

Hence, the solution set is $\{10, -2\}$.

Check We replace x by 10 and then by -2 in the given equation and get $10^2 = 8(10) + 20$, each member of which is 100; and $(-2)^2 = 8(-2) + 20$, and here each member is equal to 4.

EXAMPLE 5 Solve $1 - 2x = 3x^2$.

Solution

$$1 - 2x = 3x^2 \qquad \text{given equation}$$

Adding $-3x^2 - 1$ to each member gives

$$-3x^2 - 2x = -1$$

Dividing each member by -3, we get

$$x^2 + \tfrac{2}{3}x = \tfrac{1}{3}$$

and adding $[\frac{1}{2}(\frac{2}{3})]^2 = \frac{1}{9}$ to each member leads to

$$\begin{aligned} x^2 + \tfrac{2}{3}x + \tfrac{1}{9} &= \tfrac{1}{3} + \tfrac{1}{9} = \tfrac{4}{9} \\ (x + \tfrac{1}{3})^2 &= \tfrac{4}{9} \\ x + \tfrac{1}{3} &= \pm \tfrac{2}{3} & \text{by (1)} \\ x &= -\tfrac{1}{3} + \tfrac{2}{3} = \tfrac{1}{3} \end{aligned}$$

and

$$x = -\tfrac{1}{3} - \tfrac{2}{3} = -1$$

Thus, the solution set is $\{\tfrac{1}{3}, -1\}$.

Check The roots $\tfrac{1}{3}$ and -1 can be checked by replacing each of them for x in the given equation. This will be left as an exercise for the student.

EXAMPLE 6 Solve the equation $2x^2 - 2x - 13 = 0$.

Solution We shall show the steps in solving the above equation, but we shall omit most of the explanatory steps.

$2x^2 - 2x - 13 = 0$ given equation
$2x^2 - 2x = 13$
$x^2 - x = \frac{13}{2}$

Adding $[\frac{1}{2}(-1)]^2 = \frac{1}{4}$ to each member gives

$x^2 - x + \frac{1}{4} = \frac{13}{2} + \frac{1}{4} = \frac{27}{4}$

$(x - \frac{1}{2})^2 = \frac{27}{4}$

$x - \frac{1}{2} = \pm\sqrt{\frac{27}{4}} = \pm\frac{3\sqrt{3}}{2}$

$x = \frac{1}{2} \pm \frac{3\sqrt{3}}{2} = \frac{1}{2}(1 \pm 3\sqrt{3})$

and the solution is $\{\frac{1}{2}(1 + 3\sqrt{3}), \frac{1}{2}(1 - 3\sqrt{3})\}$.

EXAMPLE 7 Solve the equation $4x^2 + 2x + 1 = 0$.

Solution The steps in the solution process are shown below without explanation:

$4x^2 + 2x + 1 = 0$

$4x^2 + 2x = -1$

$x^2 + \frac{1}{2}x = -\frac{1}{4}$

$x^2 + \frac{1}{2}x + \frac{1}{16} = -\frac{1}{4} + \frac{1}{16}$

$(x + \frac{1}{4})^2 = -\frac{3}{16}$

$x + \frac{1}{4} = \pm\sqrt{-\frac{3}{16}} = \pm\frac{\sqrt{3}\,i}{4}$

$x = -\frac{1}{4} \pm \frac{\sqrt{3}\,i}{4} = \frac{1}{4}(-1 \pm \sqrt{3}\,i)$

Hence, the solution set is $\{\frac{1}{4}(-1 + \sqrt{3}\,i), \frac{1}{4}(-1 - \sqrt{3}\,i)\}$.

EXERCISE 11.2

Some of the steps in the solution of a quadratic equation by completing the square are indicated. Fill the blanks and find the solutions.

1. $$x^2 = 4x + 12$$
    $$x^2 - 4x = 12$$
    $$x^2 - 4x + 2^2 = 12 + 4$$
    $$(x - 2)^2 =$$
    $$x - 2 = \pm$$
    $$x = \quad \pm$$
    $$x =$$
    Solution set:

2. $$x^2 - 27 = -6x$$
    $$x^2 + 6x = 27$$
    $$x^2 + 6x + (\quad)^2 = 27 +$$
    $$(x + 3)^2 =$$
    $$x + 3 = \pm$$
    $$x = \quad \pm$$
    $$x =$$
    Solution set:

3. $$3x^2 = 4x - 1$$
    $$3x^2 - 4x = -1$$
    $$x^2 - \tfrac{4}{3}x =$$
    $$x^2 - \tfrac{4}{3}x + (\quad)^2 = \quad +$$
    $$(x - \quad)^2 =$$
    $$x - \quad = \pm$$
    $$x = \quad \pm$$
    $$x =$$
    Solution set:

4. $$6x^2 = 7x + 3$$
    $$6x^2 - 7x = 3$$
    $$x^2 - \tfrac{7}{6}x =$$
    $$x^2 - \tfrac{7}{6}x + (\quad)^2 = \quad +$$
    $$(x \quad)^2 =$$
    $$x - \quad =$$
    $$x =$$
    Solution set:

Solve each equation.

5. $x^2 - 3x + 2 = 0$

6. $x^2 - 5x + 6 = 0$

7. $x^2 - 8x + 15 = 0$

8. $x^2 - x - 6 = 0$

9 $x^2 + x - 2 = 0$

10 $x^2 + 6x + 8 = 0$

11 $x^2 + 3x - 10 = 0$

12 $x^2 + x - 12 = 0$

13 $2x^2 - 5x + 2 = 0$

14 $3x^2 - 11x + 6 = 0$

15 $4x^2 - 23x + 15 = 0$

16 $3x^2 + x - 2 = 0$

17 $15x^2 - 31x + 10 = 0$

18 $20x^2 - 37x + 15 = 0$

19 $14x^2 - 53x = -14$

20 $6x^2 - 21 = 5x$

21 $15x^2 + 19x = -6$

22 $30x^2 - 30 = -11x$

23 $6x^2 = 7x + 5$

24 $8x^2 = 26x - 21$

25 $x^2 - 2x - 1 = 0$

26 $x^2 - 4x + 1 = 0$

27 $x^2 - 6x + 4 = 0$

28 $x^2 - 8x + 14 = 0$

29 $x^2 - 8x + 20 = 0$

30 $x^2 - 6x + 10 = 0$

31 $x^2 - 10x + 34 = 0$

32 $x^2 + 4x + 13 = 0$

Solve the following equations. Obtain each root to three decimal places by using a table of square roots or a calculator.

33 $x^2 - 2x - 1 = 0$

34 $x^2 - 6x + 7 = 0$

35 $x^2 + 8x + 13 = 0$

36 $x^2 + 10x + 22 = 0$

37 $2x^2 + 2x - 1 = 0$

38 $9x^2 - 6x - 4 = 0$

39 $2x^2 - 6x + 1 = 0$

40 $9x^2 + 12x + 1 = 0$

11.4 THE QUADRATIC FORMULA

We shall now derive a formula for use in solving the quadratic equation

$$ax^2 + bx + c = 0 \qquad (11.1)$$

by solving it by the method of completing the square as given in Sec. 11.3. We thus begin by dividing through by a, which gives

$$x^2 + \frac{b}{a}x + \frac{c}{a} = 0$$

$$x^2 + \frac{b}{a}x = -\frac{c}{a} \qquad \text{transposing } \frac{c}{a}$$

Now, adding the square of half the coefficient of x to each member, we have

$$x^2 + \frac{b}{a} + \left(\frac{b}{2a}\right)^2 = -\frac{c}{a} + \frac{b^2}{4a^2}$$

$$\left(x + \frac{b}{2a}\right)^2 = \frac{b^2 - 4ac}{4a^2}$$

Hence, since $\sqrt{4a^2} = 2a$,

$$x + \frac{b}{2a} = \pm \frac{\sqrt{b^2 - 4ac}}{2a}$$

Now, transposing $b/2a$, we get

$$x = -\frac{b}{2a} \pm \frac{\sqrt{b^2 - 4ac}}{2a}$$

Since the two denominators in the right-hand member are the same, we have

$$x = \frac{-b \pm \sqrt{b^2 - 4ac}}{2a} \qquad \text{quadratic formula}$$

$$(11.3)$$

The statement (11.3) is called the quadratic formula and can be used to obtain the roots of any quadratic equation *that is in standard form*. If the given equation is not in standard form, we first obtain an equivalent equation that is in standard form and then we apply the formula.

SOLVING A QUADRATIC EQUATION In solving a quadratic equation,

1. First try to factor it (need $b^2 - 4ac$ to be a perfect square).
2. If you cannot factor the equation, use the quadratic formula since it will work every time.

EXAMPLE 1 In order to use (11.3) to get the roots of $2x^2 = 3 - 5x$, we must convert the equation to an equivalent equation that has the form (11.1). For this purpose, we shall add $-3 + 5x$ to each member, arrange terms, and get

$$2x^2 + 5x - 3 = 0$$

Now comparing the coefficients of x^2, x, and the constant term with those in (11.1), we see that $a = 2$, $b = 5$, and $c = -3$. Hence, if we replace a, b, and c in (11.3) by 2, 5, and -3, respectively, we get

$$x = \frac{-5 \pm \sqrt{5^2 - 4(2)(-3)}}{2(2)}$$

$$= \frac{-5 \pm \sqrt{25 + 24}}{4} = \frac{-5 \pm \sqrt{49}}{4} = \frac{-5 \pm 7}{4}$$

Using the plus sign and then the minus sign gives

$$x = \frac{2}{4} \quad \text{and} \quad \frac{-12}{4}$$

$$= \frac{1}{2} \quad \text{and} \quad -3$$

Hence, the solution set is $\{\frac{1}{2}, -3\}$.

Check Replacing x in $2x^2 = 3 - 5x$ by $\frac{1}{2}$ and then by -3 and simplifying, we get, with $x = \frac{1}{2}$,

$$2(\tfrac{1}{2})^2 = 3 - \tfrac{5}{2} \quad \text{or} \quad \tfrac{1}{2} = \tfrac{1}{2}$$

and with $x = -3$,

$$2(-3)^2 = 3 - 5(-3) \quad \text{or} \quad 18 = 18$$

EXAMPLE 2 Solve $12x^2 = 6 - x$ by use of the quadratic formula.

Solution We first add $-6 + x$ to each member of the above equation, arrange terms, and get

$$12x^2 + x - 6 = 0$$

Now, comparing the coefficients of x^2 and x, and the constant terms with those in (11.1), we see that $a = 12$, $b = 1$, and $c = -6$. Therefore, if we substitute these values in (11.3), we get

$$x = \frac{-1 \pm \sqrt{1^2 - 4(12)(-6)}}{2(12)} = \frac{-1 \pm \sqrt{1 + 288}}{24}$$

$$= \frac{-1 \pm \sqrt{289}}{24} = \frac{-1 \pm 17}{24}$$

$$= \tfrac{16}{24} \quad \text{and} \quad -\tfrac{18}{24}$$

$$= \tfrac{2}{3} \quad \text{and} \quad -\tfrac{3}{4}$$

Therefore, the solution set is $\{\tfrac{2}{3}, -\tfrac{3}{4}\}$.

Check Replacing x in $12x^2 = 6 - x$ by $\frac{2}{3}$ and then by $-\frac{3}{4}$, we get

$12(\frac{2}{3})^2 = 6 - \frac{2}{3}$ or $\frac{16}{3} = \frac{16}{3}$

and

$12(-\frac{3}{4})^2 = 6 - (-\frac{3}{4})$ or $\frac{27}{4} = \frac{27}{4}$

EXAMPLE 3 Solve the equation $2x = 4 + 3x^2$ by use of the quadratic formula.

Solution

$2x = 4 + 3x^2$ given equation

Adding $-3x^2 - 4$ to each member and arranging terms gives

$-3x^2 + 2x - 4 = 0$

Comparing with (11.1), we have

$a = -3 \quad b = 2 \quad c = -4$

Then substituting in (11.3) gives

$$x = \frac{-2 \pm \sqrt{2^2 - 4(-3)(-4)}}{2(-3)}$$

$$= \frac{-2 \pm \sqrt{4 - 48}}{-6} = \frac{-2 \pm \sqrt{-44}}{-6}$$

$$= \frac{-2 \pm 2i\sqrt{11}}{-6} = \tfrac{1}{3}(1 \mp i\sqrt{11})$$

Thus the solution set is

$\{\tfrac{1}{3}(1 - i\sqrt{11}), \tfrac{1}{3}(1 + i\sqrt{11})\}$

Check We replace x in $2x = 4 + 3x^2$ by the roots $\tfrac{1}{3}(1 \mp i\sqrt{11})$ and get

$2[\tfrac{1}{3}(1 \mp i\sqrt{11})] = 4 + 3[\tfrac{1}{3}(1 \mp i\sqrt{11})]^2$

Simplifying each member, we have

$\tfrac{2}{3}(1 \mp i\sqrt{11}) = 4 + 3[\tfrac{1}{9}(1 \mp 2i\sqrt{11} - 11)]$

or

$\tfrac{2}{3} \mp \tfrac{2i\sqrt{11}}{3} = 4 + \tfrac{1}{3} \mp \tfrac{2i\sqrt{11}}{3} - \tfrac{11}{3}$

$$= \frac{12 + 1 - 11}{3} \mp \frac{2i\sqrt{11}}{3}$$

$$= \tfrac{2}{3} \mp \frac{2i\sqrt{11}}{3}$$

Many equations can be simplified and then reduced to a quadratic equation in standard form. The next two examples are called **fractional equations**. Solutions should be checked to make sure there is no division by 0. Fractional equations that lead to linear equations were treated in Chap. 7.

EXAMPLE 4 Solve

$$\frac{x + 2}{2x - 1} = \frac{2x + 1}{3x - 2}$$

Solution Multiplying both members by the lcd $(2x - 1)(3x - 2)$ gives

$(x + 2)(3x - 2) = (2x + 1)(2x - 1)$
$3x^2 + 4x - 4 = 4x^2 - 1$ multiplying
$0 = x^2 - 4x + 3$ collecting like terms

Now $a = 1, b = -4, c = 3$, and thus the solutions are

$$x = \frac{4 \pm \sqrt{16 - 4(3)}}{2} = \frac{4 \pm \sqrt{4}}{2}$$

$$= \frac{4 \pm 2}{2} = 2 \pm 1 = 3, 1$$

We could also have solved the quadratic by factoring.

EXAMPLE 5 Solve

$$\frac{x + 1}{2x + 1} = \frac{3x + 2}{2x - 3}$$

Solution

$(x + 1)(2x - 3) = (3x + 2)(2x + 1)$ multiplying by lcd
$2x^2 - x - 3 = 6x^2 + 7x + 2$ multiplying
$0 = 4x^2 + 8x + 5$

Here $a = 4, b = 8, c = 5$, so

$$x = \frac{-8 \pm \sqrt{64 - 4(4)(5)}}{2(4)} = \frac{-8 \pm \sqrt{-16}}{8}$$

$$= \frac{-8 \pm 4i}{8} = -1 \pm \frac{i}{2}$$

EXERCISE 11.3

Put the quadratic equation in standard form and identify a, b, and c.

1 $x^2 + 5x - 1 = 0$

2 $2x^2 - 5x - 2 = 0$

3 $-3x^2 - 5x = 8$

4 $2x^2 + 12 = 7x$

Solve by use of the quadratic formula.

5 $x^2 - 4x + 3 = 0$

6 $x^2 - 7x + 10 = 0$

7 $x^2 - 6x + 8 = 0$

8 $x^2 + 5x - 6 = 0$

9 $5x^2 - 17x + 6 = 0$

10 $3x^2 - 7x + 2 = 0$

11 $7x^2 - 17x + 6 = 0$

12 $2x^2 - 9x - 5 = 0$

13 $12x^2 - 17x + 6 = 0$

14 $20x^2 - 31x + 12 = 0$

15 $35x^2 - 43x + 12 = 0$

16 $21x^2 - 23x + 6 = 0$

17 $20x^2 - 7x - 6 = 0$

18 $12x^2 + x - 6 = 0$

19 $6x^2 + 7x + 2 = 0$

20 $25x^2 + 25x + 6 = 0$

21 $x^2 - 4x - 1 = 0$

22 $x^2 - 6x + 7 = 0$

23 $4x^2 - 8x + 1 = 0$

24 $9x^2 - 18x + 7 = 0$

25 $x^2 - 4x + 13 = 0$

26 $x^2 - 6x + 13 = 0$

27 $2x^2 - 6x + 5 = 0$

28 $9x^2 - 12x + 5 = 0$

29 $x^2 - x + 1 = 0$

30 $3x^2 - 4x + 3 = 0$

31 $9x^2 - 18x + 11 = 0$

32 $4x^2 - 16x + 23 = 0$

33. $\dfrac{3x-1}{x+3} = \dfrac{2x+3}{4x-1}$

34. $\dfrac{3x-1}{2x+3} = \dfrac{4x-5}{x+7}$

35. $\dfrac{3x+2}{6x+1} = \dfrac{3x+6}{9x+4}$

36. $\dfrac{2x+1}{4x-1} = \dfrac{6x-5}{4x+5}$

37. $\dfrac{2x-1}{x+3} = \dfrac{x-2}{x+4}$

38. $\dfrac{2x-1}{x+5} = \dfrac{x+3}{x-4}$

39. $\dfrac{3x-1}{2x+5} = \dfrac{x-4}{2x+3}$

40. $\dfrac{3x+4}{x+5} = \dfrac{2x-1}{3x+4}$

Find the solutions of the following equations to three decimal places.

41 $x^2 - 4x + 1 = 0$

42 $x^2 - 6x + 7 = 0$

43 $x^2 - 10x + 20 = 0$

44 $x^2 - 4x - 3 = 0$

45 $4x^2 - 12x + 7 = 0$

46 $9x^2 - 12x + 1 = 0$

47 $8x^2 - 12x + 1 = 0$

48 $9x^2 - 30x + 17 = 0$

11.5 RADICAL EQUATIONS

An equation that contains a radical involving the variable is a **radical equation.** In this section we shall explain the method for solving radical equations in which the radicals involved are of the second order.

The method depends upon the fact that if two numbers are equal, their squares are equal. This follows from axiom (2.11), which states that if $A = B$, then $AC = BC$. If we replace C by A we have $A^2 = BA$, and, since $A = B$, $BA = B \cdot B = B^2$. Consequently, $A^2 = B^2$, and we have the following conclusion:

If a given equation is satisfied by some replacement for the variable, the equation obtained by equating the squares of the members of the given equation is satisfied by the same replacement. (11.4)

For example, we can readily verify that the number 2 satisfies

$$x - 1 = 2x - 3 \quad (1)$$

and

$$(x - 1)^2 = (2x - 3)^2 \quad (2)$$

ROOTS NEED TO BE CHECKED The *converse of statement (11.4) is not true* since, as you can verify, $\frac{4}{3}$ satisfies (2) but does not satisfy (1).
The equation

$$\sqrt{5x - 11} + \sqrt{x - 3} = 4 \quad (3)$$

is an example of a radical equation. We shall illustrate the method for obtaining the roots of a radical equation by solving Eq. (3).

PROCEDURE FOR SOLVING A RADICAL EQUATION The procedure consists of using (11.4) to obtain an equation that contains no radicals and at least one of whose roots is a root of (3). The computation is less complicated if we obtain an equation that is equivalent to (3) in which one member contains one radical and no other term. Therefore, we add $-\sqrt{x - 3}$ to each member of (3) and get

$$\sqrt{5x - 11} = 4 - \sqrt{x - 3} \quad (4)$$

Now we equate the squares of the members of (4) and get

$$5x - 11 = 16 - 8\sqrt{x - 3} + x - 3$$

or

$$5x - 11 = 13 + x - 8\sqrt{x - 3} \quad (5)$$

Equation (5) contains only one radical. We obtain an equivalent equation, in which one member contains only the radical, by adding $-13 - x$ to each member and simplifying. Thus, we get

$$4x - 24 = -8\sqrt{x - 3}$$

Now, dividing each member by 4, we have

$$x - 6 = -2\sqrt{x - 3} \quad (6)$$

Next, we equate the squares of the members of (6) and get

$$x^2 - 12x + 36 = 4(x - 3)$$

or

$$x^2 - 12x + 36 = 4x - 12 \quad (7)$$

Equation (7) is a quadratic equation, and we solve it as follows. Adding $-4x + 12$ to each member of (7), we get

$$x^2 - 16x + 48 = 0 \quad (8)$$
$$(x - 12)(x - 4) = 0 \quad \text{factoring left member}$$
$$x - 12 = 0$$
$$x = 12$$
$$x - 4 = 0$$
$$x = 4$$

Consequently, the roots of (8) are 12 and 4.

CHECK ALL ROOTS According to the statement (11.4), any root of (3) is also a root of (8), but as Eqs. (1) and (2) illustrate, the converse is not true. Hence it is *necessary* to replace x in (3) by 12 and 4 in order to ascertain whether or not these numbers are roots. If we replace x by 12, the left member of (3) becomes

$$\sqrt{5(12) - 11} + \sqrt{12 - 3} = \sqrt{60 - 11} + \sqrt{9}$$
$$= \sqrt{49} + 3 = 7 + 3$$
$$= 10$$

Therefore, since the right member of (3) is 4, 12 is not a root of the equation. However, when we replace x by 4, the left member of (3) becomes

$$\sqrt{5(4) - 11} + \sqrt{4 - 3} = \sqrt{20 - 11} + 1$$
$$= \sqrt{9} + 1 = 3 + 1 = 4$$

Therefore, 4 is a root of Eq. (3), so the solution set is $\{4\}$.

Equation (8) is called the *rationalized* form of Eq. (3), and the process of obtaining (8) from (3) is called **rationalizing the equation.**

As illustrated by the above example, the procedure for solving a radical equation in which the radicals are of the second order consists of the following steps:

1. By adding the same expression to each member of the given equation, obtain an equivalent equation in which one member contains one radical and no other term. This is called **isolating a radical.**
2. Equate the squares of the members of the equation in step 1.
3. If the equation obtained in step 2 contains radicals, repeat the process until an equation free of radicals is obtained. The equation thus obtained is called the rationalized equation.
4. If the rationalized equation is linear or quadratic, we solve it either by the method of Chap. 7 or by the methods presented in this chapter.
5. Substitute the roots obtained in step 4 in the given equation in order to ascertain those that are the required roots.

The procedure is further explained by Example 1, below.

EXAMPLE 1 Solve the equation

$$\sqrt{x+1} + \sqrt{2x+3} - \sqrt{8x+1} = 0$$

Solution

$$\sqrt{x+1} + \sqrt{2x+3} - \sqrt{8x+1} = 0 \quad \text{given}$$

Isolating $\sqrt{8x+1}$ leads to

$$\sqrt{x+1} + \sqrt{2x+3} = \sqrt{8x+1}$$

and squaring each member gives

$$x + 1 + 2\sqrt{(x+1)(2x+3)} + 2x + 3 = 8x + 1$$

Then isolating the radical, we have

$$2\sqrt{(x+1)(2x+3)} = 8x + 1 - x - 1 - 2x - 3$$

Combining terms and simplifying the radicand gives

$$2\sqrt{2x^2 + 5x + 3} = 5x - 3$$

Squaring each member leads to

$$4(2x^2 + 5x + 3) = 25x^2 - 30x + 9$$

and simplifying the left member leads to

$$8x^2 + 20x + 12 = 25x^2 - 30x + 9$$

Finally adding $-25x^2 + 30x - 9$ to each member, combining terms, and dividing by -1 gives

$$17x^2 - 50x - 3 = 0$$

If we solve the above equation by the quadratic formula, we get $x = 3$ and $x = -\frac{1}{17}$.

Check We check by substituting each of these values of x in the original equation. By substituting $x = 3$, we get

$$\sqrt{3+1} + \sqrt{6+3} - \sqrt{24+1}$$
$$= \sqrt{4} + \sqrt{9} - \sqrt{25} = 2 + 3 - 5 = 0$$

Therefore, 3 is a root. By substituting $x = -\frac{1}{17}$, we get

$$\sqrt{\frac{-1}{17} + 1} + \sqrt{\frac{-2}{17} + 3} - \sqrt{\frac{-8}{17} + 1}$$
$$= \sqrt{\frac{-1+17}{17}} + \sqrt{\frac{-2+51}{17}} - \sqrt{\frac{-8+17}{17}}$$
$$= \sqrt{\frac{16}{17}} + \sqrt{\frac{49}{17}} - \sqrt{\frac{9}{17}}$$
$$= \frac{4}{\sqrt{17}} + \frac{7}{\sqrt{17}} - \frac{3}{\sqrt{17}}$$
$$= \frac{8}{\sqrt{17}} \neq 0$$

Thus, $-\frac{1}{17}$ is not a root. Consequently, the solution set of the given equation is $\{3\}$.

Readers curious about how $-\frac{1}{17}$ is related to the above problem can verify that $-\frac{1}{17}$ is a root of $\sqrt{x+1} - \sqrt{2x+3} + \sqrt{8x+1} = 0$ and that the rationalized form of this equation is also $17x^2 - 50x - 3 = 0$.

EXERCISE 11.4

Solve the following equations.

1. $\sqrt{2x+5} = 3$

2. $\sqrt{5x-1} = 2$

3. $2 + \sqrt{3x-2} = 1$

4. $1 + \sqrt{4x-3} = 4$

5. $\sqrt{3x+10} = x+4$

6. $\sqrt{9+4x} = 3+2x$

7. $2x - 1 + \sqrt{x+4} = 3x - 3$

8. $2 - 5x + \sqrt{x+3} = -4x - 1$

9. $\sqrt{x+6} = \sqrt{2x+3}$

10. $\sqrt{3x+5} = -\sqrt{5x+4}$

11. $\sqrt{5x+1} - \sqrt{6x-2} = 0$

12. $\sqrt{x+3} - \sqrt{3x+7} = 0$

13 $\dfrac{\sqrt{3x-2}}{\sqrt{x-1}} = 2$

14 $\dfrac{\sqrt{4x-7}}{\sqrt{x-3}} = 3$

15 $\dfrac{\sqrt{2-x}}{\sqrt{x+3}} = 2$

16 $\dfrac{\sqrt{5x+9}}{\sqrt{3-x}} = 1$

17 $\sqrt{x^2 + 2x - 2} = \sqrt{x + 4}$

18 $\sqrt{x^2 + 5x + 9} = \sqrt{x + 5}$

19 $\sqrt{x^2 - 3x + 5} = \sqrt{x + 5}$

20 $\sqrt{x^2 - 5x + 2} = \sqrt{x + 9}$

21 $\sqrt{x + 2} = 3x - 4$

22 $\sqrt{5x + 9} = 4x + 2$

23 $\sqrt{x^2 + 2x + 8} = 3x - 2$

24 $\sqrt{2x^2 - 7x + 5} = 2x - 5$

25 $\sqrt{x+2} + \sqrt{2x+5} = 5$

26 $\sqrt{4x-3} - \sqrt{x+1} = 1$

27 $\sqrt{2x+3} - \sqrt{x+1} = 1$

28 $\sqrt{2x+6} - \sqrt{x-1} = 2$

29 $\dfrac{\sqrt{2x+7}+1}{\sqrt{5x-1}+2} = 1$

30 $\dfrac{\sqrt{3x+4}+2}{\sqrt{4x+1}+1} = 2$

31 $\dfrac{\sqrt{4x+1}+3}{\sqrt{x+2}+1} = 2$

32 $\dfrac{\sqrt{2x+4}+2}{\sqrt{x-2}+1} = 2$

33 $\sqrt{x+2} + \sqrt{3x+3} = \sqrt{12x+1}$

34 $\sqrt{x+9} - \sqrt{x+4} = \sqrt{x+1}$

35 $\sqrt{2x-1} - \sqrt{3x-2} = \sqrt{x-1}$

36 $\sqrt{2x+3} + \sqrt{x+10} = \sqrt{3-13x}$

37 $\sqrt{3x-2} + \sqrt{x-1} - \sqrt{4x+1} = 0$

38 $\sqrt{3x-5} - \sqrt{x-2} - \sqrt{3x-8} = 0$

39 $\sqrt{2x+3} + \sqrt{4x+5} = \sqrt{3-x}$

40 $\sqrt{2x+4} + \sqrt{3x+1} = \sqrt{x+9}$

11.6
SOLUTION OF FORMULAS

A **formula** is a rule or a relationship expressed in algebraic symbols. For example, from geometry we have this rule: "To get the area A of a triangle, take one-half of the product of the base b and the altitude h." Expressed as a formula, this rule becomes $A = \frac{1}{2}bh$. A formula usually expresses one variable in terms of one or more others. If we know the value, or the replacement number, of every variable in a formula except one, then we can use the formula to get the value of the unknown. A formula is an equation, and thus the methods for solving equations can be used for solving it for any one of the variables involved.

EXAMPLE 1 Solve $l = a + (n - 1)(d)$ for n.

Solution

$l = a + (n - 1)(d)$ given formula
$l = a + dn - d$ by distributive axiom

Adding $-dn - l$ to each member, dividing by -1, and arranging terms gives

$dn = l - a + d$

Dividing each member by d leads to

$n = \dfrac{l - a + d}{d}$

EXAMPLE 2 Solve $A = 2\pi rh + \pi r^2$ for r.

Solution This is a quadratic equation in r which in standard form is $\pi r^2 + (2\pi h)r - A = 0$. Using the quadratic formula with $a = \pi$, $b = 2\pi h$, $c = -A$ gives

$r = \dfrac{-2\pi h \pm \sqrt{(2\pi h)^2 - 4(\pi)(-A)}}{2\pi}$

$= \dfrac{-2\pi h \pm \sqrt{4\pi^2(h^2 + A/\pi)}}{2\pi}$ factoring out $4\pi^2$

$= -h \pm \sqrt{h^2 + \dfrac{A}{\pi}}$ simplifying

11.7
EQUATIONS IN QUADRATIC FORM

We can solve each of the equations

$(3x - 2)^2 - (3x - 2) - 6 = 0$ (1)

$(x^2 - 6)^2 - (x^2 - 6) - 6 = 0$ (2)

$\dfrac{1}{x^4} - \dfrac{1}{x^2} - 6 = 0$ (3)

by using a substitution to convert it to a quadratic equation. If we let $y = 3x - 2$ in the first equation, $y = x^2 - 6$ in the second, and $y = 1/x^2$ in the third, they become

$y^2 - y - 6 = 0$ (4)
$(y - 3)(y + 2) = 0$ factoring
$y = 3$ or $y = -2$ solutions to (4)

We solve (1) by letting

$3x - 2 = 3$ or $3x - 2 = -2$
$x = \frac{5}{3}$ $x = 0$

We solve (2) by letting

$x^2 - 6 = 3$ or $x^2 - 6 = -2$
$x^2 = 9$ $x^2 = 4$
$x = \pm 3$ $x = \pm 2$

We solve (3) by letting

$\dfrac{1}{x^2} = 3$ or $\dfrac{1}{x^2} = -2$

$x^2 = \frac{1}{3}$ or $x^2 = -\frac{1}{2}$

$x = \pm\sqrt{\dfrac{1}{3}} = \pm\dfrac{\sqrt{3}}{3}$ $x = \pm\sqrt{-\dfrac{1}{2}} = \pm\dfrac{i\sqrt{2}}{2}$

The equations

$0 = (2x)^6 - 28(2x)^3 + 27 = (8x^3)^2 - 28(8x^3) + 27$

and

$4\left(\dfrac{2x + 5}{1 - 3x}\right)^2 - 5\dfrac{2x + 5}{1 - 3x} + 3 = 0$

are both in quadratic form.

The equation $2x + 20/x = 13$ can be put in quadratic form by multiplying both members by x.

11.8
THE PYTHAGOREAN THEOREM

The *pythagorean theorem* states that if a and b are the legs of a right triangle whose hypotenuse is c (Fig. 11.1), then

$c^2 = a^2 + b^2$ (11.5)

Either leg may be called a, and the other one is then b.

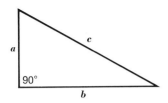

Fig. 11.1

EXAMPLE 1 If one leg of a right triangle is 7 and the hypotenuse is 25, then we can use (11.5) with $a = 7$ and $c = 25$. Thus

$$25^2 = 7^2 + b^2$$
$$625 = 49 + b^2$$
$$576 = b^2$$
$$24 = b$$

EXAMPLE 2 Find the legs of the right triangle whose hypotenuse is 10 if the difference of the legs is 2.

Solution Since $b - a = 2$, we have $b = a + 2$ and (11.5) becomes

$$a^2 + (a + 2)^2 = 10^2$$
$$a^2 + a^2 + 4a + 4 = 100 \quad \text{squaring } a + 2$$
$$2a^2 + 4a - 96 = 0 \quad \text{combining like terms}$$
$$a^2 + 2a - 48 = 0 \quad \text{dividing by 2}$$
$$(a + 8)(a - 6) = 0 \quad \text{factoring}$$
$$a = -8 \quad \text{or} \quad a = 6 \quad \text{solutions}$$

Since a length is not negative, one leg is $a = 6$, the other is $b = a + 2 = 6 + 2 = 8$.

EXERCISE 11.5

Solve the following formulas for the letter indicated. Assume all numbers represented are positive.

1. $s = \dfrac{n}{2}(a + l)$, for n

2. $s = \dfrac{n}{2}(a + l)$, for a

3. $\dfrac{1}{r} = \dfrac{1}{a} + \dfrac{1}{b}$, for b

4. $F = \dfrac{gm_1 m_2}{r^2}$, for m_1

5. $h = \tfrac{1}{2}gt^2$, for t

6. $A = \pi r^2$, for r

7. $p = \sqrt{\dfrac{2l}{g}}$, for l

8. $D = \sqrt{b^2 - 4ac}$, for a

9. $y = vt - 16t^2$, for t

10. $v^2 = w^2 h + 2g$, for w

11 $3x_0 = \dfrac{x_0}{\sqrt{1 - v^2/c^2}}$, for v

12 $3s - 4s^3 = 0$, for s

13 $A = \tfrac{1}{4}nl^2c$, for l

14 $A = \dfrac{\pi r^2 \theta}{360}$, for r

15 $A = \dfrac{\pi r^2 \theta}{360} - \dfrac{r^2 s}{2}$, for r

16 $V = \dfrac{\pi h}{3}(r_1^2 + r_1 r_2 + r_2^2)$, for r_1

Solve each of the following equations by first putting it in quadratic form.

17 $x^4 - 5x^2 + 4 = 0$

18 $x^4 + 12x^2 - 64 = 0$

19 $36x^4 - 13x^2 + 1 = 0$

20 $100x^4 - 29x^2 + 1 = 0$

21 $x^6 - 9x^3 + 8 = 0$

22 $x^6 - 35x^3 + 216 = 0$

23 $\dfrac{36}{x^4} - \dfrac{25}{x^2} + 4 = 0$

24 $\dfrac{400}{x^4} - \dfrac{289}{x^2} + 36 = 0$

25 $(x^2 - 5)^2 - 3(x^2 - 5) - 4 = 0$

26 $(x^2 - 7)^2 + (x^2 - 7) - 6 = 0$

27 $(x^2 - 3x)^2 - 2(x^2 - 3x) - 8 = 0$

28 $(x^2 + 5x)^2 + 10(x^2 + 5x) + 24 = 0$

29 $\left(\dfrac{3x+1}{x+1}\right)^2 - 2\left(\dfrac{3x+1}{x+1}\right) - 8 = 0$

30 $2\left(\dfrac{2x+3}{x-1}\right)^2 - 5\left(\dfrac{2x+3}{x-1}\right) - 3 = 0$

31 $\left(\dfrac{2x-1}{x+3}\right)^2 - 4\left(\dfrac{2x-1}{x+3}\right) + 3 = 0$

32 $\left(\dfrac{3x+2}{2x+1}\right)^2 - 3\left(\dfrac{3x+2}{2x+1}\right) + 2 = 0$

In the following problems, two sides of a right triangle are given. Find the remaining side exactly or to three decimal places.

33 $a = 9, b = 12$

34 $a = 9, b = 40$

35 $a = 9, b = 3$

36 $a = 4, b = 5$

37 $a = 15, c = 25$

38 $a = 7, c = 25$

39 $b = 23, c = 25$

40 $b = 43, c = 44$

41 One leg of a right triangle is 4 cm more than the other. Find the legs if the hypotenuse is 20 cm.

42 One leg of a right triangle is 6 m more than twice the other leg. Find the legs if the hypotenuse is 39 m.

43 The infield of a baseball diamond is a square, 90 ft on each side. How far is it from first base to third base?

44 Two people on bicycles start from the same place at the same time, John goes north and Jennifer west. Jennifer rides 3 mi/h faster than John, and after 2 h they are 30 mi apart. How fast does each one ride?

The pythagorean theorem can be used to show that if d is the distance between the two points $P_1(x_1,y_1)$ and $P_2(x_2,y_2)$ in the plane, then

$$d = \sqrt{(x_2 - x_1)^2 + (y_2 - y_1)^2}$$

Use this to find the distance between the following pairs of points.

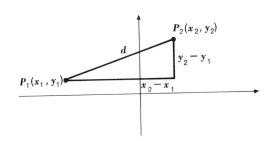

45 (1,4) and (5,7)

46 (2,5) and (−4,3)

47 (1,−8) and (−3,−2)

48 (−2,6) and (3,−1)

11.9 QUADRATIC INEQUALITIES

There are two basic methods for solving a quadratic inequality, for example

$$x^2 + 3x - 4 < 0 \qquad (1)$$

In order to find the **graphical solution** of (1) we first sketch the graph of $y = x^2 + 3x - 4$. A table of values follows.

x	-5	-4	-3	-2	-1	0	1	2
y	6	0	-4	-6	-6	-4	0	6

The graph in Fig. 11.2 is obtained by plotting and then connecting the points. The lowest point on this graph occurs at $x = -\frac{3}{2}$, $y = -\frac{25}{4}$ and is called the **vertex**. Since (1) is the same as $y = x^2 + 3x - 4 < 0$, we look at the graph and see that $y < 0$ if $-4 < x < 1$.

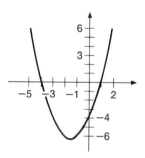

Fig. 11.2

We can find the **algebraic solution** of (1) by factoring first. Thus (1) is the same as

$$(x + 4)(x - 1) < 0 \qquad (2)$$

This means that the product of two numbers is negative, which can happen only if the numbers have opposite signs. The two possibilities are thus

$$x + 4 < 0 \quad \text{and} \quad x - 1 > 0 \qquad (3)$$
$$\text{or} \quad x + 4 > 0 \quad \text{and} \quad x - 1 < 0 \qquad (4)$$

The individual solutions of (3) are $x < -4$ and $x > 1$. However, there is no value of x which is both less than -4 and larger than 1; hence there are no solutions to (3). The individual solutions of (4) are $x > -4$ and $x < 1$, which can be combined into $-4 < x < 1$. Since the solutions of (1) are those of (3) together with those of (4), the solution to (1) is $-4 < x < 1$ (see Fig. 11.3).

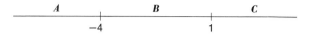

Fig. 11.3

Another way to solve (1) algebraically is to begin by replacing the inequality by the equality

$$x^2 + 3x - 4 = 0 \qquad (5)$$

Now solving (5) by any method gives $x = -4$ and $x = 1$. We now put -4 and 1 on a number line as in Fig. 11.3. The number line is now divided up into three sets A, B, and C. We simply take any number in A, any number in B, and any number in C and see whether (1) is true for these numbers. For example, take -6, 0, and 4. Then

For A: $(-6)^2 + 3(-6) - 4 = 36 - 18 - 4 = 14$
  (1) is false
For B: $0^2 + 3(0) - 4 = -4$ (1) is true
For C: $4^2 + 3(4) - 4 = 16 + 12 - 4 = 24$
  (1) is false

The solution is thus B, which is $-4 < x < 1$.

EXAMPLE 1 Solve $3x^2 - x - 8 > 0$. $\qquad (6)$

Solution If we solve $3x^2 - x - 8 = 0$ by the quadratic formula, we get $x = (1 \pm \sqrt{97})/6$, approximately 1.8 and -1.5. Using -2, 0, and 2, since $-2 < -1.5 < 0 < 1.8 < 2$, gives

$3(-2)^2 - (-2) - 8 = 12 + 2 - 8 = 6$ (6) is true
$3(0)^2 - 0 - 8 = -8$ (6) is false
$3(2)^2 - 2 - 8 = 12 - 10 = 2$ (6) is true.

The solution of (6) is therefore $x < (1 - \sqrt{97})/6 = -1.5$ together with $x > (1 + \sqrt{97})/6 = 1.8$. In set notation the solution set is

$$\left\{ x \,\middle|\, x < \frac{1 - \sqrt{97}}{6} \right\} \cup \left\{ x \,\middle|\, x > \frac{1 + \sqrt{97}}{6} \right\}$$

The solution of (6) graphically would have given only approximate values.

EXAMPLE 2 Solve $2x^2 - 10x + 13 < 0$. $\qquad (7)$

Solution The quadratic formula shows that solutions of $2x^2 - 10x + 13 = 0$ are

$$x = \frac{10 \pm \sqrt{100 - 4(2)(13)}}{4} = \frac{10 \pm 2i}{4} = \frac{5 \pm i}{2}$$

which are complex numbers. There are no real solutions. The solution to (7) thus is either all real numbers or the empty set. Checking any real number will suffice, and using $x = 0$ gives

$2(0)^2 - 10(0) + 13 = 13$ and (7) is false

Hence (7) has no solutions.

EXERCISE 11.6

Solve the following inequalities algebraically.

1. $x^2 - 5x + 4 < 0$

2. $x^2 - 3x - 4 < 0$

3. $x^2 - 4x - 12 < 0$

4. $x^2 + 4x + 3 < 0$

5. $x^2 - x - 2 > 0$

6. $x^2 - 11x + 24 > 0$

7. $x^2 - 4x - 5 > 0$

8. $x^2 + 9x + 18 > 0$

9 $2x^2 - x - 10 < 0$

10 $6x^2 - 19x - 7 < 0$

11 $12x^2 - x - 6 > 0$

12 $15x^2 - 46x + 16 > 0$

Solve the following inequalities geometrically.

13 $x^2 - 4x - 21 < 0$

14 $x^2 + 6x + 5 < 0$

15 $x^2 - 13x + 30 > 0$

16 $x^2 + 6x - 16 > 0$

17 $6x^2 + 2x - 3 > 9x + 7$

18 $4x^2 + 13x - 10 > -4x + 5$

19 $4x^2 - 8x - 5 < 3x - 2$

20 $3x^2 - 8x - 4 < 2x + 4$

21 $9x^2 - 3x - 4 < -2$

22 $12x^2 + 12x + 1 < 15 - x$

Chapter 11: Quadratic Equations

23 $6x^2 - 17x + 2 > -3$

24 $30x^2 + 30x + 6 > 1 - x$

Solve the following inequalities by first converting them to equivalent quadratic inequalities. Such inequalities were treated in Sec. 8.2 by another method.

25 $\dfrac{x-2}{x+1} < 0$ *Hint:* Since $(x+1)^2 > 0$, the given inequality is equivalent to $\dfrac{x-2}{x+1}(x+1)^2 < 0(x+1)^2$, or $(x-2)(x+1) < 0$.

26 $\dfrac{x-4}{x+3} < 0$

27 $\dfrac{x-8}{x-4} > 0$

28 $\dfrac{x+3}{x+12} > 0$

11.10
CHAPTER SUMMARY

A quadratic equation is one of the form

$$ax^2 + bx + c = 0 \qquad a \neq 0$$

where a, b, and c are real numbers. Two special cases, when $c = 0$ and when $b = 0$, can be solved easily. The solutions of

$$ax^2 + bx = 0$$

are $x = 0$ and $x = -b/a$, which are found by writing $ax^2 + bx = x(ax + b) = 0$ and setting each factor equal to 0. The solutions of

$$ax^2 + c = 0$$

are $x = \pm\sqrt{-c/a}$, which are found by setting $x^2 = -c/a$ and taking square roots. The roots are real if $-c/a$ is positive and imaginary if $-c/a$ is negative.

Quadratic equations can sometimes be solved by factoring, but we must be sure that the product of the factors is 0 before setting each factor equal to 0.

There are two methods that always work. One is completing the square. When using it be sure to divide by a so that the coefficient of x^2 is 1.

The other method that always works is use of the quadratic formula, which is derived by completing the square. According to it, the roots of $ax^2 + bx + c = 0$ are

$$x = \frac{-b \pm \sqrt{b^2 - 4ac}}{2a}$$

An equation that involves radicals (square roots) can often be reduced to a quadratic equation by isolating a radical, squaring, and repeating if necessary. All roots of the quadratic found this way are possible roots of the given equation and must be checked individually in the given equation.

An equation may sometimes be put in quadratic form by proper choice of the unknown. Formulas that are quadratic can be solved for one variable in terms of the others; this is done by using the quadratic formula with a, b, and c not as numbers but as letters, or variables.

The pythagorean theorem states that

$$a^2 + b^2 = c^2$$

if a and b are the legs of a right triangle and c is its hypotenuse. It can be used to prove the distance formula

$$d^2 = (x_2 - x_1)^2 + (y_2 - y_1)^2$$

where d is the distance between the points (x_1, y_1) and (x_2, y_2) in the plane.

Quadratic inequalities are solved by both graphical and algebraic methods.

STUDENT'S NOTES

EXERCISE 11.7 REVIEW

Solve the following quadratic equations by factoring or taking square roots.

1 $x^2 - 7x = 0$

2 $2x^2 + 5x = 0$

3 $x^2 - 7 = 0$

4 $2x^2 + 5 = 0$

5 $9x^2 - 14 = 0$

6 $x^2 - 2x - 15 = 0$

7 $2x^2 - 11x - 21 = 0$

8 $8x^2 + 34x - 9 = 0$

9 $x^2 - 3x = 28$

10 $2x^2 = 6 - x$

11 $15x^2 + 2x - 24 = 0$

12 $12x^2 - x - 35 = 0$

13 $6x^2 + 11x = 35$

14 $8x^2 - 45 = -2x$

Solve the following quadratic equations by completing the square.

15 $x^2 - 2x - 8 = 0$

16 $x^2 - x = 20$

17 $9x^2 = 10 - 9x$

18 $3x^2 = 5x - 2$

19 $6x^2 + 3 = 11x$

20 $x^2 = 2x + 4$

21 $x^2 = 10x - 26$

22 $25 + x^2 = 6x$

Solve the following quadratic equations by using the quadratic formula.

23 $x^2 + x - 30 = 0$

24 $x^2 - 4x - 21 = 0$

25 $6x^2 + x = 35$

26 $20 = 9x^2 - 3x$

27 $x^2 - 4x = -1$

28 $x^2 - 10x + 18 = 0$

29 $x^2 + 6x + 10 = 0$

30 $x^2 + 20 = 8x$

31 $9x^2 + 34 = 18x$

32 $20x = 25x^2 + 9$

33 $2c^2x^2 = 3bcx + 2b^2$

34 $x^2 - (2r - 2t)x + r^2 = 2rt + 3t^2$

Solve the following equations. Get each result to three decimals by using a table or a calculator.

35 $2x^2 + 6x = 9$

36 $3x^2 + 12x = 5$

37 $2x^2 - 7x = 3$

38 $5x^2 - 4x = 2$

Solve the following fractional equations.

39 $\dfrac{x+4}{2x-1} = \dfrac{3x+2}{2x-3}$

40 $\dfrac{x+3}{2x-1} = \dfrac{2x+3}{x+5}$

Solve the following radical equations.

41 $\sqrt{5x^2 - 4x + 3} - x = 1$

42 $3 + \sqrt{x^2 - 2x + 6} = 2x$

43 $\sqrt{2x-1} - \sqrt{x-1} = 1$

44 $\sqrt{2x+3} - \sqrt{x+1} = 1$

45 $\sqrt{x+3} + \sqrt{2x-1} = \sqrt{7x+2}$

46 $\sqrt{3x+1} - \sqrt{2x-1} = \sqrt{x-4}$

Solve the following equations in quadratic form.

47 $4x^4 + 4 = 17x^2$

48 $(x^2+6)^2 - 17(x^2+6) + 70 = 0$

Find the remaining side of the right triangle with the two given sides.

49 $a = 14, b = 48$

50 $a = 5, b = 8$

51 $a = 10, c = 26$

52 $b = 17, c = 19$

The quadratic equation $ax^2 + bx + c = 0$ has two roots. Their sum is $-b/a$, and their product is c/a. Verify this fact for the following equations.

53 $3x^2 - 8x - 4 = 0$

54 $5x^2 - 14x + 10 = 0$

55 $5x^2 - 14x + 5 = 0$

56 $5x^2 + 18x + 5 = 0$

Solve the following inequalities.

57 $3x^2 - 14x - 5 > 0$

58 $18x^2 - 91x + 45 > 0$

59 $3x^2 + 4x - 15 < 0$

60 $12x^2 + 43x + 10 < 0$

NAME _____ DATE _____ SCORE _____

EXERCISE 11.8 CHAPTER TEST

1 Solve $25x^2 + 36 = 0$.

Solve the following by factoring.

2 $25x^2 - 36 = 0$

3 $25x^2 - 36x = 0$

4 $x^2 + 3x = 18$

5 $10x^2 - 10 = -21x$

Solve by completing the square.

6 $x^2 - 2x - 2 = 0$

7 $2x^2 + 2x + 5 = 0$

Solve by the quadratic formula.

8 $4x^2 - 8x - 3 = 0$

9 $9x^2 - 6x + 17 = 0$

Solve the following equations.

10 $\sqrt{2x+7} + 1 = \sqrt{3x+9}$

11 $\sqrt{4x+1} = \sqrt{19x+7} - \sqrt{7x-6}$

12 $(x^2 + 3x)^2 - 14(x^2 + 3x) + 40 = 0$

13 $\left(\dfrac{x+1}{x-1}\right)^2 - 2\left(\dfrac{x+1}{x-1}\right) - 3 = 0$

14 Solve the inequality $6x^2 - 19x - 7 < 0$ algebraically.

15 Solve the inequality $x^2 - 13x + 30 > 0$ graphically.

16 Show that the only right triangle whose sides are x, $x+1$, and $x+2$ is the 3-4-5 right triangle.

17 Find the set of all points which are the same distance from $(1,2)$ and $(3,6)$.

CHAPTER 12

Logarithms

12.1 INTRODUCTION

If Frances has a brother named Bill and a son named Sam, she is related to both Bill and Sam. Furthermore, there is also a definite relationship between Bill and Sam, namely uncle and nephew.

In algebra we have a similar situation in the statement

$$3^2 = 9 \qquad (1)$$

The number 2 is related to both 3 and 9. The number 2 is called the **exponent** of 3, while 2 is called the **logarithm** of 9. The number 3 is called the base. Furthermore there is a definite relationship between 3 and 9 also, expressed by Eq. (1). We now define logarithm as follows:

The logarithm L to the base b of any positive number N is the exponent that indicates the power to which b must be raised in order to obtain N.

Consequently, we say that in statement (1) the logarithm of 9 to the base 3 is 2.

The statement "the logarithm to the base b of N" is abbreviated to $\log_b N$.

Four examples of the definition follow.

EXAMPLE 1 Since $2^3 = 8$, the logarithm to the base 2 of 8 is 3, or $\log_2 8 = 3$.

EXAMPLE 2 Since $7^2 = 49$, $\log_7 49 = 2$.

EXAMPLE 3 Since $5^4 = 625$, $\log_5 625 = 4$.

EXAMPLE 4 Since $16^{1/2} = \sqrt{16} = 4$, $\log_{16} 4 = \frac{1}{2}$.

419

LOGARITHMIC AND EXPONENTIAL FORMS

The definition of a logarithm can be restated in the following shorter form:

$\log_b N = L$ if and only if $b^L = N$ (12.1)

Remember that a logarithm is an exponent. The left equation in (12.1) is called the **logarithmic form** of the equation. The right equation is the **exponential form**. Both forms express the same relationship between L, b, and N. Equation (12.1) also gives the result that, by definition

$$b^{\log_b N} = N$$

If two of the letters b, N, and L in (12.1) are known, the third can often be found by inspection.

EXAMPLE 5 If $\log_5 25 = L$, then $5^L = 25$ and $L = 2$.

EXAMPLE 6 If $\log_{16} 4 = L$, then $16^L = 4$ and $L = \frac{1}{2}$.

EXAMPLE 7 If $\log_9 N = 2$, then $N = 9^2 = 81$.

EXAMPLE 8 If $\log_{27} N = \frac{1}{3}$, then $N = 27^{1/3} = \sqrt[3]{27} = 3$.

EXAMPLE 9 If $\log_b 36 = 2$, then $b^2 = 36$ and $b = 6$.

EXAMPLE 10 If $\log_b 7 = \frac{1}{2}$, then $b^{1/2} = 7$ and $b = 7^2 = 49$.

Logarithms are used extensively in theoretical mathematics and in the many applications of mathematics. In Sec. 12.8 we shall show that logarithms are very efficient in the process of obtaining products, quotients, powers, and roots.

EXAMPLE 11 Change each logarithmic equation to an exponential equation.

Solution

Logarithmic	Exponential
$\log_2 8 = 3$	$2^3 = 8$
$\log_2 \frac{1}{8} = -3$	$2^{-3} = \frac{1}{8}$
$\log_9 3 = \frac{1}{2}$	$9^{1/2} = 3$
$\log_{\sqrt{3}} 3 = 2$	$(\sqrt{3})^2 = 3$

EXAMPLE 12 Change each exponential equation to a logarithmic equation.

Solution

Exponential	Logarithmic
$5^2 = 25$	$\log_5 25 = 2$
$25^{1/2} = 5$	$\log_{25} 5 = \frac{1}{2}$
$4^{3/2} = 8$	$\log_4 8 = \frac{3}{2}$
$(\sqrt[3]{5})^3 = 5$	$\log_{\sqrt[3]{5}} 5 = 3$
$(\sqrt[3]{5})^6 = 25$	$\log_{\sqrt[3]{5}} 25 = 6$

Problems 1 to 28 of Exercise 12.1 may be done now.

EXAMPLE 13 To solve the equation $\log_b 81 = 4$, we write it in the exponential form

$$b^4 = 81$$

Now $81 = 9^2 = (3^2)^2 = 3^4$, so we must have

$$b^4 = 3^4$$

We conclude that $b = 3$ since the exponents, 4, are the same.

The following lists of powers will be helpful to you.

$2^0 = 1$	$2^1 = 2$	$2^2 = 4$
$2^3 = 8$	$2^4 = 16$	$2^5 = 32$
$2^6 = 64$	$2^7 = 128$	$2^8 = 256$
$2^9 = 512$	$2^{10} = 1024$	
$3^0 = 1$	$3^1 = 3$	$3^2 = 9$
$3^3 = 27$	$3^4 = 81$	$3^5 = 243$
$3^6 = 729$		
$4^1 = 4$	$4^2 = 2^4 = 16$	$4^3 = 2^6 = 64$
$4^4 = 2^8 = 256$	$4^5 = 2^{10} = 1024$	
$5^1 = 5$	$5^2 = 25$	$5^3 = 125$
$5^4 = 625$		
$6^1 = 6$	$6^2 = 36$	$6^3 = 216$
$7^1 = 7$	$7^2 = 49$	$7^3 = 343$
$8^1 = 8$	$8^2 = 64$	$8^3 = 2^9 = 512$
$9^1 = 9$	$9^2 = 81$	$9^3 = 3^6 = 729$

Since a logarithm is by definition an exponent, we can get a better understanding of logarithms by graphing logarithmic and exponential equations. To graph $y = \log_2 x$, we may use the preceding list of powers to obtain (logarithms here are to base 2)

$\log 1 = \log 2^0 = 0$
$\log 2 = \log 2^1 = 1$
$\log 4 = \log 2^2 = 2$
$\log 8 = \log 2^3 = 3$
$\log 16 = \log 2^4 = 4$
$\log \frac{1}{2} = \log 2^{-1} = -1$
$\log \frac{1}{4} = \log 2^{-2} = -2$

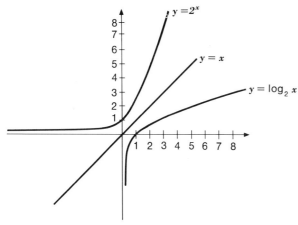

Fig. 12.1

The graph is shown in Fig. 12.1. The graph of $y = 2^x$ is also drawn in Fig. 12.1; it is the reflection of $y = \log_2 x$ through the line $y = x$. Notice that (8,3) is on the graph of $y = \log_2 x$, while (3,8) is on the graph of $y = 2^x$.

The same situation holds for $y = \log_b x$ and $y = b^x$ for any $b > 1$.

STUDENT'S NOTES

EXERCISE 12.1

Fill in the spaces in the following problems with the appropriate words or numbers.

1. Since $4^3 = 64$, $\log_4 64 = $ _____.

2. Since $6^3 = 216$, $\log_6 216 = $ _____.

3. Since $3^5 = 243$, $\log_3 243 = $ _____.

4. Since $2^9 = 512$, $\log_2 512 = $ _____.

5. In the equation $7^2 = 49$, 2 is the _____ of 7 and the _____ of 49.

6. In the equation $5^{-3} = \frac{1}{125}$, -3 is the _____ of $\frac{1}{125}$ and the _____ of 5.

7. In the equation $121^{1/2} = 11$, the logarithm of 11 is _____ and the base is _____.

8. In the equation $125^{-1/3} = \frac{1}{5}$, the logarithm of $\frac{1}{5}$ is _____ and the base is _____.

9. The number $\log_2 8$ is the exponent to which _____ must be raised in order to give _____.

10. The number $\log_5 8$ is the exponent to which _____ must be raised in order to give _____.

11. If $4^x = 32$, then $x = $ _____.

12. If $3^x = 15$, then $x = $ _____.

Change each exponential equation to a logarithmic equation.

13. $2^6 = 64$
14. $6^2 = 36$
15. $6^{-3} = \frac{1}{216}$
16. $3^{-6} = \frac{1}{729}$
17. $64^{1/3} = 4$
18. $36^{1/2} = 6$
19. $9^{3/2} = 27$
20. $64^{-2/3} = \frac{1}{16}$

Change each logarithmic equation to an exponential equation.

21. $\log_7 49 = 2$
22. $\log_{49} 7 = \frac{1}{2}$
23. $\log_5 \frac{1}{625} = -4$
24. $\log_{625} 5 = \frac{1}{4}$
25. $\log_5 5 = 1$
26. $\log_3 1 = 0$
27. $\log_8 32 = \frac{5}{3}$
28. $\log_{32} 8 = \frac{3}{5}$

Solve each equation.

29 $\log_4 256 = x$

30 $27^x = 9$

31 $\log_9 27 = x$

32 $9^x = \frac{1}{3}$

33 $\log_x 512 = 9$

34 $\log_x 125 = \frac{3}{2}$

35 $x^{1/4} = 5$

36 $\log_x 16 = -\frac{2}{3}$

37 $\log_3 x = 4$

38 $\log_9 x = -2$

39 $\log_{1/6} x = 2$

40 $\log_8 x = 0$

41 $5^x = 10$

42 $8^x = 12$

43 $\log_6 6^{1.7} = x$

44 $\log_{4^x} 4^{3.2} = 2$

Sketch the graphs on the same set of axes.

45 $y = \log_3 x,\ y = 3^x$

46 $y = \log_5 x,\ y = 5^x$

47 $y = \log_8 x,\ y = 8^x$

48 $y = \log_{10} x,\ y = 10^x$

12.2 PROPERTIES OF LOGARITHMS

The chief purpose of this chapter is to explain the application of logarithms to numerical computation. The procedure depends upon three properties of logarithms which we shall state and prove in this section. We shall first illustrate these three properties for a special case and show how they are used in computation. For this purpose we use the equation

$$N = 3^x \qquad (1)$$

and tabulate below the corresponding values of N and x for $x = 0, 1, 2, 3, \ldots, 10$. By the definition of logarithms, we have

$$x = \log_3 N \qquad (2)$$

N	1	3	9	27	81	243	729	2187	6561	19,683	59,049
$x = \log_3 N$	0	1	2	3	4	5	6	7	8	9	10

Now we shall show how to use the above information to obtain the following: First, $(27)(729)$; second, $19{,}683/2187$; and third, 9^5.

1. In order to get the product of 27 and 729, we first obtain $\log_3 (27)(729)$, and then from the table, we get the product. Since $27 = 3^3$ and $729 = 3^6$, it follows that

$$\begin{aligned}\log_3 (27)(729) &= \log_3 (3^3)(3^6) \\ &= \log_3 3^{3+6} \quad \text{by (2.29),} \\ & \qquad\qquad\qquad a^m a^n = a^{m+n} \\ &= 3 + 6 \quad \text{by (12.1),} \\ & \qquad\qquad \text{definition} \\ & \qquad\qquad \text{of logarithm} \\ &= 9\end{aligned}$$

Since
$$\begin{aligned}\log_3 (27)(729) &= 9 \\ (27)(729) &= 3^9 \quad \text{by (12.1)} \\ &= 19{,}683 \quad \text{by table}\end{aligned}$$

Notice here that the logarithm of the product is equal to the sum of the logarithms of the two factors.

2. By a method similar to that used in paragraph 1, we have

$$\begin{aligned}\log_3 \frac{19{,}683}{2187} &= \log_3 \frac{3^9}{3^7} \quad \text{by table} \\ &= \log_3 3^{9-7} \quad \text{by (2.35),} \\ & \qquad\qquad\qquad a^m/a^n = a^{m-n} \\ &= 9 - 7 \quad \text{by (12.1)} \\ &= 2\end{aligned}$$

Hence,
$$\frac{19{,}683}{2187} = 3^2 = 9$$

We call attention to the fact that in the above problem, the logarithm of the quotient is equal to the logarithm of the numerator minus the logarithm of the denominator.

3. To compute 9^5, we first get $\log_3 9^5$ and proceed as below:

$$\begin{aligned}\log_3 9^5 &= \log_3 (3^2)^5 \quad \text{since } 9 = 3^2 \\ &= \log_3 3^{2 \times 5} \quad (a^m)^n = a^{mn} \\ &= 2(5) = 10 \quad \text{by (12.1), definition of} \\ & \qquad\qquad\qquad \text{logarithm}\end{aligned}$$

Also,
$$\begin{aligned}5 \log_3 9 &= 5 \log_3 3^2 \\ &= 5(2) = 10\end{aligned}$$

Hence, $\log_3 9^5 = 5 \log_3 9$ and

$$\begin{aligned}9^5 &= 3^{10} \quad \text{by (12.1)} \\ &= 59{,}049 \quad \text{by table}\end{aligned}$$

In the above problem we see that the logarithm of a power is equal to the exponent of the power times the logarithm of the base.

The general forms of the three properties used in the above three problems are

1. The **logarithm of the product** of two numbers is the sum of the logarithms of the numbers; that is,

$$\log_b MN = \log_b M + \log_b N \qquad (12.2)$$

2. The **logarithm of the quotient** of two numbers is equal to the logarithm of the dividend minus the logarithm of the divisor; that is,

$$\log_b \frac{M}{N} = \log_b M - \log_b N \qquad (12.3)$$

3. The **logarithm of the power** of a number is equal to the exponent of the power times the logarithm of the number; that is,

$$\log_b M^n = n(\log_b M) \qquad (12.4)$$

We can use (12.4) to obtain the **logarithm of the root** of any number if we first convert the indicated root to exponential form, as follows:

$$\log_b \sqrt[n]{M} = \log_b M^{1/n} = \frac{1}{n} (\log_b M) \qquad (12.5)$$

In order to prove the above properties, we shall let

$$\log_b M = x \quad \text{and} \quad \log_b N = y \qquad (3)$$

Then, by definition of logarithm (12.1), we have

$$M = b^x \quad \text{and} \quad N = b^y \qquad (4)$$

PROOF OF (12.2)

$$\begin{aligned}
\log_b MN &= \log_b (b^x)(b^y) & \text{by (4)} \\
&= \log_b b^{x+y} & a^m a^n = a^{m+n} \\
&= x + y & \text{by (12.2), definition of logarithm} \\
&= \log_b M + \log_b N & \text{by (3)}
\end{aligned}$$

PROOF OF (12.3)

$$\begin{aligned}
\log_b \frac{M}{N} &= \log_b \frac{b^x}{b^y} & \text{by (4)} \\
&= \log_b b^{x-y} & a^m/a^n = a^{m-n} \\
&= x - y & \text{by (12.1)} \\
&= \log_b M - \log_b N & \text{by (3)}
\end{aligned}$$

PROOF OF (12.4)

$$\begin{aligned}
\log M^n &= \log_b (b^x)^n & \text{by (4)} \\
&= \log_b b^{nx} & (a^m)^n = a^{mn} \\
&= nx & \text{by (12.1)} \\
&= n(\log_b M) & \text{by (3)}
\end{aligned}$$

The above properties are illustrated by Examples 1 to 5 in which we use the following facts: $\log_2 512 = 9$, $\log_2 4096 = 12$, and $\log_2 32 = 5$.

EXAMPLE 1

$$\log_2 [512(4096)] = \log_2 512 + \log_2 4096$$
$$\text{by (12.2)}$$
$$= 9 + 12 = 21$$

EXAMPLE 2

$$\log_2 \frac{4096}{32} = \log_2 4096 - \log_2 32 = 12 - 5 = 7$$
$$\text{by (12.3)}$$

EXAMPLE 3

$$\log_2 \sqrt{4096} = \tfrac{1}{2}(\log_2 4096) = \tfrac{1}{2}(12) = 6$$
$$\text{by (12.5)}$$

EXAMPLE 4

$$\log_2 32^6 = 6(\log_2 32) = 6(5) = 30 \qquad \text{by (12.4)}$$

EXAMPLE 5

$$\log_2 \frac{512(32)}{4096} = \log_2 512 + \log_2 32 - \log_2 4096$$
$$\text{by (12.2) and (12.3)}$$
$$= 9 + 5 - 12 = 2$$

EXERCISE 12.2

Fill in the blanks.

1. Since $\log_3 81 = 4$ and $\log_3 27 = 3$, $\log_3 [81(27)] =$ _____ + _____ = _____.

2. Since $\log_2 512 = 9$ and $\log_2 64 = 6$, $\log_2 \frac{512}{64} =$ _____ − _____ = _____.

3. Since $\log_5 25 = 2$, $\log_5 25^4 = ($ $)($ $) =$ _____.

4. Since $\log_4 8 = \frac{3}{2}$ and $\log_4 16 = 2$, $\log_4 [8(16)] =$ _____ + _____ = _____.

5. Since $\log_6 1 = 0$ and $\log_6 6 = 1$, $\log_6 \frac{1}{6} =$ _____.

6. Since $\log_7 49 = 2$, $\log_7 49^{1/3} =$ _____.

7. Since $\log_7 \frac{1}{49} = -2$ and $\log_7 343 = 3$, $\log_7 [\frac{1}{49}(343)] =$ _____.

8. Since $\log_8 16 = \frac{4}{3}$ and $\log_8 \frac{1}{4} = -\frac{2}{3}$, $\log_8 (16/\frac{1}{4}) =$ _____.

Express as a single logarithm.

9. $\log_5 16 + \log_5 3$

10. $\log_4 135 - \log_4 27$

11. $4 \log_4 6$

12. $\log_7 0.44 + 2 \log_7 0.8$

13. $\log_{10} x + \log_{10} 2y$

14. $\log_9 4x - \log_9(y/2)$

15. $\frac{1}{3} \log_8 xy^2$

16. $3 \log_7 x - \log_7 x^2$

Chapter 12: Logarithms

Use the following data to find the required logarithm:

$3^2 = 9$ $\quad 3^3 = 27$ $\quad 3^4 = 81$
$3^5 = 243$ $\quad 3^6 = 729$ $\quad 3^7 = 2187$
$3^8 = 6561$ $\quad 3^9 = 19{,}683$ $\quad 3^{10} = 59{,}049$

17 $\log_3 [243(6561)]$

18 $\log_3 [27(729)]$

19 $\log_3 \dfrac{19{,}683}{729}$

20 $\log_3 \dfrac{6561}{27}$

21 $\log_3 \dfrac{81(729)}{59{,}049}$

22 $\log_3 \dfrac{6561(9)}{2187}$

23 $\log_3 \dfrac{19{,}683(729)}{27(59{,}049)}$

24 $\log_3 \dfrac{59{,}049}{243(81)(27)}$

25 $\log_3 \sqrt[3]{6561}$

26 $\log_3 \sqrt[3]{243^2}$

27 $\log_3 \dfrac{6561}{\sqrt{729}}$

28 $\log_3 \dfrac{\sqrt{81}\,\sqrt[3]{729}}{\sqrt{243}}$

If $\log_7 2 = x$, $\log_7 3 = y$, and $\log_7 5 = z$, find the required logarithm.

29 $\log_7 8$

30 $\log_7 18$

31 $\log_7 24$

32 $\log_7 30$

33 $\log_7 105$

34 $\log_7 \tfrac{21}{20}$

35 $\log_7 \sqrt{45}$

36 $\log_7 \sqrt[3]{70}$

12.3 COMMON, OR BRIGGS, LOGARITHMS

A set of logarithms of all numbers to a given base is called a **system** of logarithms. The system usually used in elementary mathematics is the **common, or Briggs, logarithms,** in which the base is 10. It is customary to omit the symbol for the base when referring to common logarithms. For example, the statement $\log a = n$ means $\log_{10} a = n$ or that the common logarithm of a is n, and thus $a = 10^n$.

If a number is an integral power of 10, we can obtain its common logarithm by inspection. For example, $\log 10{,}000 = \log 10^4 = 4$, $\log 1 = \log 10^0 = 0$, and $\log 0.001 = \log 10^{-3} = -3$. However, if a number is not an integral power of 10, its common logarithm is often an irrational number and cannot be computed by elementary methods. We can obtain an approximation to the common logarithms of such numbers by use of tables. Table 12.1 (Sec. 12.6) enables us to get the common logarithm of any number to four digits, and in the next four sections we shall explain how this is done.

In the discussion we shall frequently refer to the significant digits in an approximate number. The concepts of accuracy, precision, and significant digits were discussed in Sec. 1.8.

> The **significant digits** in an approximate number are the first nonzero digit in it and all other digits to the right of the first up to and including the digit that indicates the precision of the number.

Usually, the digits from 1 to 9 in a number as well as any 0 or 0s that appear between two such digits are significant. In a decimal between 0 and 1 such as 0.0031 or 0.023, the first 0s are never significant.

The final 0 or 0s in an approximate number may or may not be significant. This ambiguity does not appear if the number is expressed in scientific notation, as discussed in the next section.

12.4 SCIENTIFIC NOTATION

Some of the material from Sec. 1.12 will be repeated here for convenience.

In the definition of the scientific notation for a number and also in the procedure for finding the common logarithm of a number, we employ the concept of the reference position for the decimal point defined below.

> The position immediately to the right of the first nonzero digit in a number is the **reference position** for the decimal point.

The reference position for the decimal point in the following numbers is indicated by a caret: $3_\wedge 20.6$, $0.1_\wedge 23$, $0.003_\wedge 62$.

If a number is between 1 and 10, the decimal point is in reference position, as in 2.36, 8.012, and 9.86.

In order to illustrate the concepts involved, we next show the scientific notation for six numbers. The reference position for the decimal point in each is indicated by a caret.

Number	Scientific notation
$3_\wedge 12$	3.12×10^2
$3_\wedge 4.61$	3.461×10^1
$4_\wedge 235.1$	4.2351×10^3
$0.06_\wedge 21$	6.21×10^{-2}
$3_\wedge 20$ (0 significant)	3.20×10^2
$1_\wedge 600$ (0s not significant)	1.6×10^3

The above tabulation illustrates the definition below. The *scientific notation* for a number N is $N' \times 10^c$, where

1. Only the significant digits in N appear in N'.
2. N' is a number between 1 and 10, and therefore the decimal point in N' is in the reference position.
3. c stands for the number of digits in N that are between the decimal point and the reference position.
4. c is positive or negative according as the decimal point is to the right or to the left of the reference position.

EXAMPLE 1 To express $4_\wedge 16.2$ in scientific notation, we first note that $N' = 4.162$. Furthermore, $c = 2$, since there are two digits between the decimal point and the reference position and the former is to the right of the latter. Therefore $416.2 = 4.162 \times 10^2$.

EXAMPLE 2 $0.04_\wedge 23 = 4.23 \times 10^{-2}$, since the decimal point is two places to the left of the reference position.

430 | **Chapter 12: Logarithms**

EXAMPLE 3 $3_\wedge 2640$ (0 significant) = 3.2640×10^4.

EXAMPLE 4 $2_\wedge 13600$ (0s not significant) = 2.136×10^5. Note that $N' = 2.136$, since the 0s in 213,600 are not significant.

Problems 9 to 20 of Exercise 12.3 may be done now.

12.5 THE CHARACTERISTIC AND MANTISSA

If N is an integral power of 10, say $N = 10^c$, then $\log N = \log_{10} N = c$. Thus $\log 1 = \log 10^0 = 0$, $\log 1000 = \log 10^3 = 3$, and $\log 0.01 = \log 10^{-2} = -2$.

We now show that if N is not an integral power of 10, $\log N$ is equal to an integer plus a positive pure decimal fraction. The scientific notation for N is

$$N = N' \times 10^c \qquad \begin{array}{l} 1 \le N' < 10 \\ c \text{ an integer} \end{array} \qquad (1)$$

Consequently,

$$\begin{aligned} \log N &= \log (N' \times 10^c) \\ &= \log N' + \log 10^c \quad \text{by (12.2)} \\ &= \log N' + c \quad \text{since } \log 10^c = c \end{aligned} \qquad (2)$$

By (1), $1 \le N' < 10$, and so

$$\log 1 \le \log N' < \log 10$$
$$0 \le \log N' < 1$$

Therefore since c is an integer, we have the following conclusion:

If N is positive and is not an integral power of 10, then the common logarithm of N is equal to an integer plus a positive decimal fraction. If N is an integral power of 10, the logarithm of N is an integer.

When $\log N$ is expressed in the form described in the above statement, the integer is called the **characteristic** of the logarithm and the positive decimal fraction is the **mantissa**. If N is a power of 10, the mantissa of the logarithm is 0.

The characteristic of $\log N$ is the number c in (2), and c is also the exponent of 10 in the scientific notation for N. Hence, we see that the following rule can be used for obtaining the characteristic:

The **characteristic** of $\log N$ is numerically equal to the number of digits between the decimal point in N and the reference position and is positive or negative according as the former is to the right or to the left of the latter. If N is between 1 and 10, the characteristic of $\log N$ is zero.

The application of this rule is illustrated in the following tabulation:

N	Position of decimal point relative to reference position	Characteristic
462	2 places to right	2
46.2	1 place to right	1
462,000	5 places to right	5
4.62	Decimal point is in reference position	0
0.00462	3 places to left	-3

The scientific notation for each of the above values of N is $N' \times 10^c$, where $N' = 4.62$, and in each case c is the characteristic shown in the tabulation. Hence, $\log N = c + \log 4.62$. Therefore, in each case, the mantissa of $\log N$ is $\log 4.62$. Now if $\log 4.62 = 0.6646$, then the mantissa of $\log N$ is .6646. Therefore,

$$\log 462 = 2 + .6646 = 2.6646$$
$$\log 46.2 = 1 + .6646 = 1.6646$$
$$\log 462{,}000 = 5 + .6646 = 5.6646$$
$$\log 4.62 = 0 + .6646 = 0.6646$$
$$\log 0.00462 = -3 + .6646 \ne -3.6646$$

In the first four of the above logarithms, both the characteristic and the mantissa are positive, so their sum is written as a single number in which the decimal portion is the mantissa.

In the case of $\log 0.00462$, we have a different situation; $\log 0.00462 = -3 + .6646 = -2.3354$, which is correct; but neither the characteristic nor mantissa can be read off immediately from -2.3354 since $-2.3354 = -2 - .3354$. A mantissa is positive, whereas $-.3354$ is negative.

In order to be able to read off the mantissa directly, we write $\log 0.00462 = -3 + .6646 = 7 - 10 + .6646 = 7.6646 - 10$. We have merely written -3 as $7 - 10$, and this is the usual approach. Sometimes it is advantageous to write $-3 + .6646$ as $17.6646 - 20$ or $47.6646 - 50$, or even $5.6646 - 8$. Similarly, we have $\log 0.462 = -1 + .6646 = 9.6646 - 10$.

The **mantissa** of $\log N$ is positive or 0 and is determined solely by the significant digits in N.

The **characteristic** of $\log N$ is an integer which is positive, 0, or negative and is determined solely by the position of the decimal point in N.

EXERCISE 12.3

Which three numbers have logarithms with the same characteristic?

1 (a) 412.5   (b) 371   (c) 4810   (d) 987.22

2 (a) 0.12   (b) 0.214   (c) 0.32555   (d) 0.0246

3 (a) 0.23   (b) 4.17   (c) 5.822   (d) 8.5

4 (a) 0.002   (b) 0.00272   (c) 0.000100   (d) 0.0051

Which three numbers have logarithms with the same mantissa?

5 (a) 3.65   (b) 30.65   (c) 36.5   (d) 0.000365

6 (a) 0.0272   (b) 0.2720   (c) 2,720,000   (d) 293

7 (a) 333   (b) 3033   (c) 3330   (d) 0.0333

8 (a) 543   (b) 0.543   (c) 345   (d) 5.43

Express the numbers in scientific notation.

9   4673

10 47.3

11 0.892

12 0.091

13 807

14 4106

15 0.0507

16 0.9003

17 850, zero significant

18 850, zero not significant

19 3600, last 0 not significant

20 3600, neither 0 significant

Write the characteristic of the logarithm of the number.

21 38.1

22 3280

23 471.3

24 8.422

25 18,481

26 8472.3

27 201,002.44

28 1,000,000

29 0.4192

30 0.00312

31 0.0000043

32 0.0005

Use log 2 = 0.30 and log 3 = 0.48 to find the logarithm.

33 log 20

34 log 0.3

35 log 0.020

36 log 300

37 log 60

38 log 6000

39 log $\frac{2}{30}$

40 log $\frac{1}{200}$

41 log 40

42 log 900

43 log 8000

44 log 270

45 log $\sqrt{2}$

46 log 0.05

47 log $\sqrt[3]{12}$

48 log $\frac{1}{18}$

12.6 USE OF TABLES TO OBTAIN MANTISSA

In this section and the next we shall use the letter N to represent any positive number, and we shall show how to obtain the mantissa of log N. We shall discuss the procedure for the following four possibilities:

1. N contains only three significant digits.
2. N contains only one or only two significant digits.
3. N contains only four significant digits.
4. N contains more than four significant digits.

The student should now look at Table 12.1 and notice that the column headed by N at the left side of each of the two parts contains the numbers from 10 to 99, and that each of the other columns is headed by one of the integers from 0 to 9. The letter N at the head of the left column is printed in roman type and should not be confused with the italicized N that is used to represent the number under consideration.

Since *the position of the decimal point in N affects only the characteristic of log N, we shall disregard it* in the process of finding the mantissa.

CASE 1: N CONTAINS THREE DIGITS We shall illustrate the procedure for this case by explaining each step in the process of finding the mantissa of log 463. The steps are

1. Find the number composed of the first two digits in 463, that is, 46, in the column headed by N.
2. Follow the horizontal line that contains 46 to the right across the table to the column headed by 3, the third digit in 463, and there find 6656.
3. Place a decimal point to the left of 6656 and obtain .6656, which is the desired mantissa.

CASE 2: N CONTAINS ONE OR TWO DIGITS In this case we mentally annex zeros on the right of N until we have a three-digit number and then

Table 12.1 Common Logarithms

N	0	1	2	3	4	5	6	7	8	9	N	0	1	2	3	4	5	6	7	8	9
10	0000	0043	0086	0128	0170	0212	0253	0294	0334	0374	55	7404	7412	7419	7427	7435	7443	7451	7459	7466	7474
11	0414	0453	0492	0531	0569	0607	0645	0682	0719	0755	56	7482	7490	7497	7505	7513	7520	7528	7536	7543	7551
12	0792	0828	0864	0899	0934	0969	1004	1038	1072	1106	57	7559	7566	7574	7582	7589	7597	7604	7612	7619	7627
13	1139	1173	1206	1239	1271	1303	1335	1367	1399	1430	58	7634	7642	7649	7657	7664	7672	7679	7686	7694	7701
14	1461	1492	1523	1553	1584	1614	1644	1673	1703	1732	59	7709	7716	7723	7731	7738	7745	7752	7760	7767	7774
15	1761	1790	1818	1847	1875	1903	1931	1959	1987	2014	60	7782	7789	7796	7803	7810	7818	7825	7832	7839	7846
16	2041	2068	2095	2122	2148	2175	2201	2227	2253	2279	61	7853	7860	7868	7875	7882	7889	7896	7903	7910	7917
17	2304	2330	2355	2380	2405	2430	2455	2480	2504	2529	62	7924	7931	7938	7945	7952	7959	7966	7973	7980	7987
18	2553	2577	2601	2625	2648	2672	2695	2718	2742	2765	63	7993	8000	8007	8014	8021	8028	8035	8041	8048	8055
19	2788	2810	2833	2856	2878	2900	2923	2945	2967	2989	64	8062	8069	8075	8082	8089	8096	8102	8019	8116	8122
20	3010	3032	3054	3075	3096	3118	3139	3160	3181	3201	65	8129	8136	8142	8149	8156	8162	8169	8176	8182	8189
21	3222	3243	3263	3284	3304	3324	3345	3365	3385	3404	66	8195	8202	8209	8215	8222	8228	8235	8241	8248	8254
22	3424	3444	3464	3483	3502	3522	3541	3560	3579	3598	67	8261	8267	8274	8280	8287	8293	8299	8306	8312	8319
23	3617	3636	3655	3674	3692	3711	3729	3747	3766	3784	68	8325	8331	8338	8344	8351	8357	8363	8370	8376	8382
24	3802	3820	3838	3856	3874	3892	3909	3927	3945	3962	69	8388	8395	8401	8407	8414	8420	8426	8432	8439	8445
25	3979	3997	4014	4031	4048	4065	4082	4099	4116	4133	70	8451	8457	8463	8470	8476	8482	8488	8494	8500	8506
26	4150	4166	4183	4200	4216	4232	4249	4265	4281	4298	71	8513	8519	8525	8531	8537	8543	8549	8555	8561	8567
27	4314	4330	4346	4362	4378	4393	4409	4425	4440	4456	72	8573	8579	8585	8591	8597	8603	8609	8615	8621	8627
28	4472	4487	4502	4518	4533	4548	4564	4579	4594	4609	73	8633	8639	8645	8651	8657	8663	8669	8675	8681	8686
29	4624	4639	4654	4669	4683	4698	4713	4728	4742	4757	74	8692	8698	8704	8710	8716	8722	8727	8733	8739	8745
30	4771	4786	4800	4814	4829	4843	4857	4871	4886	4900	75	8751	8756	8762	8768	8774	8779	8785	8791	8797	8802
31	4914	4928	4942	4955	4969	4983	4997	5011	5024	5038	76	8808	8814	8820	8825	8831	8837	8842	8848	8854	8859
32	5051	5065	5079	5092	5105	5119	5132	5145	5159	5172	77	8865	8871	8876	8882	8887	8893	8899	8904	8910	8915
33	5185	5198	5211	5224	5237	5250	5263	5276	5289	5302	78	8921	8927	8932	8938	8943	8949	8954	8960	8965	8971
34	5315	5328	5340	5353	5366	5378	5391	5403	5416	5428	79	8976	8982	8987	8993	8998	9004	9009	9015	9020	9025
35	5441	5453	5465	5478	5490	5502	5514	5527	5539	5551	80	9031	9036	9042	9047	9053	9058	9063	9069	9074	9079
36	5563	5575	5587	5599	5611	5623	5635	5647	5658	5670	81	9085	9090	9096	9101	9106	9112	9117	9122	9128	9133
37	5682	5694	5705	5717	5729	5740	5752	5763	5775	5786	82	9138	9143	9149	9154	9159	9165	9170	9175	9180	9186
38	5798	5809	5821	5832	5843	5855	5866	5877	5888	5899	83	9191	9196	9201	9206	9212	9217	9222	9227	9232	9238
39	5911	5922	5933	5944	5955	5966	5977	5988	5999	6010	84	9243	9248	9253	9258	9263	9269	9274	9279	9284	9289
40	6021	6031	6042	6053	6064	6075	6085	6096	6107	6117	85	9294	9299	9304	9309	9315	9320	9325	9330	9335	9340
41	6128	6138	6149	6160	6170	6180	6191	6201	6212	6222	86	9345	9350	9355	9360	9365	9370	9375	9380	9385	9390
42	6232	6243	6253	6263	6274	6284	6294	6304	6314	6325	87	9395	9400	9405	9410	9415	9420	9425	9430	9435	9440
43	6335	6345	6355	6365	6375	6385	6395	6405	6415	6425	88	9445	9450	9455	9460	9465	9469	9474	9479	9484	9489
44	6435	6444	6454	6464	6474	6484	6493	6503	6513	6522	89	9494	9499	9504	9509	9513	9518	9523	9528	9533	9538
45	6532	6542	6551	6561	6571	6580	6590	6599	6609	6618	90	9542	9547	9552	9557	9562	9566	9571	9576	9581	9586
46	6628	6637	6646	6656	6665	6675	6684	6693	6702	6712	91	9590	9595	9600	9605	9609	9614	9619	9624	9628	9633
47	6721	6730	6739	6749	6758	6767	6776	6785	6794	6803	92	9638	9643	9647	9652	9657	9661	9666	9671	9675	9680
48	6812	6821	6830	6839	6848	6857	6866	6875	6884	6893	93	9685	9689	9694	9699	9703	9708	9713	9717	9722	9727
49	6902	6911	6920	6928	6937	6946	6955	6964	6972	6981	94	9731	9736	9741	9745	9750	9754	9759	9763	9768	9773
50	6990	6998	7007	7016	7024	7033	7042	7050	7059	7067	95	9777	9782	9786	9791	9795	9800	9805	9809	9814	9818
51	7076	7084	7093	7101	7110	7118	7126	7135	7143	7152	96	9823	9827	9832	9836	9841	9845	9850	9854	9859	9863
52	7160	7168	7177	7185	7193	7202	7210	7218	7226	7235	97	9868	9872	9877	9881	9886	9890	9894	9899	9903	9908
53	7243	7251	7259	7267	7275	7284	7292	7300	7308	7316	98	9912	9917	9921	9926	9930	9934	9939	9943	9948	9952
54	7324	7332	7340	7348	7356	7364	7372	7380	7388	7396	99	9956	9961	9965	9969	9974	9978	9983	9987	9991	9996

proceed as in Case 1. For example, the mantissas of log 5 and log 500 are the same. Also, the mantissa of log 32 is equal to the mantissa of log 320.

If N has more than three digits, the table can be used to approximate log N.

CASE 3: N CONTAINS FOUR DIGITS If the last digit of a four-digit number is 0, we can get the mantissa of the logarithm directly from the table. For example, the mantissa of log 6320 = the mantissa of log 632.

If the fourth digit in N is not 0, we use a process called **interpolation** to obtain the mantissa of the logarithm. We shall explain the process by using it to get the mantissa of log 6238. Since 6230 < 6238 < 6240, mantissa of log 6230 < mantissa of log 6238 < mantissa of log 6240. Therefore,

Mantissa of log 6238
    = mantissa of log 6230 + some number c

We shall now show how to find the number c by interpolation. The process is diagrammed below, and the following steps refer to the diagram:

1. Record the numbers 6230, 6238, and 6240 in the column headed by *number*. Note that the smallest of the three numbers is on the last line.
2. Turn to Table 12.1 and find that the mantissa of log 6240 = .7952 and the mantissa of log 6230 = .7945. Write these in the column headed by *Mantissa*. Designate the mantissa of log 6238 by m and record it in the diagram.
3. Find the differences 6240 − 6230 = 10 and 6238 − 6230 = 8 and record them in the positions indicated at the left of the diagram.
4. Find the differences .7952 − .7945 = .0007 and m − .7945 = c and record them on the right side of the diagram in the position indicated.

The remainder of the interpolation process is shown below the diagram.

$$10\begin{bmatrix}\text{Number} & \text{Mantissa}\\ 6240 & .7952\\ 8\begin{bmatrix}6238 & m\end{bmatrix} & \\ 6230 & .7945\end{bmatrix}\begin{matrix}.0007\\ = .7952\\ - .7945\\ c =\\ m - .7945\end{matrix}$$

Within proper limits the differences at the left and right of the brackets are proportional. That is

$$\frac{c}{.0007} = \frac{8}{10}$$

Hence,

$$c = \frac{8(.0007)}{10} = .00056 = .0006 \quad \text{rounded off to four decimal places}$$

Therefore,

$m = .7945 + .0006 = .7951$

Thus, we have log 6238 = 3.7951, since the decimal point in 6238 is three places to the right of the reference position.

CASE 4: N CONTAINS MORE THAN FOUR DIGITS If N contains more than four significant digits, we round it off to the nearest four-digit number and then proceed as in Case 3.

EXAMPLE 1 log 2.47 = .3927, as found in the table by locating 24 on the left and 7 at the top. It follows that log 2470 = log 2.47 × 10^3 = 3.3927 and log .00247 = 7.3927 − 10.

EXAMPLE 2 log 9.2 = log 9.20 = .9638 since we find 92 to the left and 0 at the top of .9638. Thus log 92,000 = 4.9638 and log (9.2 × 10^{-6}) = 4.9638 − 10.

EXAMPLE 3 To find m = log 2.173, the essential steps are these. First we find log 2.17 = .3365 and log 2.18 = .3385. Then we calculate the differences 2173 − 2170 = 3 and 2180 − 2170 = 10, and also $c = m$ − .3365 and .3385 − .3365 = .0020.

$$10\begin{bmatrix}\text{Number} & \text{Mantissa}\\ 2.180 & .3385\\ 3\begin{bmatrix}2.173 & m\end{bmatrix} & \\ 2.170 & .3365\end{bmatrix}\begin{matrix}\\ \\ c\end{matrix}.0020$$

We now equate the corresponding ratios 3/10 = c/.0020 and solve for $c = \frac{3}{10}(.0020) = .0006$. Hence log 2.173 = $m = c$ + .3365 = .3371. Thus also log 217.3 = 2.3371 and log 2.173 × 10^{-3} = 7.3371 − 10.

There are many calculators now which calculate logarithms as well as other things. They are very useful for their intended functions, i.e., to compute and to serve as tables. However, the person using a calculator must still think through and understand the situation. The numbers in any logarithm table, from a calculator or this book, are decimal approximations, so you shouldn't be surprised or upset if answers obtained in different ways are slightly different. Normally a calculator will give log 0.0135, for example, as -1.8696662. Notice (1) that the log is negative, so neither the characteristic nor the mantissa can be read off directly (you must add and subtract 10 to find that log $0.0135 = -1.8696662 = 8.1303338 - 10$, hence the mantissa is .1303338) and (2) even though eight significant digits are given by the calculator, only about four of them are accurate if 0.0135 was accurate to only three significant digits. The study of logarithms in this chapter includes properties given in Sec. 12.5, which are important for both calculation and theoretical work.

STUDENT'S NOTES

EXERCISE 12.4

Find the common logarithm.

1. log 728 =
2. log 83.1 =
3. log 0.387 =
4. log 0.621 =
5. log 0.0491 =
6. log 0.00138 =
7. log 31 =
8. log 0.42 =
9. log 0.056 =
10. log 5 =
11. log 8 =
12. log 4 =
13. log 0.07 =
14. log 0.0009 =
15. log 0.002 =
16. log 100 =
17. log 1000 =
18. log 100,000 =
19. log 3270 =
20. log 4100 =
21. log 613,000 =
22. log 861 =
23. log 74.2 =
24. log 9.36 =
25. log 0.138 =
26. log 0.0248 =
27. log 0.00421 =
28. log 2.60 × 10^2 =
29. log 3.8 × 10^4 =
30. log 3 × 10^5 =
31. log 3.46 × 10^{-1} =
32. log 5.270 × 10^{-4} =
33. log 5.00 × 10^{-6} =
34. log 127 =
35. log 482 =
36. log 312 =
37. log 0.782 =
38. log 0.0139 =
39. log 0.000486 =
40. log 519 =
41. log 6.12 =
42. log 0.814 =
43. log 831 =
44. log 749 =

Find the logarithm of the numbers by interpolation. Show your work in the space below the problem and enter the logarithm at the right of the equality sign.

45. log 0.0062438 =
46. log 0.00051738 =

47 $\log 1.628 \times 10^2 =$

48 $\log 2.469 \times 10^4 =$

49 $\log 7.8260 \times 10^6 =$

50 $\log 5.178 \times 10^{-1} =$

51 $\log 2.139 \times 10^{-3} =$

52 $\log 9.148 \times 10^{-5} =$

In the following problems do as much of the interpolation process as you can mentally. Do the necessary pencil work in the space below the problem and enter the logarithm at the right of the equality sign.

53 $\log 6914 =$

54 $\log 81.24 =$

55 $\log 931.7 =$

56 $\log 2.817 =$

57 $\log 528.9 =$

58 $\log 3783 =$

59 $\log 0.1863 =$

60 $\log 0.06783 =$

12.7 GIVEN LOG N, TO FIND N

If we know the value of log N, we can use the table to obtain the first four digits in N. If a higher degree of accuracy than this is desired, more comprehensive tables must be used. The position of the decimal point in N is determined by the characteristic of log N.

FINDING N IF MANTISSA OF LOG N IS IN THE TABLE We shall again let m represent the mantissa of log N. If m can be found in the body of the table:

1. The number in the N column that is on the same horizontal line with m contains the first two digits in N.
2. The integer at the head of the column that contains m is the third digit in N.
3. If the characteristic of log N indicates that N contains more than three digits, all digits after the third one are 0.

EXAMPLE 1 If log $N = 4.8768$, find N.

Solution The mantissa .8768 is found in line with 75 (in the N column) and in the column headed by 3. Hence, three digits in N are 7, 5, and 3 in that order. The characteristic of log N is 4. Hence, the decimal point is four places to the right of the reference position. Thus, $N = 75,300$.

EXAMPLE 2 If log $N = 8.6201 - 10$, find N.

Solution The mantissa .6201 is in line with 41 and in the column headed by 7. The characteristic of log N is -2. Hence, $N = .0417$.

If the mantissa m of log N is not listed in the table, we find the mantissa in the table that is *nearest to and less than* m. The number corresponding to the latter mantissa contains the first three digits in N. The fourth digit d in N is found by **interpolation.** We shall illustrate the process by obtaining N when log $N = 2.8535$.

In the discussion that follows, we frequently refer to a four-digit number that corresponds to a given mantissa. In such cases we assume that the decimal point follows the fourth digit. We shall let N' represent the four-digit number corresponding to the mantissa .8535.

The mantissa in the table that is nearest to and less than .8535 is .8531, and the next greater mantissa in the table is .8537. The four-digit numbers corresponding to the mantissas .8531 and .8537 are 7130 and 7140, respectively. Since $.8531 < .8535 < .8537$, we have $7130 < N' < 7140$, and we obtain the fourth digit d by the interpolation method diagrammed below.

$$.0006\left[\begin{array}{cc} \text{Mantissa} & \text{Number} \\ .8537 & 7140 \\ .0004\left[\begin{array}{c}.8535 \\ .8531\end{array}\right. & \left.\begin{array}{c}N' \\ 7130\end{array}\right]d \end{array}\right]10$$

As in the previous discussion of interpolation, the differences at the right and left are in proportion. Hence,

$$\frac{d}{10} = \frac{.0004}{.0006}$$

Therefore,

$$d = 10 \cdot \frac{.0004}{.0006} = \frac{.0040}{.0006} = 7 \quad \text{rounded off to nearest integer}$$

Hence,

$$N' = 7130 + 7 = 7137$$

The characteristic of log N is 2; therefore $N = 713.7$.

EXAMPLE 3 Find N if log $N = 2.8212$.

Solution In the tables we find that

$$.0006\left[\begin{array}{cc} \text{Mantissa} & \text{Number} \\ .8215 & 6630 \\ .0003\left[\begin{array}{c}.8212 \\ .8209\end{array}\right. & \left.\begin{array}{c}N' \\ 6620\end{array}\right]d \end{array}\right]10$$

Equating corresponding ratios gives

$$\frac{.0003}{.0006} = \frac{d}{10}$$

and so $d = 10(\frac{3}{6}) = 5$. Thus $N' = 6620 + 5 = 6625$, and hence $N = 662.5$. It follows that if log $M = 4.8212$, then $M = 66,250$, and if log $P = 4.8212 - 10$, then $P = 0.000006625$.

If a calculator is used to find N in the above example, we can use the definition $N = 10^{\log N}$ to calculate

$$N = 10^{2.8212} = 662.5$$

The term **antilog** is sometimes used in this connection. In general, *the antilog of log N is N* and it can be found by the formula

$N = 10^{\log N}$

EXAMPLE 4 Find N if $\log N = 7.2487 - 10$.

Solution From the table

$$.0024 \begin{bmatrix} & Mantissa & Number & \\ & .2504 & 1780 & \\ .0007 \begin{bmatrix} .2487 & N' \\ .2480 & 1770 \end{bmatrix} d & \end{bmatrix} 10$$

Equating corresponding ratios gives

$$\frac{.0007}{.0024} = \frac{d}{10}$$

and hence $d = 10(\frac{7}{24}) = 3$. Thus $N' = 1770 + 3 = 1773$, and so $N = 0.001773$.

After you are thoroughly familiar with interpolation, you will mentally find $d = 10(\frac{7}{24})$ almost automatically.

To use antilogs and a calculator, antilogs are computed using the key 10^x on some calculators and INV LOG on some others. In Example 4, to find N, just press the buttons 7.2487 $\boxed{-}$ 10 $\boxed{=}$ $\boxed{\text{INV}}$ $\boxed{\text{LOG}}$ or 7.2487 $\boxed{-}$ 10 $\boxed{=}$ $\boxed{10^x}$ and read on the display 0.00177296, which we round to 0.001773 since the original number was accurate to only four decimal places.

EXERCISE 12.5

Find the value of n.

1. $\log n = 1.1644$
   $n =$

2. $\log n = 3.7723$
   $n =$

3. $\log n = 4.5328$
   $n =$

4. $\log n = 2.8976$
   $n =$

5. $\log n = 2.4814$
   $n =$

6. $\log n = 3.8825$
   $n =$

7. $\log n = .6180$
   $n =$

8. $\log n = 1.9405$
   $n =$

9. $\log n = 9.2455 - 10$
   $n =$

10. $\log n = 7.7938 - 10$
    $n =$

11. $\log n = 6.5988 - 10$
    $n =$

12. $\log n = 9.9258 - 10$
    $n =$

13. $\log n = 8.6702 - 10$
    $n =$

14. $\log n = 9.9609 - 10$
    $n =$

15. $\log n = 7.6222 - 10$
    $n =$

16. $\log n = 8.5922 - 10$
    $n =$

Use the method of interpolation diagrammed in Sec. 12.7 to find the value of n. Show your work below the problem and enter the value of n in the space provided for it.

17. $\log n = .4255$
    $n =$

18. $\log n = 1.5723$
    $n =$

19. $\log n = 2.7390$
    $n =$

20. $\log n = 5.8601$
    $n =$

21 $\log n = 9.8365 - 10$
$n =$

22 $\log n = 7.3734 - 10$
$n =$

23 $\log n = 8.2233 - 10$
$n =$

24 $\log n = 7.8077 - 10$
$n =$

Find the value of n. Do as much of the interpolation process as you can mentally.

25 $\log n = 1.7408$
$n =$

26 $\log n = .7814$
$n =$

27 $\log n = 3.4860$
$n =$

28 $\log n = 5.4889$
$n =$

29 $\log n = 9.8930 - 10$
$n =$

30 $\log n = 7.6320 - 10$
$n =$

31 $\log n = 6.3380 - 10$
$n =$

32 $\log n = 9.3683 - 10$
$n =$

12.8 LOGARITHMIC COMPUTATION

In this section we shall explain the use of logarithms in numerical computation. Since we shall be using a four-place logarithm table in the examples that follow, we shall in no case obtain a result to more than four significant digits. If the numbers in a problem contain only three significant digits, we shall carry the result to three significant digits.

The first step in making a logarithmic computation is to make an outline of the solution and leave blanks in which the logarithms can be entered and at the same time arrange the outline so that the computations can be made conveniently.

EXAMPLE 1 Use logarithms to find the value of

$$N = \frac{713(5.62)}{16.3(2.78)}$$

Solution By (12.2) and (12.3), we know that

$\log N = (\log 713 + \log 5.62)$
$\qquad - (\log 16.3 + \log 2.78)$

This suggests the outline that follows for the solution. Note that N is a fraction whose numerator is 713(5.62) and whose denominator is 16.3(2.78):

$\log 713 =$ $\qquad$ $\log 16.3 =$
$(+) \log 5.62 =$ $\qquad$ $(+) \log 2.78 =$
$\log \text{numerator} =$ $\qquad$ $\log \text{denominator} =$
$\log \text{denominator} =$ $\quad (-)$
$\log N =$
$N =$

We now enter the characteristics at the right of the equality signs in the first two lines, then turn to the tables and obtain the mantissas. When the latter are entered in the proper places, the computation can be completed as below:

$\log 713 = 2.8531$ $\qquad$ $\log 16.3 = 1.2122$
$(+) \log 5.62 = 0.7497$ $\qquad$ $(+) \log 2.78 = 0.4440$
$\log \text{numerator} = 3.6028$ $\qquad$ $\log \text{denominator} = 1.6562$
$\log \text{denominator} = 1.6562 \; (-)$
$\log N = 1.9466$
$N = 88.4$

Note 1: The mantissa of $\log N$ is between .9465 and .9469 and is nearer the former than the latter. Hence we take the number corresponding to .9465 as the three digits in N, without interpolation.

Note 2: The + and − signs appearing in the outline above indicate the operations to be performed.

EXAMPLE 2 Use logarithms to find $N = 4.17/849$.

Solution

$\log 4.17 = 0.6201$
$\log 849 = \underline{2.9289} \; (-)$
$\log N =$

Now we see that we must subtract a number from another number that is smaller, and this would lead to a negative result. To avoid the difficulty, we add 10 to and subtract 10 from the characteristic of 0.6201, as at the end of Sec. 12.5. Thus, we obtain

$\log 4.17 = 10.6201 - 10$
$\log 849 = \underline{2.9289} \qquad (-)$
$\log N = 7.6912 - 10$
$N = .00491$

Note: The mantissa .6912 is between .6911 and .6920 and is nearer the former. Hence we take 491 as the three digits in N without interpolation.

If we attempt to obtain a root of a pure decimal fraction, we meet a difficulty similar to that in Example 2, and we use a similar device for dealing with it. Example 3 illustrates the situation.

EXAMPLE 3 Use logarithms to obtain $N = \sqrt[3]{.0817}$.

Solution

$\log N = \dfrac{\log .0817}{3}$

$ = \dfrac{28.9122 - 30}{3}$

$ = 9.6374 - 10$
$N = 0.434$

Example 4 includes most of the processes usually found in a computation problem. Study it carefully.

EXAMPLE 4 Use logarithms to find

$$N = \sqrt[3]{\frac{43.6^2 \sqrt{5.72}}{1.35(25.9)^4}}$$

Solution In making the outline, we have provided for the denominator first:

$$\log 1.35 = 0.1303$$
$$\log 25.9^4 = 4(\log 25.9)$$
$$= 4(1.4133) = 5.6532 \ (+)$$
$$\log \text{denominator} = 5.7835$$
$$\log 43.6^2 = 2(\log 43.6)$$
$$= 2(1.6395) = 3.2790$$
$$\log \sqrt{5.72} = \tfrac{1}{2}(\log 5.72)$$
$$= \tfrac{1}{2}(0.7574) = 0.3787 \ (+)$$
$$\log \text{numerator} = 13.6577 - 10$$
$$\log \text{denominator} = \underline{5.7835} \ (-)$$
$$\log \text{radicand} = 7.8742 - 10$$
$$= 27.8742 - 30$$
$$\log N = \frac{27.8742 - 30}{3} = 9.2914 - 10$$
$$N = 0.196$$

The logarithm of the numerator was 3.6577, but we added $10 - 10$ to it so that we could subtract 5.7835. Since it was necessary to divide the logarithm of the radicand $7.8742 - 10$ by 3, we added $20 - 20$ to it, obtaining $27.8742 - 30$. Then the division could be performed without difficulty.

Problems 1 to 28 of Exercise 12.6 may be done now.

All numbers in Example 5 contain four digits, and we shall obtain the result to four significant digits. Interpolation must be used to obtain all logarithms. The details of the interpolation are not shown, but you should verify each logarithm and the computation.

EXAMPLE 5 Use logarithms to obtain

$$N = \frac{436.2 \times 69.21}{5.186 \times 32.59}$$

Solution

$$\log 436.2 = 2.6397 \qquad \log 5.186 = 0.7148$$
$$(+) \log 69.21 = \underline{1.8402} \qquad (+) \log 32.59 = \underline{1.5131}$$
$$\log \text{numerator} = 4.4799 \qquad \log \text{denominator} = 2.2279$$
$$\log \text{denominator} = \underline{2.2279} \ (-)$$
$$\log N = 2.2520$$
$$N = 178.6$$

Problems 29 to 40 of Exercise 12.6 may be done now.

We know that if $3^{2x+1} = 27$, we can write 27 as 3^3 and have

$$3^{2x+1} = 3^3$$
$$2x + 1 = 3 \qquad \text{equating exponents}$$
$$\qquad\qquad\qquad \text{(or taking logs to base 3)}$$
$$x = 1 \qquad \text{solving for } x$$

The next example shows how to solve a related problem.

EXAMPLE 6 Solve $3^{2x+1} = 25$.

Solution If we take logarithms of both members of the equation, we have

$$\log (3^{2x+1}) = \log 25$$
$$(2x + 1) \log 3 = \log 25 \qquad \text{by (12.4)}$$
$$2x + 1 = \frac{\log 25}{\log 3} \qquad \text{dividing by } \log 3$$
$$x = \frac{1}{2}\left(\frac{\log 25}{\log 3} - 1\right) \qquad \text{solving for } x$$

In order to calculate x, we let $x = \tfrac{1}{2}(y - 1)$, where

$$y = \frac{\log 25}{\log 3} = \frac{1.3979}{0.4771} \qquad \text{using tables}$$

$$y = \frac{1.398}{0.4771} \qquad \text{using four digits}$$
$$\log y = \log 1.398 - \log 0.4771 \qquad \text{by (12.3)}$$
$$= 0.1455 - (9.6786 - 10) \qquad \text{using tables}$$
$$= 0.4669$$
$$y = 2.93 \qquad \text{tables}$$
$$x = \tfrac{1}{2}(2.93 - 1) = 0.965$$

EXERCISE 12.6

Find n to three significant figures by using logarithms.

1 $n = 85.2(0.712)$

2 $n = 4.83(0.744)$

3 $n = 813(0.000512)$

4 $n = 0.348(0.0291)$

5 $n = \dfrac{81.2}{47.3}$

6 $n = \dfrac{4.78}{53.9}$

7 $n = \dfrac{846}{47.5}$

8 $n = \dfrac{0.0138}{0.00742}$

9 $n = \dfrac{91.2(4.73)}{225}$

10 $n = \dfrac{8.42(7.55)}{141}$

11 $n = \dfrac{48.2}{37.5(16.2)}$

12 $n = \dfrac{8.55}{16.3(41.5)}$

13 $n = \dfrac{15.5(4.69)}{83.2(0.117)}$

14 $n = \dfrac{(42.5)^2}{18.9(242)}$

15 $n = \dfrac{9.12(82.7)}{(14.7)^3}$

16 $n = \dfrac{4.92(5.03)}{16.8(43.2)}$

17 $n = \sqrt{471}$

18 $n = \sqrt[3]{0.822}$

19 $n = \sqrt[4]{0.0532}$

20 $n = \sqrt[5]{686}$

21 $n = (4.75)^3 \sqrt[4]{81.5}$

22 $n = \sqrt{51.6}\,(0.812)^4$

23 $n = \dfrac{\sqrt[3]{744}}{(1.15)^7}$

24 $n = \dfrac{\sqrt{222}}{\sqrt[3]{333}}$

25 $n = \dfrac{41.6\sqrt{21.5}}{\sqrt[3]{68.2}}$

26 $n = \sqrt{\dfrac{58.5}{279}}$

27 $n = \sqrt[3]{\dfrac{6.43}{1.25(2.41)}}$

28 $n = \sqrt[4]{\dfrac{\sqrt{16.5}}{7.59}}$

Find n to four significant figures.

29 $n = 41.41(3.552)$

30 $n = 6.522(0.01187)$

31 $n = \dfrac{6588}{29.92}$

32 $n = \dfrac{8.431}{747.8}$

33 $n = \sqrt{7.461}$

34 $n = \sqrt[5]{0.8243}$

35 $n = \sqrt[8]{0.001825}$

36 $n = \sqrt[4]{0.08569}$

37 $n = \sqrt{\dfrac{45.59}{0.09091}}$

38 $n = \dfrac{\sqrt[3]{8603}}{50.93}$

39 $n = \dfrac{0.08555}{\sqrt[4]{6.509}}$

40 $n = \dfrac{\sqrt{4.848}}{\sqrt[5]{20.06}}$

Solve the following equations to three digits.

41 $2^x = 7$

42 $7^x = 2$

43 $5^x = 28$

44 $14^x = 191$

45 $4^{2x+1} = 59^x$

46 $16^{4x-3} = 11^x$

47 $5^{3x-4} = 3^{2x-1}$

48 $15^{2x-5} = 4^{x-1}$

Do the following calculations with a calculator if one is available.

49 $48.2(3.71) - \dfrac{15.3}{0.0120} =$

50 $\dfrac{5.62}{3.48 + 6.55(0.392)} =$

51 $\dfrac{417(312)}{689(5.22)(3.48)} =$

52 $\dfrac{684(712)\sqrt{555}}{(33.8)(972)} =$

53 $\dfrac{\sqrt{612}\ \sqrt[3]{4100}}{\log 1260} =$

54 $\dfrac{\sqrt{588} + \sqrt[4]{392}}{\sqrt{649}} =$

55 $\dfrac{2^{0.6} + 3^{0.7}}{4^{0.8}} =$

56 $\dfrac{3^{1.4} + \sqrt{48.4}}{41.2} =$

12.9 CHAPTER SUMMARY

If N is a positive number and b is positive but $b \neq 1$, then the logarithm of N to the base b is the exponent L satisfying the equation

$$b^L = N \quad \text{or} \quad L = \log_b N \tag{12.1}$$

These two equations express the same relationship between b, L, and N. By definition

$$b^{\log_b N} = N \quad \text{and} \quad \log_b b^L = L$$

The basic properties of logarithms are

$$\log_b MN = \log_b M + \log_b N \tag{12.2}$$

$$\log_b \frac{M}{N} = \log_b M - \log_b N \tag{12.3}$$

$$\log_b M^n = n \log_b M \tag{12.4}$$

$$\log_b \sqrt[n]{M} = \log_b M^{1/n} = \frac{1}{n} \log_b M \tag{12.5}$$

Common logarithms are logarithms to the base 10, and they are the ones used most often in computations. Any common logarithm is equal to its characteristic plus its mantissa. The characteristic is an integer and the mantissa is a nonnegative decimal fraction less than 1. Common logarithms can be found in a table, by interpolation, or by calculator.

STUDENT'S NOTES

EXERCISE 12.7 REVIEW

1. Since $3^4 = 81$, $\log_3 81 = $ _____.

2. Since $25^{3/2} = 125$, $\log_{25} 125 = $ _____.

3. In $4^{-5/2} = \frac{1}{32}$, $-5/2$ is the _____ of 4 and the _____ of $\frac{1}{32}$.

4. In $27^{-5/3} = \frac{1}{243}$, $-5/3$ is the _____ of $\frac{1}{243}$ and the _____ of 27.

5. The number $\log_4 7$ is the exponent to which _____ must be raised in order to get _____.

6. In order to get 5, we must raise 7 to the power _____.

7. The equation $(\frac{1}{4})^{3/2} = \frac{1}{8}$ is equivalent to the logarithmic equation _____.

8. The equation $\log_{1/2} 16 = -4$ is equivalent to the exponential equation _____.

Solve the following equations.

9. $\log_5 x = 3$

10. $\log_{1/3} x = 3$

11. $\log_8 4 = x$

12. $\log_4 8 = x$

13. $\log_x \frac{1}{9} = -\frac{2}{3}$

14. $\log_x \frac{1}{64} = -\frac{3}{2}$

Express the following as one logarithm.

15. $\log_4 152 - \log_4 8 =$

16. $\log_5 6 + \log_5 8 - \log_5 12 =$

17. $\log_9 30 + \log_9 20 - \log_9 150 =$

18. $3 \log_8 2 + 2 \log_8 3 =$

19 $\frac{1}{2}\log_5 16 - \frac{3}{2}\log_5 2 =$

20 $\log_3 4 + \frac{1}{2}\log_3 8 =$

If $\log_7 2 = 0.36$, $\log_7 3 = 0.56$, and $\log_7 5 = 0.83$, find the following logarithms.

21 $\log_7 42$

22 $\log_7 \sqrt[4]{6}$

23 $\log_7 \frac{15}{4}$

24 $\log_7 54$

Find the characteristic and mantissa of the following common logarithms.

25 $\log 472$

26 $\log 0.00307$

27 $\log 0.0124$

28 $\log 3^{17}$

29 $\log \sqrt[5]{0.333}$

30 $\log 0.925^{15}$

How many digits are there in each of the following numbers?

31 13^{13}

32 5^{50}

33 429^{85}

Use interpolation to find the following logarithms.

34 log 4.156

35 log 21.32

36 log 0.01477

37 log 8414

Chapter 12: Logarithms

Perform the following calculations.

38 $5.41(38.0) =$

39 $4.13(7.56)(8.12) =$

40 $\dfrac{80.5}{47.1} =$

41 $\dfrac{6.58(0.713)}{12.6} =$

42 $\sqrt[3]{0.916} =$

43 $\sqrt[5]{8.72} =$

44 $\dfrac{44.4\sqrt{653}}{2470} =$

45 $\dfrac{\sqrt[8]{47.8}}{6.89\sqrt{10.9}} =$

46 $8448\sqrt{0.007531} =$

47 $\dfrac{64.68}{\sqrt[9]{7992}} =$

It is true that

$$\log_b N = \dfrac{\log_{10} N}{\log_{10} b}$$

Use this formula to find the following values.

48 $\log_2 66$

49 $\log_9 8$

Common logarithms use base 10. The other base ordinarily used (in calculus and higher courses) is called e and is approximately 2.72. The number e is the limiting value of $(1 + 1/n)^n$ as n gets larger and larger. Calculate

$$\left(1 + \frac{1}{n}\right)^n = \left(\frac{n+1}{n}\right)^n$$

for the following values of n.

50 20

51 64

52 512

Use a calculator, if one is available, to find the following numbers.

53 $\sqrt{514} \log 37.8 =$

54 $\dfrac{\log 709}{4.71 + \sqrt{5.01}} =$

55 $\dfrac{(\log 616)^3}{\log \sqrt{353}} =$

56 $\dfrac{(51.6)^{1.7}}{\sqrt{840} \sqrt[3]{1170}} =$

57 $\dfrac{\sqrt[5]{21.2} + \log 903}{\sqrt{2.3} - \log 2.3} =$

58 $\dfrac{652}{1.01 - 4.31^{2.16} + 6.11^{1.98}} =$

NAME _____ DATE _____ SCORE _____

EXERCISE 12.8 CHAPTER TEST

Find the value of the unknown.

1. $\log_2 x = -3$

2. $x = 9^{3/2}$

3. $\log_4 \frac{1}{8} = x$

4. $25^x = \frac{1}{5}$

5. $\log_x 32 = \frac{5}{2}$

6. $x^{8/5} = 256$

Use the powers of 2 to find the following logarithms.

7. $\log_2 [64(16)]$

8. $\log_2 \frac{1024}{128}$

9. $\log_2 \frac{512(128)}{64}$

10. $\log_2 \sqrt[4]{256}$

11. $\log_2 \frac{\sqrt[3]{512}}{\sqrt{64}}$

Find the logarithm of the following numbers.

12. 86.3

13. 40,700

14. 0.007

15. 0.01307

16. Find n if $\log n = 2.3324$.

17. Find n if $\log n = 1.7427$.

18. Find n if $\log n = 9.9607 - 10$.

Calculate the following numbers.

19 $n = 27.1(3.88)$

20 $n = \dfrac{41.2}{6500}$

21 $n = \dfrac{5.41(83.2)}{51.8(6.83)}$

22 $n = 0.752^4$

23 $n = \sqrt[4]{615}(0.521)^4$

24 $n = \sqrt{\dfrac{8.16(47.2)}{\sqrt{659}}}$

25 $n = \dfrac{\sqrt[3]{8.502}}{5.911}$

CHAPTER 13

Numerical Trigonometry

13.1
INTRODUCTION

Trigonometry is a branch of mathematics that deals with ratios associated with angles. The methods that have been developed in this field enable us to find the unknown parts of a triangle when we know three parts including a side. These methods also enable us to find distances that cannot be measured directly, such as the height of a flagpole or the distance between two mountain peaks. Trigonometry is useful in any vocation that employs geometry, and it is essential for the development of most branches of mathematics and of any science based on mathematics. In this section we shall define and explain some of the terms that we shall use.

A **half line** or **ray** is the portion of a straight line that originates at a fixed point and extends indefinitely in one direction.

If two rays originate at the same point, an **angle** is formed. For example, in Fig. 13.1 the rays r and s originate at A and are the sides of the **angle** P_1AP that has the point A as the **vertex.** We shall consider r as a stationary ray and s as a ray that revolves in a plane about the point A. If P is a point on s and s starts revolving from a position coincident with r, then P coincides with P_1, and as s revolves, P traces an arc of a circle. If s makes a complete revolution, the arc is a circle. Although the arc (or circle) is not a part of the angle, we use

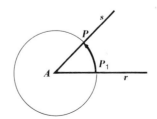

Fig. 13.1

it for measuring the angle. Suppose the circumference is divided into 360 equal parts with one of the points of division on the ray r. Then as P moves over one of these subdivisions, s generates an angle of 1 degree, denoted by $1°$. Therefore if the arc P_1P in Fig. 13.1 is of length $n/360$ of the circumference, we say that the measure of the angle P_1AP is n degrees, or $A = n°$. Furthermore, the arc P_1P is said to subtend the angle P_1AP at the center. We therefore define a **degree** as an angle which, if placed with it's vertex at the center of a circle, subtends an arc that is $\frac{1}{360}$ of the circumference. Of course, units other than the degree are used for angular measurement. They include the radian, which is used extensively in calculus, the mil, which is used in artillery firing, and the grad. Each of these units is defined by means of an arc intercepted by the sides of an angle on a circle with the center at the vertex. In this chapter, we use only the degree.

One degree ($1°$) is divided into sixty **minutes** ($60'$), and one minute is equal to sixty **seconds** ($60''$). Hence one degree contains 3600 seconds.

In an angle generated by revolution, the stationary ray is called the **initial side,** and the rotating ray is called the **terminal side.** The angle is positive or negative according as the direction of revolution is counterclockwise or clockwise.

In the discussion that follows, an angle of a triangle will be designated by a capital letter at its vertex, and the side opposite the angle will be designated by the lowercase form of the same letter. This practice is illustrated in Fig. 13.2. A right triangle is a triangle in which one of the angles is a right angle ($=90°$), and in such a triangle the right angle is usually designated by C. The capital letters that designate the angles are also used to refer to the vertices of the triangle and therefore to the ends of the sides.

In a right triangle the two line segments that form the right angle are called the **sides** of the triangle, and the third line segment is called the **hypotenuse.** In Fig. 13.2 the angle A is formed by the hypotenuse c and the side b, and side a is drawn across from the angle. We refer to b as the **adjacent side** of A and to a as the **opposite side.** Likewise, a is the adjacent side of angle B, and b is the opposite side of B.

Problems 1 to 4 in Exercise 13.1 may be done now.

13.2
TRIGONOMETRIC RATIOS OF AN ACUTE ANGLE

In geometry it is proved that the ratio of two sides of a triangle is equal to the ratio of the two corresponding sides of a similar triangle. We shall use this fact to show that we can associate six ratios with any given angle and that the value of each ratio can be determined when the magnitude of the angle is known.

The six ratios that can be formed from the sides of a right triangle are called *trigonometric ratios*. Each has a definite value, which is the same regardless of the lengths of the line segments forming the triangle, and each has a name, stated and defined below.

In order to define the trigonometric ratios (see Fig. 13.3) of the acute angle A, we choose a point on one side of the angle and construct a perpendicular from that point to the ray that forms the other side of the angle, thus forming a right triangle.

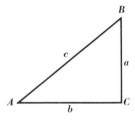

Fig. 13.3

Then the **trigonometric ratios** of the angle A are defined as follows:

$$\text{Sine of } A = \frac{\text{side opposite } A}{\text{hypotenuse}} = \frac{a}{c}$$

$$\text{Cosine of } A = \frac{\text{side adjacent to } A}{\text{hypotenuse}} = \frac{b}{c}$$

$$\text{Tangent of } A = \frac{\text{side opposite } A}{\text{side adjacent to } A} = \frac{a}{b}$$

$$\text{Cotangent of } A = \frac{\text{side adjacent to } A}{\text{side opposite } A} = \frac{b}{a}$$

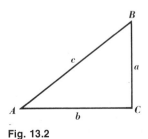

Fig. 13.2

Secant of $A = \dfrac{\text{hypotenuse}}{\text{side adjacent to } A} = \dfrac{c}{b}$

Cosecant of $A = \dfrac{\text{hypotenuse}}{\text{side opposite } A} = \dfrac{c}{a}$

We ordinarily use the following abbreviations for the trigonometric ratios:

Ratio	Abbreviation	Ratio	Abbreviation
Sine of A	sin A	Cosecant of A	csc A
Cosine of A	cos A	Secant of A	sec A
Tangent of A	tan A	Cotangent of A	cot A

The trigonometric ratios can be divided into the following reciprocal pairs, also called **coratios**: sin A, csc A; cos A, sec A; and tan A, cot A. Also, from Fig. 13.3 and from the definition of the trigonometric ratios, we see that

$$\sin B = \dfrac{b}{c} \quad \cos B = \dfrac{a}{c} \quad \tan B = \dfrac{b}{a}$$

$$\csc B = \dfrac{c}{b} \quad \sec B = \dfrac{c}{a} \quad \cot B = \dfrac{a}{b}$$

Thus,

$$\sin A = \cos B = \dfrac{a}{c} \quad \cos A = \sin B = \dfrac{b}{c}$$

$$\tan A = \cot B = \dfrac{a}{b} \quad \cot A = \tan B = \dfrac{b}{a}$$

$$\sec A = \csc B = \dfrac{c}{b} \quad \csc A = \sec B = \dfrac{c}{a}$$

TRIGONOMETRIC RATIOS OF COMPLEMENTARY ANGLES Since $A + B = 90°$, it follows that A and B are complementary and we see that *any trigonometric ratio of an angle is equal to the corresponding coratio of the complement of the angle.* Accordingly, cos 41° = sin 49° and tan 33° = cot 57°.

Problems 5 to 16 in Exercise 13.1 may be worked now.

13.3
USE OF TABLES AND CALCULATORS

By use of advanced methods we can calculate the values of the trigonometric ratios to any desired degree of accuracy. These values have been calculated to four, five, six, seven, and even more places and have been collected and arranged in tables. Table 13.1 gives the values to four decimal places. Calculators are another source of these values. We can use the table to obtain to four places the value of the sine, cosine, tangent, and cotangent of any acute angle. It is the purpose of this section to explain how this can be done. Later we shall show how these tables can be used to obtain the values of the trigonometric ratios of larger angles.

FEATURES OF TABLE 13.1 The student should now turn to Table 13.1 pages 464 and 465.

1. The table appears in three parts, and each part is divided into six columns.
2. The word angle appears at the top and bottom of the right and left columns on each part. The second, third, fourth, and fifth columns have the abbreviations for sine, cosine, tangent, and cotangent, respectively, at the top and the same words, with sine and cosine interchanged and tangent and cotangent interchanged, at the bottom.
3. If we start at the top of the left column of the first part of the table and read *down* through the table to the last part, we find the angles at intervals of 10' from 0 to 45°; then if we move to the right column of the last part and read *up* through the table to the first part, we find the angles from 45 to 90°. For example, the angle 25°40' is found in the left column of the table *below* 25°, and the angle 65°50' is found in the right column *above* 65°.
4. The names of the trigonometric ratios at the top of each column are associated with the angles in the left column of the table, and the names of the ratios at the bottom of each column are associated with the angles in the right column.

We shall explain the method of using the table and a calculator with the following examples.

EXAMPLE 1 In order to find the value of sin 35°40', we first find 35° in the left column of the table, then read *down* to 40'. Now we move to the right across the table to the column headed by sin, and we find .5831. Hence sin 35°40' = .5831.

USE OF TABLE FOR ANGLES LESS THAN 45° (*a*) 35°40' is *less* than 45°, so we find the angle in the *left* column of the table. (*b*) Since the angle is

Table 13.1 Trigonometric Functions

Angle	sin	cos	tan	cot	Angle	sin	cos	tan	cot	Angle	sin	cos	tan	cot	Angle
0° 00'	.0000	1.0000	.0000	∞	90° 00'	15° 00'	.2588	.9659	.2679	3.7321	75° 00'	30° 00'	.5000	.8660	.5774	1.7321	60° 00'
10	.0029	1.0000	.0029	343.77	50	10	.2616	.9652	.2711	3.6891	50	10	.5025	.8646	.5812	1.7205	50
20	.0058	1.0000	.0058	171.89	40	20	.2644	.9644	.2742	3.6470	40	20	.5050	.8631	.5851	1.7090	40
30	.0087	1.0000	.0087	114.59	30	30	.2672	.9636	.2773	3.6059	30	30	.5075	.8616	.5890	1.6977	30
40	.0116	.9999	.0116	85.940	20	40	.2700	.9628	.2805	3.5656	20	40	.5100	.8601	.5930	1.6864	20
50	.0145	.9999	.0145	68.750	10	50	.2728	.9621	.2836	3.5261	10	50	.5125	.8587	.5969	1.6753	10
1° 00'	.0175	.9998	.0175	57.290	89° 00'	16° 00'	.2756	.9613	.2867	3.4874	74° 00'	31° 00'	.5150	.8572	.6009	1.6643	59° 00'
10	.0204	.9998	.0204	49.104	50	10	.2784	.9605	.2899	3.4495	50	10	.5175	.8557	.6048	1.6534	50
20	.0233	.9997	.0233	42.964	40	20	.2812	.9596	.2931	3.4124	40	20	.5200	.8542	.6088	1.6426	40
30	.0262	.9997	.0262	38.188	30	30	.2840	.9588	.2962	3.3759	30	30	.5225	.8526	.6128	1.6319	30
40	.0291	.9996	.0291	34.368	20	40	.2868	.9580	.2994	3.3402	20	40	.5250	.8511	.6168	1.6212	20
50	.0320	.9995	.0320	31.242	10	50	.2896	.9572	.3026	3.3052	10	50	.5275	.8496	.6208	1.6107	10
2° 00'	.0349	.9994	.0349	28.636	88° 00'	17° 00'	.2924	.9563	.3057	3.2709	73° 00'	32° 00'	.5299	.8480	.6249	1.6003	58° 00'
10	.0378	.9993	.0378	26.432	50	10	.2952	.9555	.3089	3.2371	50	10	.5324	.8465	.6289	1.5900	50
20	.0407	.9992	.0407	24.542	40	20	.2979	.9546	.3121	3.2041	40	20	.5348	.8450	.6330	1.5798	40
30	.0436	.9990	.0437	22.904	30	30	.3007	.9537	.3153	3.1716	30	30	.5373	.8434	.6371	1.5697	30
40	.0465	.9989	.0466	21.470	20	40	.3035	.9528	.3185	3.1397	20	40	.5398	.8418	.6412	1.5597	20
50	.0494	.9988	.0495	20.206	10	50	.3062	.9520	.3217	3.1084	10	50	.5422	.8403	.6453	1.5497	10
3° 00'	.0523	.9986	.0524	19.081	87° 00'	18° 00'	.3090	.9511	.3249	3.0777	72° 00'	33° 00'	.5446	.8387	.6494	1.5399	57° 00'
10	.0552	.9985	.0553	18.075	50	10	.3118	.9502	.3281	3.0475	50	10	.5471	.8371	.6536	1.5301	50
20	.0581	.9983	.0582	17.169	40	20	.3145	.9492	.3314	3.0178	40	20	.5495	.8355	.6577	1.5204	40
30	.0610	.9981	.0612	16.350	30	30	.3173	.9483	.3346	2.9887	30	30	.5519	.8339	.6619	1.5108	30
40	.0640	.9980	.0641	15.605	20	40	.3201	.9474	.3378	2.9600	20	40	.5544	.8323	.6661	1.5013	20
50	.0669	.9978	.0670	14.924	10	50	.3228	.9465	.3411	2.9319	10	50	.5568	.8307	.6703	1.4919	10
4° 00'	.0698	.9976	.0699	14.301	86° 00'	19° 00'	.3256	.9455	.3443	2.9042	71° 00'	34° 00'	.5592	.8290	.6745	1.4826	56° 00'
10	.0727	.9974	.0729	13.727	50	10	.3283	.9446	.3476	2.8770	50	10	.5616	.8274	.6787	1.4733	50
20	.0756	.9971	.0758	13.197	40	20	.3311	.9436	.3508	2.8502	40	20	.5640	.8258	.6830	1.4641	40
30	.0785	.9969	.0787	12.706	30	30	.3338	.9426	.3541	2.8239	30	30	.5664	.8241	.6873	1.4550	30
40	.0814	.9967	.0816	12.251	20	40	.3365	.9417	.3574	2.7980	20	40	.5688	.8225	.6916	1.4460	20
50	.0843	.9964	.0846	11.826	10	50	.3393	.9407	.3607	2.7725	10	50	.5712	.8208	.6959	1.4370	10
5° 00'	.0872	.9962	.0875	11.430	85° 00'	20° 00'	.3420	.9397	.3640	2.7475	70° 00'	35° 00'	.5736	.8192	.7002	1.4281	55° 00'
10	.0901	.9959	.0904	11.059	50	10	.3448	.9387	.3673	2.7228	50	10	.5760	.8175	.7046	1.4193	50
20	.0929	.9957	.0934	10.712	40	20	.3475	.9377	.3706	2.6985	40	20	.5783	.8158	.7089	1.4106	40
30	.0958	.9954	.0963	10.385	30	30	.3502	.9367	.3739	2.6746	30	30	.5807	.8141	.7133	1.4019	30
40	.0987	.9951	.0992	10.078	20	40	.3529	.9356	.3772	2.6511	20	40	.5831	.8124	.7177	1.3934	20
50	.1016	.9948	.1022	9.7882	10	50	.3557	.9346	.3805	2.6279	10	50	.5854	.8107	.7221	1.3848	10
6° 00'	.1045	.9945	.1051	9.5144	84° 00'	21° 00'	.3584	.9336	.3839	2.6051	69° 00'	36° 00'	.5878	.8090	.7265	1.3764	54° 00'
10	.1074	.9942	.1080	9.2553	50	10	.3611	.9325	.3872	2.5826	50	10	.5901	.8073	.7310	1.3680	50
20	.1103	.9939	.1110	9.0098	40	20	.3638	.9315	.3906	2.5605	40	20	.5925	.8056	.7355	1.3597	40
30	.1132	.9936	.1139	8.7769	30	30	.3665	.9304	.3939	2.5386	30	30	.5948	.8039	.7400	1.3514	30
40	.1161	.9932	.1169	8.5555	20	40	.3692	.9293	.3973	2.5172	20	40	.5972	.8021	.7445	1.3432	20
50	.1190	.9929	.1198	8.3450	10	50	.3719	.9283	.4006	2.4960	10	50	.5995	.8004	.7490	1.3351	10
7° 00'	.1219	.9925	.1228	8.1443	83° 00'	22° 00'	.3746	.9272	.4040	2.4751	68° 00'	37° 00'	.6018	.7986	.7536	1.3270	53° 00'
10	.1248	.9922	.1257	7.9530	50	10	.3773	.9261	.4074	2.4545	50	10	.6041	.7969	.7581	1.3190	50
20	.1276	.9918	.1287	7.7704	40	20	.3800	.9250	.4108	2.4342	40	20	.6065	.7951	.7627	1.3111	40
													sin	cos	tan	cot	Angle

13.3 Use of Tables and Calculators | 465

Angle	cos	sin	cot	tan	Angle	cos	sin	cot	tan	Angle	cos	sin	tan	cot	Angle
	.1305	.9914	7.5958	.1317		.3827	.9239	.4142	2.4142		.6088	.7934	.7673	1.3032	30
30	.1334	.9911	7.4287	.1346	30	.3854	.9228	.4176	2.3945	30	.6111	.7916	.7720	1.2954	20
40	.1363	.9907	7.2687	.1376	40	.3881	.9216	.4210	2.3750	40	.6134	.7898	.7766	1.2876	10
50					50					50					
8° 00'	.1392	.9903	7.1154	.1405	23° 00'	.3907	.9205	.4245	2.3559	38° 00'	.6157	.7880	.7813	1.2790	52° 00'
10	.1421	.9899	6.9682	.1435	10	.3934	.9194	.4279	2.3369	10	.6180	.7862	.7860	1.2723	50
20	.1449	.9894	6.8269	.1465	20	.3961	.9182	.4314	2.3183	20	.6202	.7844	.7907	1.2647	40
30	.1478	.9890	6.6912	.1495	30	.3987	.9171	.4384	2.2998	30	.6225	.7826	.7954	1.2572	30
40	.1507	.9886	6.5606	.1524	40	.4014	.9159	.4383	2.2817	40	.6248	.7808	.8002	1.2497	20
50	.1536	.9881	6.4348	.1554	50	.4041	.9147	.4417	2.2637	50	.6271	.7790	.8050	1.2423	10
9° 00'	.1564	.9877	6.3138	.1584	24° 00'	.4067	.9135	.4452	2.2460	39° 00'	.6293	.7771	.8098	1.2349	51° 00'
10	.1593	.9872	6.1970	.1614	10	.4094	.9124	.4487	2.2286	10	.6316	.7753	.8146	1.2276	50
20	.1622	.9868	6.0844	.1644	20	.4120	.9112	.4522	2.2113	20	.6338	.7735	.8195	1.2203	40
30	.1650	.9863	5.9758	.1673	30	.4147	.9100	.4557	2.1943	30	.6361	.7716	.8243	1.2131	30
40	.1679	.9858	5.8708	.1703	40	.4173	.9088	.4592	2.1775	40	.6383	.7698	.8292	1.2059	20
50	.1708	.9853	5.7694	.1733	50	.4200	.9075	.4628	2.1609	50	.6406	.7679	.8342	1.1988	10
10° 00'	.1736	.9848	5.6713	.1763	25° 00'	.4226	.9063	.4663	2.1445	40° 00'	.6428	.7660	.8391	1.1918	50° 00'
10	.1765	.9843	5.5764	.1793	10	.4253	.9051	.4699	2.1283	10	.6450	.7642	.8441	1.1847	50
20	.1794	.9838	5.4845	.1823	20	.4279	.9038	.4734	2.1123	20	.6472	.7623	.8491	1.1778	40
30	.1822	.9833	5.3955	.1853	30	.4305	.9026	.4770	2.0965	30	.6494	.7604	.8541	1.1708	30
40	.1851	.9827	5.3093	.1883	40	.4331	.9013	.4806	2.0809	40	.6517	.7585	.8591	1.1640	20
50	.1880	.9822	5.2257	.1914	50	.4358	.9001	.4841	2.0655	50	.6539	.7566	.8642	1.1571	10
11° 00'	.1908	.9816	5.1446	.1944	26° 00'	.4384	.8988	.4877	2.0503	41° 00'	.6561	.7547	.8693	1.1504	49° 00'
10	.1937	.9811	5.0658	.1974	10	.4410	.8975	.4913	2.0353	10	.6583	.7528	.8744	1.1436	50
20	.1965	.9805	4.9894	.2004	20	.4436	.8962	.4950	2.0204	20	.6604	.7509	.8796	1.1369	40
30	.1994	.9799	4.9152	.2035	30	.4462	.8949	.4986	2.0057	30	.6626	.7490	.8847	1.1303	30
40	.2022	.9793	4.8430	.2065	40	.4488	.8936	.5022	1.9912	40	.6648	.7470	.8899	1.1237	20
50	.2051	.9787	4.7729	.2095	50	.4514	.8923	.5059	1.9768	50	.6670	.7451	.8952	1.1171	10
12° 00'	.2079	.9781	4.7046	.2126	27° 00'	.4540	.8910	.5095	1.9626	42° 00'	.6691	.7431	.9004	1.1106	48° 00'
10	.2108	.9775	4.6382	.2156	10	.4566	.8897	.5132	1.9486	10	.6713	.7412	.9057	1.1041	50
20	.2136	.9769	4.5736	.2186	20	.4592	.8884	.5169	1.9347	20	.6734	.7392	.9110	1.0977	40
30	.2164	.9763	4.5107	.2217	30	.4617	.8870	.5206	1.9210	30	.6756	.7373	.9163	1.0913	30
40	.2193	.9757	4.4494	.2247	40	.4643	.8857	.5243	1.9074	40	.6777	.7353	.9217	1.0850	20
50	.2221	.9750	4.3897	.2278	50	.4669	.8843	.5280	1.8940	50	.6799	.7333	.9271	1.0786	10
13° 00'	.2250	.9744	4.3315	.2309	28° 00'	.4695	.8829	.5317	1.8807	43° 00'	.6820	.7314	.9325	1.0724	47° 00'
10	.2278	.9737	4.2747	.2339	10	.4720	.8816	.5354	1.8676	10	.6841	.7294	.9380	1.0661	50
20	.2306	.9730	4.2193	.2370	20	.4746	.8802	.5392	1.8546	20	.6862	.7274	.9435	1.0599	40
30	.2334	.9724	4.1653	.2401	30	.4772	.8788	.5430	1.8418	30	.6884	.7254	.9490	1.0538	30
40	.2363	.9717	4.1126	.2432	40	.4797	.8774	.5467	1.8291	40	.6905	.7234	.9545	1.0477	20
50	.2391	.9710	4.0611	.2462	50	.4823	.8760	.5505	1.8165	50	.6926	.7214	.9601	1.0416	10
14° 00'	.2419	.9703	4.0108	.2493	29° 00'	.4848	.8746	.5543	1.8040	44° 00'	.6947	.7193	.9657	1.0355	46° 00'
10	.2447	.9696	3.9617	.2524	10	.4874	.8732	.5581	1.7917	10	.6967	.7173	.9713	1.0295	50
20	.2476	.9689	3.9136	.2555	20	.4899	.8718	.5619	1.7796	20	.6988	.7153	.9770	1.0235	40
30	.2504	.9681	3.8667	.2586	30	.4924	.8704	.5658	1.7675	30	.7009	.7133	.9827	1.0176	30
40	.2532	.9674	3.8208	.2617	40	.4950	.8689	.5696	1.7556	40	.7030	.7112	.9884	1.0117	20
50	.2560	.9667	3.7760	.2648	50	.4975	.8675	.5735	1.7347	50	.7050	.7092	.9942	1.0058	10
15° 00'	.2588	.9659	3.7321	.2679	30° 00'	.5000	.8660	.5774	1.7321	45° 00'	.7071	.7071	1.0000	1.0000	45° 00'
Angle	cos	sin	cot	tan	Angle	cos	sin	cot	tan	Angle	cos	sin	cot	tan	Angle

in the left column of the table, we read *down* from 35° to 35°40′. (*c*) Again, since the angle is less than 45°, we find the value of the function in the column with sines at the *top*.

EXAMPLE 2 In order to obtain the value of cos 67°50′, we find 67° in the *right* column of the table and read *up* to 50′. Across the table to the left from 67°50′ in the column with cos at the *bottom*, we find .3773. Hence, cos 67°50′ = .3773.

USE OF TABLE FOR ANGLES BETWEEN 45° AND 90° (*a*) The angle 67°50′ is *greater* than 45°, and thus the angle is in the *right* column of the table. (*b*) We read *up* from 67° to 67°50′. (*c*) The value of the ratio is found in the column with cos at the *bottom*.

In order to find the value of a trigonometric ratio by use of a calculator, follow the manufacturer's instructions. Most calculators use angles in decimal form only. Thus 42°30′ = 42.5° and 51°20′ = 51.333°.

EXAMPLE 3 Find sin 35°40′ by use of a calculator.

Solution We convert 40′ to a decimal part of a degree and find that it is 0.667°. Therefore, we want sin 35.667°. Hence, we put 35.667 on the display and then depress the sin button. We thus get 0.5831 on the display. Consequently, sin 35°40′ = sin 35.667° = 0.5831, as was found by use of Table 13.1. Both calculators and tables give decimal *approximations*.

EXAMPLE 4 You should practice using both the table and a calculator by verifying the following statements.

sin 17°20′ = 0.2979   tan 39°50′ = 0.8342
sin 49°40′ = 0.7623   tan 52°50′ = 1.3190
cos 26°20′ = 0.8962   cot 8°30′ = 6.6912
cos 73°20′ = 0.2868   cot 81°20′ = 0.1524

We see by studying the table that as an acute angle increases, the values of the *sine* and *tangent* of the angle *increase* and the values of the *cosine* and *cotangent* of the angle *decrease*. This fact should be noted very carefully because it is important in the discussion of interpolation.

Problems 17 to 24 of Exercise 13.1 may be done now.

In order to find a trigonometric ratio of an angle that is not listed in the table, we resort to linear interpolation or use a calculator. Since the procedure is the same as was used in connection with logarithms, we shall not discuss the general procedure but give three examples.

EXAMPLE 5 Find the value of sin 27°36′.

Solution Since 27°36′ is not listed in Table 13.1, we shall resort to linear interpolation. We begin by finding sin 27°30′ = .4617 and sin 27°40′ = .4643 because 27°30′ and 27°40′ are nearer 27°36′ than any other entries in the table. We then make the usual diagram for interpolation. In this case, it is

$$10'\left[\begin{array}{c}\text{Angle}\\ 27°40'\\ 6'\left[\begin{array}{c}27°36'\\ 27°30'\end{array}\right.\end{array}\right.\quad\begin{array}{c}\text{Sine}\\ .4643\\ \\ .4617\end{array}\left.\right]c\right].0026$$

It is now clear that 27°36′ is six-tenths of the way from 27°30′ to 27°40′; hence, we assume that sin 27°36′ is six-tenths of the way from sin 27°30′ = .4617 to sin 27°40′ = .4643. Consequently, we have

$$\frac{c}{.0026} = \frac{6}{10}$$

$$c = \tfrac{6}{10}(.0026) = .0016 \quad \text{to four decimal places}$$

Therefore, sin 27°36′ = .4617 + .0016 = .4633.

Using a calculator, we have

$$\sin 27°36' = \sin(27 + \tfrac{36}{60})° = \sin 27.6°$$
$$= 0.46329604 = 0.4633$$

to four decimal places.

EXAMPLE 6 Find the value of tan 55°27′.

Solution We show the diagram and the computation below:

$$10'\left[\begin{array}{c}\text{Angle}\\ 55°30'\\ 7'\left[\begin{array}{c}55°27'\\ 55°20'\end{array}\right.\end{array}\right.\quad\begin{array}{c}\text{Tangent}\\ 1.4550\\ \\ 1.4460\end{array}\left.\right]c\right].0090$$

$$\frac{7}{10} = \frac{c}{.0090}$$

$$c = \tfrac{7}{10}(.0090) = .0063$$
$$\tan 55°27' = 1.4460 + .0063 = 1.4523$$

Note that the tangent of an angle *increases* with the angle. Therefore the value of *c* is *added* to tan 55°20′.

EXAMPLE 7 Find the value of cos 37°28'.

$$10' \left[8' \left[\begin{array}{cc} \text{Angle} & \text{Cosine} \\ 37°30' & .7934 \\ 37°28' & \\ 37°20' & .7951 \end{array} \right] c \right] .0017$$

Solution

$$\frac{8}{10} = \frac{c}{.0017}$$

$$c = \tfrac{8}{10}(.0017) = .0014 \quad \text{to four decimal places}$$

$$\cos 37°28' = .7951 - .0014 = .7937$$

Note that the cosine of an angle *decreases* as the angle increases from 0 to 90°. Thus the value of c is *subtracted* from cos 37°20'.

As with logarithms, most of the interpolation can be performed mentally, especially after you become familiar with it.

If a calculator is used, interpolation is not needed.

EXAMPLE 8 To test your ability to interpolate and to use a calculator verify the following statements:

sin 47°15' = 0.7343   tan 17°26' = 0.3140
cos 38°39' = 0.7810   cot 63°12' = 0.5052

Problems 25 to 32 in Exercise 13.1 may be done now.

The process of finding the angle when we know the value of one of its trigonometric ratios is similar to the process described in the preceding section. That is, we find the value of the given ratio in the proper column of the table and then move to the right or left across the table to get the angle. We shall illustrate the method with several examples, depending on whether the ratio name is at the top or bottom of the column.

EXAMPLE 9 Given sin A = 0.5495, find A.

Solution The values of the sines are found in the second column of each part of the table reading *down* from .0000 to .7071 and in the third column reading *up* from .7071 to 1.0000. We find .5495 on the third part of the tables in the second column (note that the word sin is at the head of this column). We move to the *left* to the angle column and find 33°20'. Hence, A = 33°20'.

EXAMPLE 10 Given sin A = 0.9696, find A.

Solution The value .9696 is found in the third column of the first part of the table, and this column has the word sin at the bottom, so we move to the *right* to the angle column and find 75°50'. Hence, A = 75°50'.

EXAMPLE 11 Given tan A = 0.2994, find A.

Solution The values of the tangents are found in the fourth column of each part of the table reading *down* from .0000 to 1.0000 and in the fifth column reading *up* from 1.0000 to ∞. We find the value .2994 in the fourth column of the second part of the table. The name of the given ratio is at the top of the column, so we move to the left to the angle column and find 16°40'. Hence, A = 16°40'.

FINDING THE ANGLE GIVEN A TRIGONOMETRIC RATIO From these examples, we can see that the steps in obtaining the angle when we know one of its trigonometric ratios are as follows:

1. Determine the column of the table in which the given value will be found.
2. Find the given value in that column.
3. Notice whether the name of the given ratio is at the top or the bottom of the column.
4. To find the required angle, move to the left across the table to the angle column if the name of the given ratio is at the top of the column and to the right across the table if the name of the ratio is at the bottom.

In order to find the angle by use of a calculator when a trigonometric ratio is given, we put the value of the ratio on the display. We then use the buttons giving "the angle whose function is" (written $\sin^{-1} x$, $\cos^{-1} x$, $\tan^{-1} x$ or arcsin x, arccos x, arctan x or inv sin x, inv cos x, inv tan x) and read the angle in decimal form on the display.

EXAMPLE 12 If cos A = 0.5495, find A.

Solution We begin by putting 0.5495 on the display. We continue by depressing the $\cos^{-1} x$ button and reading 56.667 on the display. Therefore, A = 56°40' since 0.667° = 40' to the nearest 10 minutes.

In order to test your understanding of the process and obtain further practice in it, verify the statements in Example 13.

EXAMPLE 13

$\cos A = 0.9763$, $A = 12°30'$
$\cos A = 0.3529$, $A = 69°20'$
$\cot A = 0.6128$, $A = 58°30'$
$\cot A = 2.4342$, $A = 22°20'$
$\tan A = 0.4877$, $A = 26°$
$\sin A = 0.9674$, $A = 75°20'$
$\sin A = 0.2022$, $A = 11°40'$

Problems 33 to 40 in Exercise 13.1 may be done now.

If the given value of the ratio cannot be found in the table, we use interpolation to find the angle. The principle is the same as in our previous discussion.

EXAMPLE 14 Given $\sin A = .4551$, find A.

Solution We look in the sines column of the tables for the numbers that are nearer to .4551 than any others. We find the numbers .4540 and .4566. Now we write the numbers .4540, .4551, and .4566 in the column under *sine* in the diagram with .4540 at the bottom. Then in the column under *angle* we write the corresponding angles 27°10' and 27°. Now we proceed as in other interpolation problems.

$$.0026 \left[.0011 \left[\begin{array}{cc} \textit{Sine} & \textit{Angle} \\ .4566 & 27°10' \\ .4551 & \\ .4540 & 27°00' \end{array} \right] c \right] 10'$$

$$\frac{.0011}{.0026} = \frac{c}{10'}$$

$$c = 10' \left(\frac{.0011}{.0026} \right) = 10' \left(\frac{11}{26} \right) = \frac{110'}{26} = 4'$$

to one digit

Therefore, $A = 27° + 4' = 27°4'$. Interpolation is not needed if a calculator is used. However, we must convert from decimal degrees to minutes, as in Example 12.

EXAMPLE 15 Given $\cos A = .6216$, find A.

Solution The procedure in this example is the same as in Example 14 except that the difference at the right of the right bracket is negative because the angle decreases as the cosine increases. The steps in the solution are shown below.

$$.0023 \left[.0014 \left[\begin{array}{cc} \textit{Cosine} & \textit{Angle} \\ .6225 & 51°30' \\ .6216 & \\ .6202 & 51°40' \end{array} \right] c \right] -10'$$

Note that this difference is
$51°30' - 51°40' = -10'$

$$\frac{.0014}{.0023} = \frac{c}{-10'}$$

$$c = -10' \left(\frac{.0014}{.0023} \right) = -10' \left(\frac{14}{23} \right)$$

$$= \frac{-140'}{23}$$

$$= -6' \quad \text{to one digit}$$

Therefore, $A = 51°40' - 6' = 51°34'$.

EXERCISE 13.1

Draw each of the angles.

1. $30°, 140°, -210°$
2. $45°, 160°, -250°$
3. $-72°, 129°, 320°$
4. $60°, 150°, -330°$

Find the trigonometric ratios of the smaller angle (opposite the smaller side) of a right triangle that has the following parts. Use $a^2 + b^2 = c^2$. Write answer in the order sin, cos, tan, cot, sec, csc.

5. $a = 3, b = 4, c = \sqrt{3^2 + 4^2} = 5$

6. $a = 5, b = 12$

7. $a = 24, b = 7$

8. $a = 15, b = 8$

9. $a = 10, c = 26, b = \sqrt{26^2 - 10^2} = 24$

10. $a = 14, c = 50$

11. $b = 15, c = 17$

12. $b = 36, c = 39$

13 If $b = 7$ and $\tan A = 0.27$, find a.

14 If $c = 23$ and $\sin B = 0.81$, find b.

15 If $a = 38$ and $\cos B = 0.43$, find c.

16 If $b = 73$ and $\sin B = 0.27$, find c.

By use of the table or a calculator, find the value of each trigonometric ratio; interpolate if necessary.

17 $\sin 34°20' =$

18 $\cos 29°50' =$

19 $\tan 41°10' =$

20 $\cot 15°30' =$

21 $\cos 72°50' =$

22 $\tan 58° =$

23 $\cot 84°30' =$

24 $\sin 64°40' =$

25 tan 17°37′ =

26 cot 81°24′ =

27 sin 72°56′ =

28 cos 35°29′ =

29 cot 47°13′ =

30 sin 29°32′ =

31 cos 42°47′ =

32 tan 57°48′ =

Find the value of the angle to the nearest 10 minutes.

33 cos A = 0.9948

34 tan A = 0.3476

35 cot A = 1.3432

36 sin A = 0.6799

37 $\tan A = 1.2497$

38 $\cot A = 0.4734$

39 $\sin A = 0.9483$

40 $\cos A = 0.0814$

Find the value of the angle to the nearest minute.

41 $\cot A = 1.5791$

42 $\sin A = 0.3451$

43 $\cos A = 0.7887$

44 $\tan A = 0.2763$

45 $\sin A = 0.8320$

46 $\cos A = 0.4182$

47 $\tan A = 3.4567$

48 $\cot A = 0.1439$

13.4 SOLUTION OF RIGHT TRIANGLES

We can use the trigonometric ratios to find three parts of any right triangle if we are given two parts including a side. The process is known as **solving the triangle**; we shall illustrate it with several examples. In each case we shall assume that the right triangle is lettered as in Fig. 13.2, and in the first three examples we shall use data correct to three significant digits and obtain results to only three significant digits.

GIVEN A SIDE AND AN ACUTE ANGLE Given $a = 104$ and $A = 25°20'$, solve the triangle. We first draw the figure (Fig. 13.4) approximately to scale and show the values of the known parts.

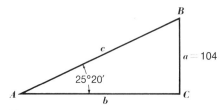

Fig. 13.4

Since $A + B = 90°$, $B = 90° - 25°20' = 64°40'$. The unknown parts are now b and c, and we shall find the value of c first. For this purpose, we select the trigonometric ratio that involves the two known parts A and a and the desired unknown part c. This ratio is $\sin A$, since $\sin A = a/c$. By substituting the given values in this equation, we get

$$\sin 25°20' = \frac{104}{c}$$

In Table 13.1 or by use of a calculator we find that $\sin 25°20' = 0.4279$. Therefore, we have

$$0.4279 = \frac{104}{c}$$

and we solve for c as follows:

$0.4279c = 104$ multiplying each member of above equation by c

$c = \dfrac{104}{0.4279}$ dividing each member by 0.4279

$ = 243$ to three significant digits

In order to find side b, we *return to the original data* and select the trigonometric ratio that involves the two known parts a and A and the unknown b. This ratio is $\tan A$, since $\tan A = a/b$. Therefore,

$\tan 25°20' = \dfrac{104}{b}$ substituting known values of A and a

$0.4734 = \dfrac{104}{b}$ since $\tan 25°20' = 0.4734$

$0.4734b = 104$ multiplying each member by b

$b = \dfrac{104}{0.4734}$ dividing each member by 0.4734

$ = 220$ to three significant digits

We can check the above solution by using the fact that b, c, and B, whose values were computed, are connected by the relation $\sin B = b/c$. Hence $b = c(\sin B) = 243(\sin 64°40') = 243(0.9038) = 219.63 = 220$ to three significant digits. Therefore, since we found that $b = 220$ is our solution, we have a check.

GIVEN THE HYPOTENUSE AND AN ACUTE ANGLE Given $c = 250$ and $B = 42°10'$, solve the right triangle. We first construct Fig. 13.5 approximately to scale and record the given data on it.

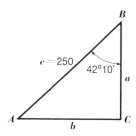

Fig. 13.5

Now we find side a, and for this purpose we select the trigonometric ratio that involves the two known parts and the side a. This ratio is $\cos B$, since $\cos B = a/c$. Therefore,

$\cos 42°10' = \dfrac{a}{250}$ substituting given values for B and c

$0.7412 = \dfrac{a}{250}$ since $\cos 42°10' = 0.7412$

$a = 250(0.7412)$ multiplying each member by 250

$ = 185$ to three significant digits

To get side b we use $\sin B$, since this ratio involves the two known parts and the desired unknown. Therefore, we have

$$\sin B = \frac{b}{c}$$

$\sin 42°10' = \dfrac{b}{250}$ substituting given values for B and c

$0.6713 = \dfrac{b}{250}$ since $\sin 42°10' = 0.6713$

$b = 250(0.6713)$ multiplying each member by 250

$ = 168$ to three significant digits

Since $A + B = 90°$, $A = 90° - 42°10' = 47°50'$; hence the solution is $a = 185$, $b = 168$, and $A = 47°50'$.

As a check, we use the fact that the three parts computed above are connected by $\tan A = a/b$. Therefore, $a = b(\tan A) = 168(\tan 47°50') = 168(1.1041) = 185$ (to three significant digits). Hence, since this was the value of a found in the original computation, we have a check.

The number of significant digits in the result of computation cannot exceed the number of significant digits that appears in any one of the numbers involved in the computation. In Examples 1 and 2 the given data involved three digits, so the required results were obtained to three significant digits.

If we are given only three significant digits in the value of the trigonometric ratios, we can obtain the value of the corresponding angle only to the nearest multiple of 10 minutes. If the value of a trigonometric ratio of an angle is given to only two significant digits, we can obtain the value of the corresponding angle to only the nearest degree.

As illustrated by Examples 1 and 2, the process of solving a right triangle consists of the following steps:

STEPS IN SOLVING A RIGHT TRIANGLE

1. Draw the figure as nearly to scale as practical and label the given sides and angles with their values.
2. Decide which unknown part is to be found first.
3. Select the trigonometric ratio that involves the two known parts and the unknown part selected in step 2. Write this ratio down and substitute the known values. This yields an equation.
4. With the help of the table or a calculator solve the equation in step 3 and express the result to the number of significant digits justified by the data.
5. If one of the acute angles is given, find the other by subtracting the given angle from 90°.

GIVEN THE HYPOTENUSE AND ONE SIDE

Given $a = 325$ and $c = 538$, solve the triangle. We first draw the figure approximately to scale as in Fig. 13.6. The side b can be found by use of the

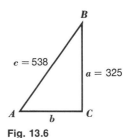

Fig. 13.6

pythagorean theorem. If a calculator is available, use it to find $\sqrt{c^2 - a^2}$ since this involves known parts (see step 3 above). However we shall find one of the acute angles first. We shall select the angle A for this computation. In order to find A, we shall use $\sin A$, since this ratio involves the two known sides and the angle A. Therefore, we have

$$\sin A = \frac{325}{538} = 0.604 \quad \text{to three significant digits}$$

Hence, $A = 37°10'$ to the nearest multiple of 10 minutes and $B = 90° - 37°10' = 52°50'$.

We can use the above value of A with either a or c to obtain b. We shall use c, and we have $\cos A = b/c$. Hence,

$$\cos 37°10' = \frac{b}{538}$$

and

$b = 538(\cos 37°10')$
$ = 538(0.7969) = 429$

Thus, the solution is $A = 37°10'$, $B = 52°50'$, $b = 429$.

We check the solution by using the fact that

$$b = \sqrt{c^2 - a^2} = \sqrt{289{,}444 - 105{,}625}$$
$$= \sqrt{183{,}819} = 429$$

Because of approximations inherent in tables and calculators, *different methods of solving the same problem may well give slightly different answers.*

As a last example, we shall use data correct to four significant digits and obtain the values of the angles to the nearest minute. This, of course, will require interpolation. We shall not show the interpolation process in the solution, but for practice you should check all results for which interpolation was required.

GIVEN TWO SIDES Given $a = 0.3162$ and $b = 0.6214$, solve the triangle. The triangle drawn approximately to scale is shown in Fig. 13.7. We

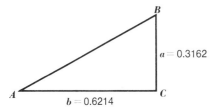

Fig. 13.7

shall first find angle A, using tan A for this purpose. Thus, we have

$$\tan A = \frac{0.3162}{0.6214} = 0.5089 \quad \text{to four significant digits}$$

We now interpolate and get $A = 26°58'$. Since $A + B = 90°$, $B = 90° - 26°58' = 63°2'$.

We shall now use sin A to get side c as follows:

$$\sin A = \frac{a}{c}$$

$$\sin 26°58' = \frac{0.3162}{c}$$

$$0.4535 = \frac{0.3162}{c} \quad \text{since } \sin 26°58' = .4535$$

$$c(0.4535) = 0.3162 \quad \text{multiplying both members by } c$$

$$c = \frac{0.3162}{0.4535} = 0.6972$$

This can be checked by the pythagorean theorem.

Problems 1 to 20 in Exercise 13.2 may be worked now.

13.5
SOME APPLICATIONS

If an observer at A, as in Fig. 13.8, is looking at an object B, then AB is the **line of sight**. If C is any point on a horizontal ray through A such that AC is in the same vertical plane as AB, then the angle CAB between the horizontal and the line of sight is called the **angle of elevation** or **angle of depression** of B according as B is above or below A.

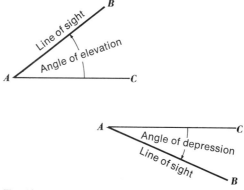

Fig. 13.8

EXAMPLE 1 A 13.4-m flagpole is standing vertically on horizontal ground and casts a shadow 17.5 m long. What is the angle of elevation of the top of the flagpole as seen from the end of the shadow?

Solution Figure 13.9 shows the situation described in the problem. We know the two legs and want the angle A; hence, we use tan A and have tan $A = 13.4/17.5 = 0.7657$. Therefore, by use of tables or a calculator, we find that $A = 37°30'$ to the nearest multiple of 10 minutes.

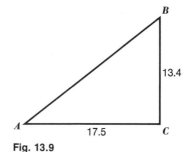

Fig. 13.9

EXAMPLE 2 From the top of a lighthouse 150 ft high, the angles of depression of two boats in the bay at the foot of the lighthouse are 42 and 46°. If the boats are in the same vertical plane as the lighthouse, find the distance between them.

476 | Chapter 13: Numerical Trigonometry

Solution The data in the problem are shown in Fig. 13.10, in which A and B represent the positions of the boats. Since the lines CB and PD are parallel, angle B = angle BPD = 42° and angle A = angle APD = 46°. We shall use right triangle CBP to find the length of CB and right triangle CAP to find the length of CA. Then the distance $AB = CB - CA$.

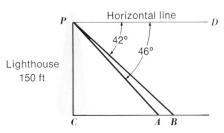

Fig. 13.10

In right triangle CBP

$$\cot B = \frac{CB}{CP}$$

Therefore,

$$\cot 42° = \frac{CB}{150} \quad \text{since } CP = \text{height of lighthouse} = 150 \text{ ft}$$

$$\begin{aligned} CB &= 150(\cot 42°) \\ &= 150(1.1106) \\ &= 167 \text{ ft} \quad \text{to three significant digits} \end{aligned}$$

In right triangle CAP

$$\cot A = \frac{AC}{CP}$$

Therefore,

$$\cot 46° = \frac{AC}{150}$$

$$\begin{aligned} AC &= 150(\cot 46°) \\ &= 150(0.9657) \\ &= 145 \text{ ft} \quad \text{to three significant digits} \end{aligned}$$

Hence, the distance $AB = 167 \text{ ft} - 145 \text{ ft} = 22 \text{ ft}$.

DIRECTIONS IN SURVEYING AND MARINE NAVIGATION In surveying and in marine navigation, the direction or bearing of a line through a given point is specified by giving the acute angle at which the direction of the angle is to the east or west of the north-south line through the points. Accordingly, the bearing of OA in Fig. 13.11 is N40°E since it is 40° east of north. Similarly the

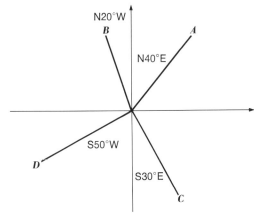

Fig. 13.11

bearing of OB is N20°W, of OC is S30°E, and of OD is S50°W.

DIRECTION IN AIR NAVIGATION In air navigation, the direction of flight or course is specified by stating the clockwise angle the ray makes with the due north direction. In Fig. 13.12, the directions of OA, OB, OC, and OD are 50, 120, 300, and 240°, respectively.

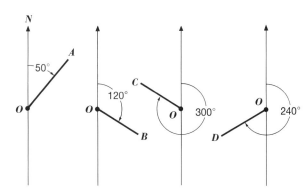

Fig. 13.12

EXAMPLE 3 An airplane flew at 340 mi/h for 2 h in the direction of 115°, made a stop, then flew in the direction of 205° at 360 mi/h for $1\frac{1}{2}$ h. In what direction must it fly in order to return to its starting point? How far is it from the starting point?

Solution A sketch of the situation is shown in Fig. 13.13. The angle BFS is 90° since XBF is 25° as is TFS also. Hence, $\tan S = \frac{680}{540} = 1.2593$. Therefore, $S = 51.55°$ by use of a calculator and

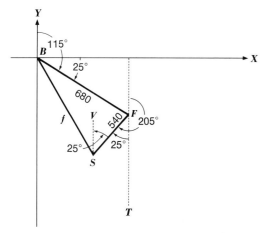

Fig. 13.13

51°30' to the nearest 10 minutes by use of tables. Consequently, VSB is $51°30' - 25° = 26°30'$, and the plane must fly at 333°30' to return to the starting point. We can find f by use of $\sin 51°30' = 680/f$ since then

$$f = \frac{680}{\sin 51°30'}$$
$$= \frac{680}{0.7826} = 869$$

Therefore, it is 869 mi from the starting point.

STUDENT'S NOTES

EXERCISE 13.2

Solve the right triangles that have the parts given.

1 $A = 28°20'$, $c = 573$

2 $A = 53°40'$, $c = 7.28$

3 $B = 64°10'$, $c = 82.9$

4 $B = 39°50'$, $c = 437$

5 $a = 224$, $B = 38°30'$

6 $a = 115$, $B = 52°20'$

7 $b = 323$, $A = 55°40'$

8 $b = 198$, $A = 50°10'$

9 $a = 5.61$, $A = 67°50'$

10 $a = 113$, $A = 32°30'$

11 $b = 71.8$, $B = 61°20'$

12 $b = 407$, $B = 44°50'$

13 $b = 359$, $c = 727$

14 $b = 5.03$, $c = 9.26$

15 $a = 63.1$, $c = 91.3$

16 $a = 48.6$, $c = 84.9$

17 $a = 21.33$, $b = 34.75$

18 $a = 72.19$, $b = 62.27$

19 $a = 88.25$, $b = 47.76$

20 $a = 17.73$, $b = 80.36$

21 From the top of a lighthouse 75 ft high, the angle of depression of a boat is 3°. Assume that the base of the lighthouse is on the shoreline and find the distance from the boat to the base.

22 At a point 500 ft from the base of a building the angles of elevation of the top and bottom of a flagpole on the roof of the building are 24 and 22°, respectively. Find the height of the building and the length of the flagpole.

23 A pilot in a helicopter flying at an altitude of 1000 ft above a highway observes a speeding car directly below him. At the same time his instrument indicated that the angle of depression of a highway intersection ahead of the car was 18°40′, and 25 s later the car passed through the intersection. Find the speed of the car in feet per second.

24 An observer on a boat approaching a bridge over a river found that the angle of elevation of the bottom of the bridge was 10°. A short time later the angle of elevation was 11°20′. If the bridge was known to be 150 ft above the river, how far apart were the points of observation?

25 Airfield B is 260 mi due south of field A. Field C is due east of A and in the direction N58°40′E from B. A pilot flew from A to B and then to C in 4 h flying time. Find his average speed.

26 In order to find the width of a river flowing due south, a surveyor set a stake A in the edge of the water due west of a stone B in the edge of the water across the river. Then at point C 600 ft south of A she measured the angle ACB and found it to be 14°00′. What was the width of the river?

27 A plane heads due east at 420 mi/h with a south wind of 22 mi/h. Find the speed and direction of flight.

28 How far apart are Mt. Janice and Mary's Peak if the former is 48.32 mi N17°31′W of an observation point and the latter is 40.37 mi N72°29′E of it?

13.6
TRIGONOMETRIC RATIOS FOR ANGLES BETWEEN 90 AND 180°

In this section we shall extend the definition of the trigonometric ratios to include angles from 90 to 180°. We use the Greek letter α (alpha) to designate an angle that varies but is always between 90 and 180° and the Greek letter θ (theta) to represent the supplement of α. Thus in our discussion $\alpha + \theta = 180°$, α is an obtuse angle, and θ is an acute angle.

α BETWEEN 90 AND 180° In order to define the trigonometric ratios of α, we shall first construct the angle DAB equal to α (Fig. 13.14). Then from

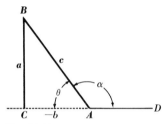

Fig. 13.14

the point B we drop a perpendicular BC to the side AD produced through the vertex A. We shall now require that distances measured to the right be positive and distances measured to the left be negative; thus the length of AD is positive and the length of AC is negative. We let $AB = c$, $CB = a$, and $AC = -b$, where a, b, and c are positive. The angle CAB is equal to θ, since it is the supplement of α. We now define the trigonometric ratios of α as follows:

$$\sin \alpha = \frac{a}{c} \qquad \csc \alpha = \frac{c}{a}$$

$$\cos \alpha = \frac{-b}{c} = -\frac{b}{c} \qquad \sec \alpha = \frac{c}{-b} = -\frac{c}{b}$$

$$\tan \alpha = \frac{a}{-b} = -\frac{a}{b} \qquad \cot \alpha = \frac{-b}{a} = -\frac{b}{a}$$

We shall show that each of the above functions can be expressed in terms of the supplement θ of α. For this purpose we shall construct the triangle $A'B'C'$ (Fig. 13.15), in which angle $C'A'B' = \theta$ and $A'B' = c$. The right triangles $C'A'B'$ and ACB are congruent, because the hypotenuses are equal and each has an angle equal to θ. Therefore

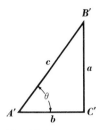

Fig. 13.15

$C'B' = a$, and $A'C'$ has the same length as AC. However, $A'C'$ is positive. Therefore, $A'C' = b$. By the definition of Sec. 13.2, we have in triangle $A'B'C'$

$$\sin \theta = \frac{a}{c} \qquad \csc \theta = \frac{c}{a}$$

$$\cos \theta = \frac{b}{c} \qquad \sec \theta = \frac{c}{b}$$

$$\tan \theta = \frac{a}{b} \qquad \cot \theta = \frac{b}{a}$$

TRIGONOMETRIC RATIOS FOR SUPPLEMENTARY ANGLES If we compare the above trigonometric ratios with those of α, we see that with $\alpha = 180° - \theta$

$$\sin \alpha = \frac{a}{c} = \sin \theta \qquad \cot \alpha = -\frac{b}{a} = -\cot \theta$$

$$\cos \alpha = -\frac{b}{c} = -\cos \theta \qquad \sec \alpha = -\frac{c}{b} = -\sec \theta$$

$$\tan \alpha = -\frac{a}{b} = -\tan \theta \qquad \csc \alpha = \frac{c}{a} = \csc \theta$$

Hence, the sine and cosecant of an angle between 90 and 180° are respectively equal to the sine and cosecant of the supplement of the angle, and all other ratios of the angle are respectively equal to the negative of the corresponding ratio of the supplement. The following examples illustrate the rule:

$$\sin 155° = \sin (180° - 155°) = \sin 25°$$
$$= 0.4226$$
$$\cos 125°40' = -\cos (180° - 125°40')$$
$$= -\cos 54°20' = -0.5831$$
$$\tan 110°10' = -\tan (180° - 110°10')$$
$$= -\tan 69°50' = -2.7228$$
$$\cot 148°20' = -\cot (180° - 148°20')$$
$$= -\cot 31°40' = -1.6212$$

Problems 1 to 12 in Exercise 13.3 may be worked now.

13.7
THE OBLIQUE TRIANGLE

An **oblique triangle** is a triangle in which no one of the angles is a right angle. A triangle has six parts, three sides and three angles, and the sum of the measures of the three angles is 180°. We shall develop formulas that will enable us to solve an oblique triangle completely if we are given three of the six parts including a side.

The necessary information, one side and two other parts of a triangle, can be

1. Two angles and any side (designated **AAS**)
2. Two sides and the angle opposite one of them (designated **SSA**)
3. Two sides and the angle between them (designated **SAS**)
4. Three sides (designated **SSS**)

In the next three sections we develop two formulas, the law of sines and the law of cosines, that will enable us to solve an oblique triangle if information in any one of the four categories is given.

13.8
THE LAW OF SINES

In order to derive the law of sines, we construct the triangle ABC (Figs. 13.16 and 13.17) and *letter the sides opposite A, B, and C with the small letters a, b, and c, respectively*. The letters A, B, and C refer to both the vertices and the angles at the vertices. From the point C we drop the perpendicular CD to the side AB in Fig. 13.16 and to AB produced in Fig. 13.17. We call the length of this perpendicular h.

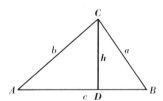

Fig. 13.16

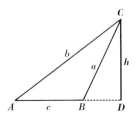

Fig. 13.17

In the right triangle ADC in each figure, $\sin A = h/b$. Therefore,

$$h = b \sin A \tag{1}$$

Furthermore, in right triangle DBC in either figure, $\sin B = h/a$. Hence,

$$h = a \sin B \tag{2}$$

Now we equate the values of h in formulas (1) and (2) and get

$$a \sin B = b \sin A \tag{3}$$

Then we divide each member of (3) by $(\sin A)(\sin B)$ and get

$$\frac{a}{\sin A} = \frac{b}{\sin B} \tag{4}$$

If we drop a perpendicular from B to AC and repeat the above argument, we get

$$\frac{a}{\sin A} = \frac{c}{\sin C} \tag{5}$$

Now, by comparing (4) and (5), we see that

$$\frac{a}{\sin A} = \frac{b}{\sin B} = \frac{c}{\sin C} \tag{13.1}$$

This relation is known as the **law of sines** and it is a concise way of writing the equations (4) and (5). Each of formulas (4) and (5) contains four of the six parts of triangle ABC. Hence, if we know three of these parts, we can solve for the fourth one.

We can use the law of sines to solve triangles included in Cases 1 (AAS) and 2 (SSA), and we shall illustrate the methods with Examples 1 to 4.

LAW OF SINES FOR AAS Given the triangle ABC, with $A = 34°20'$, $B = 46°40'$, and side $a = 210$, solve the triangle. Figure 13.18 shows the given data. In order to solve the triangle, we must find the angle C, the side b, and the side c.

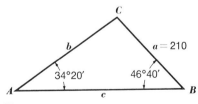

Fig. 13.18

We find angle C by use of the relation $A + B + C = 180°$. Therefore,

$C = 180° - (A + B)$
$= 180° - (34°20' + 46°40')$
$= 180° - 81° = 99°$

We shall use formula (4), which involves b and the known parts a, A, and B, to find the side b. So, starting with

$$\frac{b}{\sin B} = \frac{a}{\sin A}$$

and substituting the known parts $a = 210$, $A = 34°20'$, and $B = 46°40'$, we get

$$\frac{b}{\sin 46°40'} = \frac{210}{\sin 34°20'}$$

Hence,

$b = \dfrac{210(\sin 46°40')}{\sin 34°20'}$ multiplying each member by $\sin 46°40'$

$= \dfrac{210(0.7274)}{0.5640}$ since $\sin 46°40' = 0.7274$ and $\sin 34°20' = 0.5640$

$= 271$ to three significant digits

To get side c, we use formula (5). After substituting the known parts, we have

$$\frac{c}{\sin 99°} = \frac{210}{\sin 34°20'}$$

Therefore,

$c = \dfrac{210(\sin 99°)}{\sin 34°20'}$

$= \dfrac{210(0.9877)}{0.5640}$ $\sin 99° = \sin(180° - 99°)$
$\qquad\qquad\qquad\qquad = \sin 81° = 0.9877$

$= 368$

Problems 13 to 24 in Exercise 13.3 may be worked now.

13.9
THE AMBIGUOUS CASE

Before considering the procedure for solving a triangle with two sides and an angle opposite one of them given (SSA), we call attention to the fact that if the sine of an angle is positive and less than 1, the angle may be an acute angle, as found in tables or by use of a calculator, or it may be 180° minus that angle. Thus if $\sin A = \frac{1}{2}$, then $A = 30°$ or $180° - 30° = 150°$. Furthermore, if a, b, and $\sin B$ are given, then $\sin A = a(\sin B)/b$ is determined and may be less than 1, equal to 1, or greater than 1. If

$\sin A \begin{cases} < 1 & A \text{ from the table and} \\ & A' = 180 - A \text{ are possible angles} \\ = 1 & A = 90° \\ > 1 & \text{no angle } A \text{ since a leg is never} \\ & \text{greater than the hypotenuse} \end{cases}$

Finally, if a, b, and $\sin B$ are given, we find $\sin A$ and get all possible angles A (none, one, or two) that fit into a triangle with the given angle, noting that two angles fit into a triangle if their sum is less than 180°. SSA is called the ambiguous case since there may be more than one angle if $\sin A < 1$.

EXAMPLE 1 If $B = 29°20'$, $b = 243$, and $a = 445$, then by the law of sines

$$\sin A = \frac{445 \sin 29°20'}{243} = 0.8971$$

Consequently, $A = 63°50'$ and $A' = 180° - 63°50' = 116°10'$. Therefore

$A + B = 63°50' + 29°20' = 93°10' < 180$

and

$A' + B = 116°10' + 29°20' = 145°30' < 180°$

Hence both A and A' fit into a triangle with the given angle B. The solution of each of the triangles ABC and $A'BC'$ can now be completed by use of the law of sines.

EXAMPLE 2 If $B = 67°40'$, $a = 342$, and $b = 574$, then by the law of sines

$$\sin A = \frac{342 \sin 67°40'}{574} = 0.5511$$

$A = 33°30'$ and $A' = 180° - 33°30' = 146°30'$

Therefore

$A + B = 33°30' + 67°40' = 101°10' < 180°$

and

$A' + B = 146°30' + 67°40' = 214°10' > 180°$

Consequently, A fits into a triangle with B, but A' does not. The triangle ABC can be solved by use of the law of sines.

EXAMPLE 3 If $B = 132°20'$, $a = 270$, and $b = 213$, then by the law of sines

$$\sin A = \frac{270 \sin 132°20'}{213} = 0.9371$$

$$A = 69°30' \quad \text{and} \quad A' = 110°30'$$

Neither A nor A' fits into a triangle with B; hence, there is no triangle with the given data.

EXAMPLE 4 If $B = 48°50'$, $b = 473$, and $a = 337$, then

$$\sin B = \frac{473 \sin 48°50'}{337} = 1.06$$

There is no angle B and no solution since the sine of an angle is never greater than 1.

Problems 25 to 40 of Exercise 13.3 may be worked now.

EXERCISE 13.3

Use the relation between the ratios of an angle and those of its supplement and a table or a calculator to find the ratios.

1 $\sin 103° =$

2 $\cos 98° =$

3 $\tan 147° =$

4 $\cot 119° =$

5 $\cos 154°20' =$

6 $\tan 139°40' =$

7 $\cot 148°50' =$

8 $\sin 161°30' =$

9 $\tan 98°45' =$

10 $\cot 138°23' =$

11 $\sin 127°38' =$

12 $\cos 159°11' =$

Solve the oblique triangles that have the parts given (Case AAS).

13 $A = 38°20', B = 67°30', a = 214$

14 $A = 41°30', C = 59°20', a = 13.6$

15 $B = 68°50'$, $C = 45°20'$, $b = 70.3$

16 $B = 61°40'$, $A = 49°10'$, $b = 487$

17 $A = 96°20'$, $C = 37°50'$, $a = 787$

18 $A = 34°10'$, $B = 99°30'$, $a = 2.35$

19 $B = 22°50'$, $A = 48°20'$, $c = 80.4$

20 $B = 28°30'$, $C = 41°40'$, $a = 98.7$

21 $B = 49°36'$, $C = 71°17'$, $a = 4193$

22 $C = 111°33'$, $B = 43°26'$, $c = 8888$

23 $A = 94°38'$, $C = 42°42'$, $b = 3849$

24 $A = 36°36'$, $C = 47°47'$, $b = 73.73$

Find the number of solutions; this is Case SSA, the ambiguous case.

25 $B = 36°50'$, $b = 382$, $a = 501$

26 $B = 41°10'$, $b = 321$, $a = 239$

27 $B = 144°20'$, $b = 421$, $a = 532$

28 $B = 151°30'$, $b = 181$, $a = 405$

Find the solutions; this is Case SSA, the ambiguous case.

29 $A = 131°10'$, $a = 571$, $b = 706$

30 $C = 34°20'$, $c = 473$, $b = 193$

31 $C = 36°20'$, $c = 187$, $a = 243$

32 $A = 29°40'$, $a = 177$, $c = 113$

33 A saleswoman can arrive at C by traveling N28°40′E from A or at B by going N73°30′E from A. How far apart are A and B if C is 58.1 km from B and 31.3 km from A?

34 The angle of elevation of a kite as seen from one point on horizontal ground is 54°20′, and it is 82°30′ from another point that is 103 m from the first. How high is the kite?

35 At a point on a hill whose slope is 15° with the horizontal, the angle of elevation of the top of a tree 100 ft up the slope is 34°. Find the height of the tree to the nearest foot.

36 A motorist traveling on a state highway intended to turn off on a farm road that intersected the highway at an angle of 12° and drive 25 km to a country store. She missed her turn, however, and proceeded to a second farm road that intersected the highway at an angle of 100° and drove to the store on the latter road. How far did she go out of the way?

37 Airfield *A* is 560 mi due north of field *B*. A pilot flew in the direction 130° from *A* to *C* and then in the direction 200° to *B*. Find the total distance flown.

38 An 18-m loading ramp from the ground to the door of a warehouse made an angle of 30° with the horizontal. It was replaced by a ramp 40 m long. Find the angle the second ramp made with the horizontal.

39 In order to avoid a storm, a pilot flew 80.0 mi in a direction that was 16°21′ off course. He then changed direction and flew 72.0 mi to his destination. If his average speed was 210 mi/h, find the delay caused by the storm.

40 A girl walked 100 m due east from her camp and then turned in a direction east of south and walked 985 m to a point where she could see her camp. Her compass indicated that the direction of her camp from this point was N38°W. How far was she from her camp?

13.10
THE LAW OF COSINES

Since the law of sines does not enable us to solve a triangle when two sides and the included angle or the three sides are given, we shall derive a formula that can be used with those parts given. For that purpose, we draw oblique triangles as in Figs. 13.19 and 13.20, and in each triangle we drop a perpendicular CD from C to the line AB or to AB produced. We shall let h represent the length of

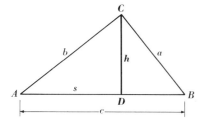

Fig. 13.19

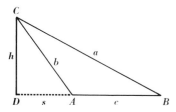

Fig. 13.20

DC and s represent the length of AD. Then in the right triangle DBC in either figure, we have

$$a^2 = h^2 + (DB)^2 \qquad (1)$$

In Fig. 13.19, $DB = c - s$, and when this is substituted in (1), we have

$$\begin{aligned}a^2 &= h^2 + (c - s)^2 \\ &= h^2 + c^2 - 2cs + s^2 \qquad \text{squaring } c - s \\ &= h^2 + s^2 + c^2 - 2cs \qquad \text{rearranging terms}\end{aligned}$$

In the right triangle ADC, $h^2 + s^2 = b^2$, and since $\cos A = s/b$, we have $s = b \cos A$. Therefore, we have

$$a^2 = b^2 + c^2 - 2bc \cos A$$

In Fig. 13.20, $DB = c + s$, and if we substitute this in (1), simplify, and rearrange terms as above, we get

$$\begin{aligned}a^2 &= h^2 + (c + s)^2 \\ &= h^2 + c^2 + 2cs + s^2 \\ &= h^2 + s^2 + c^2 + 2cs\end{aligned}$$

Again in the right triangle DAC, we have $h^2 + s^2 = b^2$, and also $s = b \cos DAC$. Therefore,

$$a^2 = b^2 + c^2 + 2bc \cos DAC$$

However, angle DAC + angle $A = 180°$ (by angle A, we mean angle BAC). Therefore, $A = 180° - DAC$ and

$$\begin{aligned}\cos A &= \cos (180° - DAC) \\ &= -\cos DAC\end{aligned}$$
$$\text{since } \cos(180° - \theta) = -\cos \theta$$

Thus,

$$\cos DAC = -\cos A$$

Hence, we again have

$$a^2 = b^2 + c^2 - 2bc \cos A \qquad \textbf{law of cosines} \qquad (13.2a)$$

Since a represented any one of the sides of the triangle and A the opposite angle, similar arguments for b and the angle B and c and the angle C will yield

$$b^2 = a^2 + c^2 - 2ac \cos B \qquad (13.2b)$$
$$c^2 = a^2 + b^2 - 2ab \cos C \qquad (13.2c)$$

If we solve each of the above formulas for the cosine of the angle, we get

$$\cos A = \frac{b^2 + c^2 - a^2}{2bc} \qquad (13.3a)$$

$$\cos B = \frac{a^2 + c^2 - b^2}{2ac} \qquad (13.3b)$$

$$\cos C = \frac{a^2 + b^2 - c^2}{2ab} \qquad (13.3c)$$

We can use (13.2a) to (13.2c) if two sides and the included angle are given and formulas (13.3a) to (13.3c) if we know the three sides. We shall illustrate the use of the law of cosines with two examples.

THE CASE SAS Given $b = 112$, $c = 215$, and $A = 35°30'$, solve the triangle. The triangle is shown in Fig. 13.21. Since b, c, and A are given,

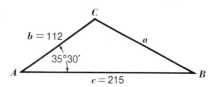

Fig. 13.21

we use formula (13.2a) to obtain the side a. If we substitute the above values in (13.2a), we get

$$a^2 = 112^2 + 215^2 - 2(112)(215)(\cos 35°30')$$

Since the data are given to three significant digits, we can obtain only three significant digits in the value of a. Therefore, we round off the values of 112^2, 215^2, and $2(112)(215)(\cos 35°30')$ to three significant digits before combining them. Thus, we obtain

$$a^2 = 12{,}500 + 46{,}200 - 39{,}200 = 19{,}500$$

and

$$a = 140 \quad \text{to three significant digits}$$

The angles B and C can be found by means of formulas (13.3b) and (13.3c). However, the computation is easier if we use the law of sines.

Remember that different methods of working the same problem may give slightly different answers.

We can gain some idea of the relative magnitude of the angles by use of the theorem in plane geometry which states that the order of magnitude of the angles of a triangle is the same as that of the opposite sides. This means that if

$$\text{side } a < \text{side } b < \text{side } c$$

then

$$\text{angle } A < \text{angle } B < \text{angle } C$$

In this problem we have $b < a < c$, and therefore it follows that $B < A < C$.

In order to find angle B, we shall use the law of sines

$$\frac{a}{\sin A} = \frac{b}{\sin B}$$

substitute the known values, and get

$$\frac{140}{\sin 35°30'} = \frac{112}{\sin B}$$

Therefore,

$$\sin B = \frac{112(\sin 35°30')}{140}$$
$$= \frac{112(0.5807)}{140} = 0.4646$$

Hence, $B = 27°40'$ and $B' = 180° - 27°40' = 152°20'$. However, since we know that B is less than $A = 35°30'$, we must choose only the former value. Thus, $B = 27°40'$.

We can find the value of C by use of the fact that $C = 180° - (A + B)$ or by a repeated use of the law of sines. We shall find C by the first method and check the result by the second. Accordingly, we have

$$C = 180° - (35°30' + 27°40')$$
$$= 180° - 63°10' = 116°50'$$

As a check, we have by use of the law of sines

$$\sin C = \frac{c \sin A}{a} = \frac{215(0.5807)}{140} = 0.8918$$
$$C = 63°10' \quad \text{or} \quad 116°50'$$

The second of the two values checks with the value of C previously obtained.

The thoughtful student may ask at this point: Suppose I had solved for C before finding the value of B; how would I know whether to choose $63°10'$ or $116°50'$, since both satisfy the condition $C > A$? The answer is that if $63°10'$ is temporarily chosen, then B must be equal to $180° - (35°30' + 63°10') = 180° - 98°40' = 81°20'$, and this contradicts the condition $B < A$. Hence, $116°50'$ must be the value chosen for C.

Problems 1 to 12 in Exercise 13.4 may be worked now.

THE CASE SSS Given $a = 125$, $b = 175$, and $c = 200$, solve the triangle. The figure drawn to scale is shown in Fig. 13.22. Each of the formulas

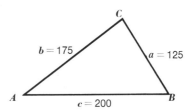

Fig. 13.22

(13.3a) to (13.3c) expresses the cosine of an angle in terms of the sides, and we shall use each of them in the solution of our problem. Since the data are given to three significant digits, we can obtain the values of the angles only to the nearest multiple of 10 minutes, and we shall round off the result in each step of the computation to three significant digits.

We shall start with formula (13.3a), substitute the given values for a, b, and c, and obtain

$$\cos A = \frac{175^2 + 200^2 - 125^2}{2(175)(200)}$$

$$= \frac{30{,}600 + 40{,}000 - 15{,}600}{70{,}000}$$

$$= \frac{55{,}000}{70{,}000} = 0.786$$

Therefore, $A = 38°10'$.

By using formula (13.3b) and substituting the given values, we get

$$\cos B = \frac{125^2 + 200^2 - 175^2}{2(125)(200)}$$

$$= \frac{15{,}600 + 40{,}000 - 30{,}600}{50{,}000}$$

$$= \frac{25{,}000}{50{,}000} = .500$$

Hence, $B = 60°$.

Finally you should check that $C = 81°50'$.

As a check, we have $38°10' + 60' + 81°50' = 179°60' = 180°$, as it should be.

STUDENT'S NOTES

EXERCISE 13.4

Find the indicated side or angle to the justified degree of accuracy.

1    $a = 17, b = 21, C = 30°, c =$

2    $a = 73, b = 67, C = 60°, c =$

3    $a = 21, c = 35, B = 150°, b =$

4    $b = 37, c = 41, A = 45°, a =$

5    $b = 39, c = 57, A = 53°, a =$

6    $b = 39, c = 57, A = 127°, a =$

7    $a = 43, b = 67, C = 130°, c =$

8    $a = 91, b = 45, C = 68°, c =$

Chapter 13: Numerical Trigonometry

9 $b = 218$, $c = 329$, $A = 78°20'$, $a =$

10 $b = 523$, $c = 481$, $A = 63°10'$, $a =$

11 $a = 631$, $b = 549$, $C = 52°40'$, $c =$

12 $a = 631$, $b = 549$, $C = 127°20'$, $c =$

13 $a = 13$, $b = 15$, $c = 19$, $A =$

14 $a = 11$, $b = 19$, $c = 23$, $C =$

15 $a = 23$, $b = 29$, $c = 31$, $B =$

16 $a = 41$, $b = 47$, $c = 73$, $C =$

17 $a = 123, b = 234, c = 345, C =$

18 $a = 175, b = 182, c = 149, B =$

19 $a = 381, b = 277, c = 219, A =$

20 $a = 257, b = 341, c = 384, B =$

21 $a = 435, b = 472, c = 327, C =$

22 $a = 797, b = 779, c = 977, A =$

23 $a = 509, b = 831, c = 648, A =$

24 $a = 871, b = 741, c = 639, B =$

25 A coast guard station received a distress signal from a ship that was 40.0 nautical miles N38°10′W of the station. How long will it take a rescue ship that is 30.0 nautical miles S80°20′W of the station to reach the distressed ship if it steams at the rate of 32 knots? (A knot is one nautical mile per hour.)

26 Points B and C are on opposite sides of a pond. Point A on the shore is 1000 m from B and 1500 m from C. If angle BAC is 60°, how far is it from B to C?

27 A pilot left airfield A on a flight to field B and 2 h later observed a storm ahead. She then changed her direction and flew 1 h to field C, due east of B and 600 mi from A. If the speed of her plane was 250 mi/h, what was the direction of her flight after observing the storm?

28 A man who planned to fence a triangular plot found that the marker at the northern corner had been destroyed. The existing markers are 450 m apart, and one is due east of the other. His deed showed that the lost marker was 200 m from the east corner and 350 m from the west. In what direction must he measure 200 m from the east corner in order to locate the north corner?

13.11
CHAPTER SUMMARY

After some definitions, including that of the trigonometric ratios, we showed how to use tables and a calculator to find angles and ratios. Our next section showed how to solve a right triangle. We concluded the chapter by presenting the law of sines and the law of cosines, which are used to solve oblique triangles.

The fundamental equations and theorems given in the chapter are

$$\sin A = \frac{\text{side opposite } A}{\text{hypotenuse}}$$

$$\cos A = \frac{\text{side adjacent to } A}{\text{hypotenuse}}$$

$$\tan A = \frac{\text{side opposite } A}{\text{side adjacent to } A}$$

Any trigonometric ratio of an acute angle is equal to the coratio of the complementary angle.
Any trigonometric ratio of an angle is numerically equal to the same function of its supplement.

$$\frac{a}{\sin A} = \frac{b}{\sin B} = \frac{c}{\sin C} \tag{13.1}$$

$$a^2 = b^2 + c^2 - 2bc \cos A \tag{13.2a}$$

$$\cos A = \frac{b^2 + c^2 - a^2}{2bc} \tag{13.3a}$$

STUDENT'S NOTES

EXERCISE 13.5 REVIEW

1 Find the trigonometric ratios of the smaller angle of a right triangle if $a = 8$ and $c = 17$.

2 $\sin 31°20' =$

3 $\tan 72°40' =$

4 $\cos 37°11' =$

5 $\cot 81°27' =$

6 $\cos 161°20' =$

7 $\sin 113°41' =$

8 If $\sin A = 0.2843$, $A =$

9 If $\cos A = 0.3241$, $A =$

10 Solve the right triangle in which $A = 41°30'$ and $b = 307$.

11 Solve the right triangle in which $A = 72°10'$ and $c = 769$.

12 Solve the right triangle in which $a = 273$ and $b = 429$.

13 Solve the oblique triangle in which $a = 238$, $B = 76°50'$, and $A = 38°50'$.

14 Find all solutions if $a = 331$, $b = 287$, and $A = 78°20'$.

15 If $a = 21$, $b = 37$, and $c = 31$, find C.

16 If $a = 222$, $c = 301$, and $B = 48°20'$, find b.

17 A rescue party found that the angle of elevation of a ledge on a vertical cliff on which two men were trapped was 65°20′. If this measurement was made at a horizontal distance of 83.8 m from the bottom of the cliff, and if from the same spot the angle of elevation of the top of the cliff at a spot directly above the ledge was 78°10′, how far below the top of the cliff was the ledge?

18 Two airplanes left their home field at the same time, one flying a course N48°20′E and the other flying a course N41°40′W. If at the end of 1 h the first plane had traveled 342 mi and the second 429 mi, how far apart were they?

NAME _____ DATE _____ SCORE _____

EXERCISE 13.6 CHAPTER TEST

Solve the right triangles.

1 $B = 29°30'$, $a = 243$

2 $A = 38°40'$, $a = 267$

3 $a = 531$, $c = 783$

Solve the oblique triangles.

4 $A = 44°40'$, $B = 58°30'$, $c = 429$

5 $a = 43$, $b = 37$, $C = 102°$

6 If $a = 391$, $b = 417$, and $c = 659$, find C.

Course Test

This course test may be used as a diagnostic test to help determine which chapters a particular student or class may omit.

CHAPTER 1

Perform the indicated operations.

1. $763 + 552$

2. $984 - 498$

3. 537×329

4. $9\overline{)785.13}$

5. Change $\frac{15}{23}$ to a decimal with three decimal places.

6. Reduce 0.28 to a fraction in lowest terms.

7. Find 27% of 309.

8. Show that $\frac{1}{2} + \frac{1}{3} + \frac{1}{4} + \frac{1}{5} + \frac{43}{60} = 2$.

9. Evaluate $\dfrac{3.20 + 5.71}{8.61 - 4.50}$.

CHAPTER 2

10. If $a = -2$ and $b = 3$, then $|a + b| - a - b =$

11. $2a^3b^2 \times 3ab^0 \times 7a^2b =$

12. Find the product of $2a^2 + 3a - 1$ and $3a - 4$.

13 Remove symbols of grouping and then collect like terms in $x\{2x + 3[x - x(x - 2) + x^2] - 8x\}$.

14 $\dfrac{35x^4y^3z^2 - 14x^2y^5z^4}{7x^2y^3z} =$ _____

15 Find the quotient and remainder if $2x^3 + x^2 - 3x + 3$ is divided by $x + 2$.

CHAPTER 3

16 Find the product of $3x + 2y$ and $2x - 5y$.

17 Factor $6ax - 8ay$.

18 Factor $6x^2 + 11xy - 10y^2$.

CHAPTER 4

19 Reduce $\dfrac{(2x - y)(3x^2 - xy - 10y^2)}{(3x + 5y)(2x^2 + 5xy - 3y^2)}$ to lowest terms.

20 $\dfrac{3x}{5y} - \dfrac{x + 2y}{2x - 5y} =$ _____

21 $\dfrac{\dfrac{2x + 1}{2x - 1} - \dfrac{x}{2x + 1}}{3 - \dfrac{4x}{2x + 1}} =$ _____

CHAPTER 5

22 Find the volume of a cone of height 12 in and base radius 5 in if the volume of one of height h and base radius r is $V = \pi r^2 h/3$.

23 If $f(x) = 3x - 5$ and $g(x) = 2x^2 - x + 4$ find $f(3)$, $f(-2)$, $g(1)$, $g(-2)$, and $f(g(0))$.

24 Is $\{(1,2),(1,3),(3,4),(4,5)\}$ a function? Why?

25 Is $\{(1,2),(2,3),(3,2),(4,5)\}$ a function? Why?

26 Sketch the graph of $y = 2x - 1$.

27 Sketch the graph of $y = x^2 + x - 2$.

28 Sketch the graph of $y = x + |x|$ for $x = -3$ to $x = 2$.

CHAPTER 6

Simplify the expression.

29 $2^4 2^3 =$

30 $\dfrac{5^7}{5^5} =$

31 $(2^3)^2 =$

32 $(2a^2 b)^3 =$

33 $\dfrac{6a^3 b^4}{18 a^2 b^5} =$

34 $(25 a^4 b^6)^{1/2} =$

35 $\sqrt{27} + \sqrt{147} - \sqrt{75} =$

36 $\dfrac{2\sqrt{3} - 3\sqrt{5}}{2\sqrt{5} - \sqrt{3}} =$

37 $(3 + 4i)(2 - 5i) =$

38 $\dfrac{3 + 7i}{2 + 3i} =$

CHAPTER 7

39 Are $3x = 4$ and $x + \tfrac{2}{3} + \tfrac{1}{4} = \tfrac{9}{4}$ equivalent equations?

40 Solve $x/a + x/b = a + b$ for x.

41 Solve $(2x + 1)(x + 3) \geq (x + 2)(x - 1)$.

42 Find two numbers whose quotient is $\tfrac{5}{13}$ and whose sum is 18.

CHAPTER 8

Solve:

43 $\dfrac{1}{x} + \dfrac{2}{3} = \dfrac{7}{6}$

44 $\dfrac{2}{x+3} - \dfrac{3}{3x+5} = \dfrac{1}{x+5}$

45 $\dfrac{5x-1}{x+7} \geq 0$

CHAPTER 9

46 Find the ratio of 7 lb to 10 oz as a fraction and reduce to lowest terms.

47 Find x if $x:42 = 2:7$.

48 If 6 sifters of flour are needed to make a cake for 15, how many people can be served by a cake that used 14 sifters of flour?

49 If $y + 5$ varies directly as $2x + 1$ and $y = 2$ for $x = 3$, find y for $x = 6$.

CHAPTER 10

50 Show that $3x - 2y = 8$ and $-6x + 4y = 16$ are inconsistent equations.

51 Solve $5x + 3y = 1$ and $7x + 2y = 8$ simultaneously.

52 Solve $x + 2y + 2z = 1$, $2x - y - 3z = 11$, and $3x + 3y + 4z = 4$ simultaneously.

CHAPTER 11

53 Solve $3x^2 + 5x - 2 = 0$ by factoring, by completing the square, and by the quadratic formula.

54 Solve $\sqrt{2x + 3} - 1 = \sqrt{x + 1}$.

55 Solve $(x^2 - x)^2 - 8(x^2 - x) + 12 = 0$.

56 Solve $2x^2 - 7x + 6 \leq 0$ graphically and then check your solution by solving algebraically.

57 Show that the only right triangle with sides x, $3x - 3$, and $2x + 3$ in increasing order is the one whose sides are 5, 12, and 13.

CHAPTER 12

Find the value of the unknown letter.

58 $\log_3 x = -2$

59 $\log_x 64 = 3$

60 $\log_4 \frac{1}{16} = x$

By use of logarithms or a calculator or both, find n.

61 $n = (38.1)(5.27)$

62 $n = \dfrac{429}{72.6}$

63 $n = \sqrt{27.52}$

64 $n = \sqrt[3]{0.8034}$

65 $\log n = 1.7185$

66 $\log n = 8.3149 - 10$

CHAPTER 13

67 Solve the right triangle in which one leg is $a = 273$ and the hypotenuse is $c = 418$.

68 Solve the oblique triangle in which $A = 49°50'$, $a = 147$, and $c = 133$.

Answers

EXERCISE 1.1

+	1	4	7	10	6	12	3	0	9	5	8	11	2
3	4	7	10	13	9	15	6	3	12	8	11	14	5
7	8	11	14	17	13	19	10	7	16	12	15	18	9
8	9	12	15	18	14	20	11	8	17	13	16	19	10
2	3	6	9	12	8	14	5	2	11	7	10	13	4
10	11	14	17	20	16	22	13	10	19	15	18	21	12
1	2	5	8	11	7	13	4	1	10	6	9	12	3
0	1	4	7	10	6	12	3	0	9	5	8	11	2
12	13	16	19	22	18	24	15	12	21	17	20	23	14
5	6	9	12	15	11	17	8	5	14	10	13	16	7
4	5	8	11	14	10	16	7	4	13	9	12	15	6
9	10	13	16	19	15	21	12	9	18	14	17	20	11
11	12	15	18	21	17	23	14	11	20	16	19	22	13
6	7	10	13	16	12	18	9	6	15	11	14	17	8

1 commutative **2** commutative **3** associative **5** associative and commutative **6** associative and commutative **7** associative and commutative **9** 7 **10** 13 **11** 13 **13** 13 **14** 12 **15** 15 **17** 13 **18** 17 **19** 14 **21** 147 **22** 157 **23** 189 **25** 1621 **26** 1659 **27** 2054 **29** 6,661 **30** 12,942 **31** 8792 **33** 1073.401 **34** 5647.18 **35** 9738.30 **37** 25 **38** 34 **39** 61 **41** 36 **42** 27 **43** 38 **45** 333 **46** 622 **47** 313 **49** 481 **50** 617 **51** 368 **53** 502.21 **54** 110.91 **55** 43.882 **57** 251.54 **58** 28.084 **59** 7.96

EXERCISE 1.2

×	7	12	5	10	3	8	0	6	11	4	9	2
2	14	24	10	20	6	16	0	12	22	8	18	4
5	35	60	25	50	15	40	0	30	55	20	45	10
9	63	108	45	90	27	72	0	54	99	36	81	18
1	7	12	5	10	3	8	0	6	11	4	9	2
7	49	84	35	70	21	56	0	42	77	28	63	14
3	21	36	15	30	9	24	0	18	33	12	27	6
11	77	132	55	110	33	88	0	66	121	44	99	22
8	56	96	40	80	24	64	0	48	88	32	72	16
6	42	72	30	60	18	48	0	36	66	24	54	12
10	70	120	50	100	30	80	0	60	110	40	90	20
4	28	48	20	40	12	32	0	24	44	16	36	8
12	84	144	60	120	36	96	0	72	132	48	108	24

1 commutative **2** commutative **3** commutative **5** associative **6** associative **7** associative **9** commutative and associative **10** commutative and associative **11** commutative and associative **13** distributive **14** distributive **15** distributive **17** 166 **18** 141 **19** 280 **21** 13,604 **22** 24,453 **23** 30,958 **25** 351,833 **26** 1,245,174 **27** 2,052,288 **29** 261 **30** 112 **31** 159 **33** 324 **34** 423 **35** 543 **37** 567, 1 **38** 852, 3 **39** 594, 3 **41** 1132, 3 **42** 1579, 13 **43** 822, 31 **45** 9.906 **46** 241.11 **47** 3.3255 **49** 162.966 **50** 3.4081 **51** 19.8576 **53** 384, 1.10 **54** 76, 7.27 **55** 20, 0.3093 **57** 1866, 0.48 **58** 2419, 1.07 **59** 2053, 0.012 **61** 11.62 **62** 2.69 **63** 46.99 **65** 7.1818··· **66** 29.1818··· **67** 4.125 **69** 0 **70** 0 **71** 0 **73** 0 **74** not a permissible operation **75** not a permissible operation **77** 7(11) **78** 2(3)(13) **79** 79 **81** 7 + 11 = 13 + 5 **82** 5 + 19 = 7 + 17 = 11 + 13 **83** 7 + 23 = 11 + 19 = 13 + 17

EXERCISE 1.3

1 proper, numerator, denominator **2** decimal, integer, fractional part **3** mixed, integral part, fractional part **5** 14 **6** 9 **7** 16 **9** 4 **10** 9 **11** 20 **13** $\frac{15}{30}, \frac{20}{30}, \frac{24}{30}$ **14** $\frac{20}{60}, \frac{45}{60}, \frac{24}{60}$ **15** $\frac{105}{140}, \frac{56}{140}, \frac{20}{140}$ **17** $\frac{10}{20}, \frac{15}{20}, \frac{8}{20}$ **18** $\frac{15}{30}, \frac{25}{30}, \frac{6}{30}$ **19** $\frac{63}{84}, \frac{70}{84}, \frac{48}{84}$ **21** $\frac{1}{3}$ **22** $\frac{2}{3}$ **23** $\frac{1}{4}$ **25** $\frac{17}{21}$ **26** $\frac{19}{23}$

27 $\frac{13}{17}$ **29** $\frac{31}{37}$ **30** $\frac{23}{29}$ **31** $\frac{71}{92}$ **33** $3\frac{1}{7}$ **34** $2\frac{2}{3}$ **35** $1\frac{2}{5}$ **37** $\frac{13}{5}$ **38** $\frac{5}{4}$ **39** $\frac{25}{9}$
41 0.656 **42** 0.419 **43** 0.671 **45** 0.636 **46** 0.360 **47** 0.875 **49** $\frac{3}{10}$ **50** $\frac{4}{5}$ **51** $\frac{7}{10}$
53 $\frac{8}{25}$ **54** $\frac{33}{200}$ **55** $\frac{59}{250}$ **57** $\frac{5}{9}$ **58** $\frac{2}{3}$ **59** 1 **61** $\frac{25}{33}$ **62** $\frac{3}{11}$ **63** $\frac{86}{333}$ **65** $\frac{21}{55}$ **66** $\frac{271}{990}$
67 $\frac{4261}{11000}$

EXERCISE 1.4

1 $\frac{4}{3}$ **2** $\frac{1}{4}$ **3** 1 **5** $\frac{7}{12}$ **6** $\frac{1}{12}$ **7** $\frac{7}{24}$ **9** $\frac{1}{3}$ **10** $\frac{1}{9}$ **11** $\frac{1}{8}$ **13** $\frac{4}{5}$ **14** 0 **15** $\frac{5}{12}$
17 $\frac{3}{7}$ **18** $\frac{4}{9}$ **19** $\frac{1}{8}$ **21** $\frac{17}{4}$ **22** $\frac{3}{2}$ **23** $\frac{13}{6}$ **25** $\frac{36}{35}$ **26** $\frac{67}{30}$ **27** $\frac{26}{9}$ **29** 4.638
30 4.139 **31** 1.083 **33** 0.481 **34** 2.061 **35** 0.573

EXERCISE 1.5

1 $\frac{1}{5}$ **2** $\frac{1}{21}$ **3** $\frac{2}{7}$ **5** $\frac{7}{8}$ **6** $\frac{11}{9}$ **7** $\frac{2}{5}$ **9** $\frac{6}{5}$ **10** $\frac{26}{11}$ **11** $\frac{21}{52}$ **13** $\frac{1}{4}$ **14** $\frac{1}{6}$ **15** $\frac{9}{4}$
17 $\frac{16}{5}$ **18** 3 **19** $\frac{1}{2}$ **21** 4 **22** $\frac{3}{4}$ **23** $\frac{49}{18}$ **25** $\frac{12}{11}$ **26** $\frac{21}{25}$ **27** $\frac{8}{3}$ **29** 4 **30** 10
31 $\frac{35}{3}$ **33** $\frac{2}{3}$ **34** $\frac{3}{2}$ **35** $\frac{6}{7}$ **37** 18.2 **38** 23.5875 **39** 4.312 **41** 28 **42** 13 **43** 0.667

EXERCISE 1.6

1 0.31 **2** 0.14 **3** 0.78 **5** 1.25 **6** 0.375 **7** 0.003 **9** $\frac{3}{4}$ **10** $\frac{1}{5}$ **11** $\frac{4}{5}$ **13** $\frac{1}{3}$
14 $\frac{3}{16}$ **15** $\frac{2}{7}$ **17** 23% **18** 41% **19** 37.4% **21** 6.1% **22** 0.47% **23** 769% **25** 20%
26 75% **27** 37.5% **29** $66\frac{2}{3}\%$ **30** $83\frac{1}{3}\%$ **31** $57\frac{1}{7}\%$ **33** 7.98 **34** 51.48 **35** 139.05
37 39.4 **38** 63.1 **39** 253.4 **41** 212.5 **42** 400 **43** 460 **45** 604.3 **46** 624.3 **47** 189.8
49 $1.00 **50** $25 **51** $1075 **53** $483 **54** gain of $9.00 **55** 20% **57** 25%, $16\frac{2}{3}\%$, 20.6%
58 $12\frac{2}{3}\%$ **59** 15% **61** $62.50 **62** $1200 **63** 21,250 people **65** $100 **66** $2800 **67** $3500

EXERCISE 1.7

1 3.87×10^2 **2** 2.53(10) **3** 4.035(10) **5** 1.37×10^{-1} **6** 2.02×10^{-2} **7** 9.009×10^{-1}
9 8.1×10^2 **10** 0.82 **11** 82 **13** 2.60×10^{-2} **14** 3.05×10^{-3} **15** 6.01×10^4 **17** 7.5
18 34 **19** 14.2 **21** 1.3×10^2 **22** 3.96×10^3 **23** 2.6×10^2 **25** 0.76 **26** 0.78 **27** 6.6
29 1.2×10^2 **30** 2.29×10^3 **31** 6.9×10^2 **33** 5.58 **34** 8.31 **35** 130.11 **37** 74.23
38 2.183 **39** 2.209 **41** 5120, 5.12 **42** 21.35, .02135 **43** .14, 140 **45** 314 **46** 15550
47 0.00876 **49** false **50** true **51** true **53** 5 **54** 55 **55** 77 **57** 253,704 in **58** 110 cups **59** 14.083 ft **61** 14 yd, 3 in **62** 853 acres 338 yd² 2 ft² **63** 0.88 lb **65** 1419.5 mL **66** 0.52 lb **67** $\frac{1}{3}$ lb shortening

EXERCISE 1.8

1 13 **2** 42 **3** 71 **5** 19.5 **6** 4.83 **7** 2.68 **9** 120 **10** 53.1 **11** 3.60×10^3
13 4.79 **14** 299 **15** 210 **17** 0.283 **18** 0.115 **19** 2.40 **21** 4.41 **22** 2.88
23 10.4 **25** 18.7 **26** 3.14 **27** 1.03 **29** .813 **30** 1.19 **31** 1.12 **33** 7.93 **34** 3.51
35 1.69 **37** 3.38 **38** 0.402 **39** 91.4

EXERCISE 1.9

1 1003 **2** 1080 **3** 632 **4** 579 **6** 606 **7** 784.57 **8** 272.1 **9** 55.73 **10** 413
11 87.7875 **12** 517.03 **13** 206.47 **14** 0 **15** 0 **16** not defined **17** 2,3,5,7 **18** $23 + 3 = 19 + 7 = 13 + 13$ **19** 23 **20** 2.390 **21** 0.41 **22** $\frac{35}{70}, \frac{28}{70}, \frac{60}{70}$ **23** $\frac{5}{8}, \frac{3}{11}$ **24** 0.591, 1.609 **25** $\frac{3}{5}, \frac{18}{25}$ **26** $\frac{8}{11}$ **27** $\frac{1}{12}$ **28** $\frac{17}{15}$ **29** $\frac{3}{14}$ **30** $\frac{1}{2}$ **31** 1.63 **32** 5.02 **33** 23.9 **34** 2.00
35 .713 **36** 47.15% **37** 82.42 **38** 113.78 **39** 3.23 **40** 100 **44** .075% **45** 18 mm
46 6400 g **47** 14 yd 1 ft 1 in **48** 4.26 L

EXERCISE 1.10

1 106 **2** 31.2 **3** $44 - \sqrt{4} - 4$ **4** $5(7)(13)$ **5** $\frac{43}{50}$ **6** 1.63 **7** 1 **8** $\frac{17}{84}$ **9** 17.52, 0.4404
10 159 **11** 3.50

EXERCISE 2.1

1 binomial **2** monomial **3** none **5** monomial **6** none **7** trinomial **17** 5 **18** 11
19 0 **21** 4 **22** -4 **23** 0 **25** -162 **26** -353 **27** 12 **29** $5a - 2a - a + 7b - 3b + 6c + 2c$ **30** $9x - 8x - 3y + 5y - 4y - 4w + 6w$ **31** $8y - 10y - 9a + 6a + 13p - 11p + 4p$
33 $3a^2b - 7a^2b + 5ab^2 + 2ab^2 - 5ab^2$ **34** $7a^2b - 3ab + 5ab - 8ab^2 + 2ab^2$ **35** $14x^3y - 17x^2y^2 + 21x^2y^2 - 13xy^3 + 9xy^3$ **37** $2a - 9c$ **38** $3t - 3u + 16m$ **39** $3w - 3h$ **41** $5x^3y + xy^3$ **42** $-pr^3 + r^4$ **43** $2pq + qr - 10p$ **45** $-2x + y - 6w$ **46** $3x + y + w$ **47** $10y + w$ **49** $7c$ **50** $a + b + c$ **51** $p + 7d + q$ **53** $2a + b$ **54** 0 **55** $2p + 2i + 2e$ **57** 6 **58** 62 **59** -16
61 3 **62** 86 **63** -75

EXERCISE 2.2

9 a^4 **10** $-a^5$ **11** a^3 **13** a^6 **14** $-a^7$ **15** $-a^9$ **17** true **18** false, $9a^6$ **19** true
21 false, $24a^9$ **22** true **23** false, $24a^9$ **25** true **26** false, $24x^4y^6$ **27** false, $6x^6y^7$ **29** $ab +$

ac **30** $ab - 2ac$ **31** $6ab - 2ac$ **33** $6r^3s^3 + 2r^2s^3$ **34** $4a^3b - 6a^2b^2$ **35** $-6x^3y^2 + 4x^2y^3$ **37** $2x^2 + 5xy - 3y^2$ **38** $6x^2 - 13xy + 6y^2$ **39** $12x^2 + xy - 6y^2$ **41** $2a^3 + 9a^2 + 16a + 15$ **42** $15x^3 - 19x^2 + x + 3$ **43** $2a^3 + 3a^2b - ab^2 - 12b^3$ **45** $6a^2 - ab - 7ac - 2b^2 + 7bc - 3c^2$ **46** $15a^2 + 13ab - 16ac - 6b^2 - 10bc + 4c^2$ **47** $3x^4 - 5x^3y + 4x^2y^2 - xy^3 - y^4$ **49** $3(x - y + w)$ **50** $-12x - y + 3w + 13$ **51** $8x - 21y + 9w$ **53** $2x^2$ **54** $-x + 3y$ **55** $8w - 5y + 9x$

EXERCISE 2.3

1 a^2 **2** a^5 **3** b^6 **5** xy^2 **6** x^2y^2 **7** b **9** true **10** true **11** false, $2x^3$ **13** false, $2c^3d^4$ **14** true **15** true **17** true **18** false, does not exist **19** true **21** $x^2y^2 - x^4y$ **22** $x^4w + x^3y^2w^3$ **23** $s^3t - 2r$ **25** $4a^2b^6c - 3b^3$ **26** $3y^3t^3 + 2xt$ **27** $5a^4b^5 - 3ab^4c^2$ **29** $x + 3$ **30** $x + 5$ **31** $3x + 1$ **33** $x^2 - 3x + 4$ **34** $3x^2 + x - 1$ **35** $3x^2 + 4x - 2$ **37** $x + 4$ **38** $2x + 1$ **39** $x + 2$ **41** $x^3 - 2x^2 + x + 3$ **42** $2x^3 + 3x - 1$ **43** $3x^3 + 4x^2 + 5$ **45** $2x^2 - x + 3, 2$ **46** $x^2 + 2x - 1, -3$ **47** $3x^2 + x + 2, -5$ **49** $2p^2 - pq - q^2, -2q^3$ **50** $2p^2 + 2pq - q^2, 4q^3$ **51** $3p^2 + pq - q^2, -5q^3$ **53** $2r^3 + 3r^2s - rs^2 + s^3, -2s^4$ **54** $2r^3 - r^2s + s^3, 7s^4$ **55** $r^3 - 2rs^2 + 3s^3, 6s^4$

EXERCISE 2.4

1 3014 **2** 600 **3** *110000000* **5** 1000 **6** 3411 **7** 33333 **9** 10,760 **10** 6560 **11** *100000000* **13** 285 **14** 13,344 **15** 25 **17** 1548 **18** 2323 **19** 122222 **21** *101* **22** *110* **23** *1011* **25** *10110* **26** *10010* **27** *111011* **29** *1111* **30** *11110* **31** *1001101* **33** *10010011* **34** *1011111* **35** *11100001* **37** *1110* **38** *1221* **39** *22112* **41** *10112* **42** *10021* **43** *12001112*

EXERCISE 2.5

4 -2 **5** $4a - b$ **6** $7a - 4b - c$ **7** $3a + 4b + 2c$ **10** x^4 **11** $-x^7$ **12** yes **13** no **14** $30x^6y^7$ **15** $6x^3y^3 - 15x^4y^5$ **16** $2a^3 - a^2 - 4a - 15$ **17** $-16x^2 + 12x$ **18** x^3y^2 **19** $5abc^4 - 3a^4b^2c$ **20** $2x^2 - x - 5, -6$ **21** $3x^2 - xy + 5y^2, -7y^3$ **22** $3x^2 - 2x + 1, 0$ **23** *100010011* **24** *1010100* **25** *11110* **26** *10112* **27** *10010* **28** *1022*

EXERCISE 2.6

2 0 **3** $30a^5b^8$ **4** $6x^3y^4 - 4x^4y^7$ **5** $6a^3 - 5a^2 - 3a + 2$ **6** $2x^3 + 9x^2 + 2x$ **7** $3x^2 - 4x + 4, -18$ **8** $3xw^3 - 2x^2y^2w$ **9** *10001011* **10** *100011* **11** *12121*

EXERCISE 3.1

1 $x^2 + 5x + 6$ **2** $x^2 + 6x + 5$ **3** $x^2 + x - 6$ **5** $x^2 - 4x - 21$ **6** $x^2 + 7x + 10$ **7** $x^2 - 7x + 12$ **9** $6x^2 + 5xy + y^2$ **10** $20x^2 + 9xy + y^2$ **11** $x^2 + 7xy + 12y^2$ **13** $x^2 - xy - 6y^2$ **14** $x^2 + 2xy - 8y^2$ **15** $10x^2 - 7xy + y^2$ **17** $6x^2 + 17xy + 12y^2$ **18** $4x^2 + 16xy + 15y^2$ **19** $12x^2 + 25xy + 12y^2$ **21** $12x^2 - 25xy + 12y^2$ **22** $20x^2 - 23xy + 6y^2$ **23** $21x^2 - 71xy + 40y^2$ **25** $8x^2 + 2xy - 15y^2$ **26** $15x^2 - 14xy - 8y^2$ **27** $42x^2 - 11xy - 20y^2$ **29** $a^2 + 6a + 9$ **30** $a^2 + 8a + 16$ **31** $4a^2 + 4a + 1$ **33** $b^2 - 4b + 4$ **34** $b^2 - 10b + 25$ **35** $4b^2 - 4b + 1$ **37** $4d^2 + 12d + 9$ **38** $9d^2 + 24d + 16$ **39** $16d^2 - 40d + 25$ **41** $4x^2 + 12xy + 9y^2$ **42** $25x^2 + 40xy + 16y^2$ **43** $16x^2 + 56xy + 49y^2$ **45** $9x^2 - 12xy + 4y^2$ **46** $16x^2 - 40xy + 25y^2$ **47** $49x^2 - 42xy + 9y^2$

EXERCISE 3.2

1 $x^2 - 9$ **2** $x^2 - 16$ **3** $x^2 - 49$ **5** $4x^2 - 1$ **6** $9y^2 - 1$ **7** $25b^2 - 1$ **9** $4x^2 - 9y^2$ **10** $25x^2 - 49y^2$ **11** $36x^2 - 25y^2$ **13** 91 **14** 396 **15** 884 **17** 1431 **18** 2379 **19** 3551 **21** $x^2 + 4y^2 + 9z^2 + 4xy + 6xz + 12yz$ **22** $4x^2 + y^2 + 25z^2 - 4xy - 20xz + 10yz$ **23** $9x^2 + 4y^2 + 16z^2 - 12xy - 24xz + 16yz$ **25** $a^2 + b^2 + c^2 + d^2 + 2ab + 2ac - 2ad + 2bc - 2bd - 2cd$ **26** $4a^2 + b^2 + 9c^2 + 16d^2 - 4ab - 12ac + 16ad + 6bc - 8bd - 24cd$ **27** $4a^2 + 9b^2 + 25c^2 + d^2 + 12ab + 20ac - 4ad + 30bc - 6bd - 10cd$ **29** $16p^2 + 24pq + 9q^2 - 4r^2$ **30** $4p^2 - 4pq + q^2 - 9r^2$ **31** $9p^2 - 25q^2 - 20qr - 4r^2$ **33** $a^2 + 2ab + b^2 - c^2 - 2cd - d^2$ **34** $p^2 + 2pr + r^2 - q^2 - 2qs - s^2$ **35** $4p^2 - 12pq + 9q^2 - 16r^2 + 40rs - 25s^2$ **37** $a^2 + 2ab + b^2 - c^2 - d^2 - f^2 + 2cd + 2cf - 2df$ **38** $4p^2 + 9q^2 + r^2 + 12pq - 4pr - 6qr - 9s^2 + 12st - 4t^2$ **39** $9b^2 + 4a^2 + 4d^2 + 12ba - 12bd - 8ad - c^2 + 6cf - 9f^2$

EXERCISE 3.3

1 $3(x + 2)$ **2** $5(x - 3)$ **3** $2(4x + 1)$ **5** $5a(x - 2y)$ **6** $3b(x + 2y)$ **7** $3r(2t - 3s)$ **9** $3xy(y + 2x)$ **10** $5ab(a + 2b)$ **11** $3a^2b^2(3b - 4a)$ **13** $3(a + 2b - 3c)$ **14** $5a^{11}(a^2 - 2a + 3)$ **15** $2xy^5(xz - 4y + 6z^2)$ **17** $p^{23}q(2pr + 3qr + 5)$ **18** $2xy^2(2xz - 3y + 4z)$ **19** $3a^2b^7(3bc^2 - 2 - c^2)$ **21** $3xyz^6(2x + 3)(x + 1)$ **22** $3y^2z^8(3x + 1)(x - 2)$ **23** $2x^2y(3z - 2)(2z + 1)$ **25** $a^{11}(3a + 2b - 5c + 4d)$ **26** $3x^4(x - 2w + 3y - 4z)$ **27** $3a(a - 2x + 3y - 4z)$ **29** $xy^6z(3x - 7y^2 - 2z + 5)$ **30** $2abc^7(ac - 3ab - 2bc + 4)$ **31** $3a^{12}b^2c^7(2a - 3b - 5c + 4)$ **33** $(x + 3)(y + z)$ **34** $(y + 1)(x - 2)$ **35** $(y + w)(x - 3s)$ **37** $(a + b)(c + a - b)$ **38** $(m - n)(p + m - 2n)$ **39** $(x - y)(2 - y)$ **41** $(a + d)(b + c)$ **42** $(x - y)(y + z)$ **43** $(a - b)(c - d)$ **45** $(2a - 3c)(a - b)$ **46** $(2x - 3w)(3x - 2y)$ **47** $(5x + 3y)(w - 3z)$

EXERCISE 3.4

1 $(x + y)(x - y)$ **2** $(a + b)(a - b)$ **3** $(w + 3)(w - 3)$ **5** $(x + 4y)(x - 4y)$ **6** $(b + 2a)(b - 2a)$ **7** $(3a + b)(3a - b)$ **9** $(a + 6b)(a - 6b)$ **10** $(q + 7r)(q - 7r)$ **11** $(r + 3s)(r - 3s)$ **13** $(2x + 5y)(2x - 5y)$ **14** $(3y + 4t)(3y - 4t)$ **15** $(4t + 7x)(4t - 7x)$ **17** $(x + y + w)(x + y - w)$ **18** $(a + b + c)(a - b - c)$ **19** $(t + 3x - a)(t - 3x + a)$ **21** $(3x + y)(3y - x)$ **22** $(a - 2b)(5a)$ **23** $(7p + 4q)(3p - 10q)$ **25** $(x^2 + y)(x^2 - y)$ **26** $(2x^2 + 3y)(2x^2 - 3y)$ **27** $(6x^2 + 5y^3)(6x^2 - 5y^3)$ **29** $(x^2 + y^2)(x + y)(x - y)$ **30** $(x^2 + 4y^2)(x + 2y)(x - 2y)$ **31** $(9a^2 + 16b^2)(3a + 4b)(3a - 4b)$ **33** $(a + c)(a^2 - ac + c^2)$ **34** $(b + d)(b^2 - bd + d^2)$ **35** $(x - w)(x^2 + xw + w^2)$ **37** $(x - 2y)(x^2 + 2xy + 4y^2)$ **38** $(x - 3y)(x^2 + 3xy + 9y^2)$ **39** $(5a + b)(25a^2 - 5ab + b^2)$ **41** $(a + b - c)(a^2 - ab + ac + b^2 - 2bc + c^2)$ **42** $(w + y + x)(w^2 + 2wy + y^2 - wx - yx + x^2)$ **43** $(a - b + 2c)(a^2 + ab - 2ac + b^2 - 4bc + 4c^2)$ **45** $(a - b)(a^2 + ab + b^2)(a^6 + a^3b^3 + b^6)$ **46** $(a - b^2)(a + b^2)(a^4 + a^2b^4 + b^8)$ **47** $(3x^4 + 2y)(9x^8 - 6x^4y + 4y^2)$ **49** $(x - y)(1 - x - y)$ **50** $(x + y)(1 + x^2 - xy + y^2)$ **51** $(a + b)(a^2 - ab + b^2 + a - b)$ **53** $(x + 2)(x - 1)$ **54** $(2x - 3y)(1 + 2x + 3y)$ **55** $(3w - 4y)(1 - 3w - 4y)$ **57** $(x + 3y)(x^2 - 3xy + 9y^2 - 2)$ **58** $(a + 2b)(2a^2 - 4ab + 8b^2 + 5)$ **59** $(a - 5b)(2a - ab - 5b^2)$

EXERCISE 3.5

1 $-,-$ **2** $-,-$ **3** $-,-$ **5** $+,+$ **6** $+,+$ **7** $+,+$ **9** $-,+$ **10** $-,+$ **11** $-,+$ **13** $4y, 2y$ **14** $3y, 4y$ **15** $5y, 4y$ **17** $(a + 3)(a + 2)$ **18** $(a + 1)(a + 4)$ **19** $(a + 3)(a + 5)$ **21** $(b - 2)(b - 5)$ **22** $(b - 2)(b - 4)$ **23** $(b - 2)(b - 7)$ **25** $(c + 4)(c - 3)$ **26** $(c - 5)(c + 2)$ **27** $(c - 3)(c + 7)$ **29** $(2x + 3y)(4x + 5y)$ **30** $(3x + 5y)(2x + 3y)$ **31** $(3x + 4y)(2x + 5y)$ **33** $(3x - 2y)(4x + 3y)$ **34** $(3x - 5y)(4x + 7y)$ **35** $(9x - 5y)(4x + 3y)$ **37** $(2a - 3b)(3a - 4b)$ **38** $(3a - 4b)(2a - 5b)$ **39** $(5a - b)(3a - 4b)$ **41** perfect square, $(2x + 5y)^2$ **42** perfect square, $(4x - 7y)^2$ **43** not a perfect square **45** not a perfect square **46** perfect square, $(3p - 4q)^2$ **47** perfect square, $(5p + 6q)^2$

EXERCISE 3.6

1 $x^2 + 11x + 28$ **2** $x^2 + 2x - 15$ **3** $x^2 - 8x + 12$ **4** $6x^2 + 5xy + y^2$ **5** $4x^2 + 13xy + 3y^2$ **6** $5x^2 + 9xy - 2y^2$ **7** $6x^2 + 11xy - 10y^2$ **8** $12x^2 + 11xy - 15y^2$ **9** $x^2 + 6x + 9$ **10** $4x^2 + 4x + 1$ **11** $9x^2 + 30xy + 25y^2$ **12** $x^2 - 49$ **13** $9x^2 - 1$ **14** $4x^2 - 81y^2$ **15** $a^2 + 4b^2 + 9c^2 - 4ab + 6ac - 12bc$ **16** $9a^2 + 4b^2 + c^2 + 16d^2 + 12ab - 6ac + 24ad - 4bc + 16bd - 8cd$ **17** $4a^2 - b^2 + 4bc - 4c^2$

18 $p^2 - 4pd + 4d^2 - 9q^2 - 24qr - 16r^2$ **19** $p^2 + 4q^2 + r^2 + 4pq - 2pr - 4qr - 9s^2 + 6st - t^2$ **20** $2(a - 4b)$ **21** $3xy(1 + 2y)$ **22** $5(x - 2y + 4w)$ **23** $3b(a^2c - 4ab + 6c^2)$ **24** $3bc(5a - 2)(a + 1)$ **25** $wx(3x + 2y)(2x + 3y)$ **26** $abc(5 - 8b - 2c + 7a)$ **27** $6x^2yw^2(y - 3x - 2yw + 5xy)$ **28** $(x + 2)(a - b)$ **29** $(x + y)(2x + y - w)$ **30** $(a - b)(b + c)$ **31** $(2r - 5s)(3x + 2y)$ **32** $(a + 5)(a - 5)$ **33** $(3x + 4y)(3x - 4y)$ **34** $(x - y)(x - 3y)$ **35** $(5a - 9b)(a + b)$ **36** $(2x^2 + 3y^2)(2x^2 - 3y^2)$ **37** $(4a^2 + 9b^2)(2a + 3b)(2a - 3b)$ **38** $(a + 3b)(a^2 - 3ab + 9b^2)$ **39** $(x + 2y - w)(x^2 + 4xy + 4y^2 + wx + 2wy + w^2)$ **40** $(2a^3 - 3b^4)(4a^6 + 6a^3b^4 + 9b^8)$ **41** $(4b^2 - 5c^5)(16b^4 + 20b^2c^5 + 25c^{10})$ **42** $(x + y)(x - y + x^2 - xy + y^2)$ **43** $(2a - 5b)(3a^2 + 2a + 5b)$ **44** $(2x + 5y)(3x - 2y)$ **45** $(3x + 7y)(2x - 3y)$ **46** perfect square, $(2x + 3y)^2$ **47** not a perfect square   **48** not a perfect square   **49** perfect square, $(3x - 4y)^2$

EXERCISE 3.7

1 $6x^2 + xy - 15y^2$ **2** $25x^2 - 40x + 16$ **3** $a^2 + 9b^2 + 4c^2 + 25d^2 - 6ab + 4ac + 10ad - 12bc - 30bd + 20cd$ **4** $9p^2 + 6ps + s^2 - 4q^2 + 12qr - 9r^2$ **5** $3a(2x - 5y)$ **6** $x(3x + 2y)(x + y)$ **7** $(3a - 4b^2)(3a + 4b^2)$ **8** $(2x + 5y)(4x^2 - 10xy + 25y^2)$ **9** $(3x - 7)(x + 1)$ **10** $(9a^2 + 4b^2)(3a + 2b)(3a - 2b)$ **11** $(2x - 3y)(4x^2 + 2x + 6xy + 3y + 9y^2)$ **12** $(2x + 5y)(3x - 7y)$ **13** $x(3x - 5y)(x + 3y)$

EXERCISE 4.1

1 $3a^2b^2$ **2** $3y^3$ **3** s **5** $-a$ **6** $y - x$ **7** $r - p$ **9** $(x - 2y)(x - y)$ **10** $(x + 3y)(x + y)$ **11** $x - y$ **13** $\dfrac{x}{y}$ **14** $\dfrac{ac}{b}$ **15** $\dfrac{y^2}{xw^2}$ **17** $\dfrac{3}{x + y}$ **18** x^2 **19** $\dfrac{3xy}{1 + 2y}$ **21** $\dfrac{x + 1}{2x + 1}$ **22** $\dfrac{x - 2y}{2x + y}$ **23** $\dfrac{x - 2y}{2x - y}$ **25** $\dfrac{2}{7}$ **26** $\dfrac{8}{17}$ **27** $\dfrac{2}{3}$ **29** $\dfrac{1}{a^3b}$ **30** $\dfrac{y^2}{x^3}$ **31** $\dfrac{x^3}{2y^3}$ **33** $\dfrac{q^3s}{pr}$ **34** $\dfrac{a^3b^2c}{d}$ **35** $\dfrac{v}{wy^3x}$ **37** $\dfrac{3c^3}{2ab^4d}$ **38** $\dfrac{z^5}{3y^3w^3}$ **39** $\dfrac{7p^2q^2}{6d^3r^4}$ **41** $\dfrac{1}{2y}$ **42** $\dfrac{x}{2y}$ **43** $\dfrac{-10x}{y}$ **45** $\dfrac{3y(x + y)}{2x^2}$ **46** $\dfrac{(a - b)b}{a^2}$ **47** $\dfrac{3y^2}{8x(x - 2y)^2}$ **49** $\dfrac{2x^3}{3y^2}$ **50** $\dfrac{pq}{6}$ **51** $\dfrac{2a^2}{b}$ **53** vt **54** $\dfrac{4b(a + b)}{a(a + 2b)}$ **55** xy^2 **57** $\dfrac{x + y}{x + 3y}$ **58** $\dfrac{x + 2y}{x - 2y}$ **59** $\dfrac{3p - 7q}{4p + 3q}$ **61** $\dfrac{a + 2b}{2a + 3b}$ **62** $\dfrac{a - 3b}{(a - b)(2a + b)}$ **63** $\dfrac{x^2(x + 4y)}{3y(x - 3y)}$ **65** $\dfrac{6}{5}$ **66** $\dfrac{1}{2}$ **67** $\dfrac{2}{3}$ **69** $\dfrac{8y}{5b}$ **70** $\dfrac{2x}{3y}$ **71** $x(x - y)$ **73** $\dfrac{xy}{(x - y)(x + y)^2}$ **74** $\dfrac{(a - b)^2}{ab}$ **75** $\dfrac{x + y}{x - y}$ **77** $\dfrac{1}{a + 3b}$ **78** $\dfrac{a}{a^2 + ab + b^2}$ **79** $\dfrac{x - 3}{x + 2}$

EXERCISE 4.2

1 $\frac{15}{30}, \frac{20}{30}, \frac{24}{30}$ **2** $\frac{77}{154}, \frac{66}{154}, \frac{112}{154}$ **3** $\frac{6}{12}, \frac{9}{12}, \frac{10}{12}$ **5** $\frac{b}{ab}, \frac{a}{ab}, \frac{1}{ab}$ **6** $\frac{2ab}{a^2b}, \frac{3b}{a^2b}, \frac{a}{a^2b}$ **7** $\frac{u^3}{u^2v^2}, \frac{2v}{u^2v^2}, \frac{uv^3}{u^2v^2}$

9 $\frac{a(a+b)}{a^2-b^2}, \frac{(a-b)^2}{a^2-b^2}, \frac{(a+b)(2b-a)}{a^2-b^2}$ **10** $\frac{2(x-y)(2x-y)}{2(x+y)(2x-y)}, \frac{2(x+y)(x+2y)}{2(x+y)(2x-y)}, \frac{(x+3y)(x+y)}{2(x+y)(2x-y)}$

11 $\frac{3(x+2y)(x-2y)}{3(x-y)(x+2y)}, \frac{3(x+y)(x-y)}{3(x-y)(x+2y)}, \frac{(x-y)(x+2y)}{3(x-y)(x+2y)}$ **13** $\frac{(a-3b)(a+3b)}{(a+b)(a-2b)(a+3b)},$

$\frac{(a+2b)(a-2b)}{(a+b)(a-2b)(a+3b)}, \frac{(a-b)(a+b)}{(a+b)(a-2b)(a+3b)}$ **14** $\frac{(a-2b)(a+2b)}{(a-b)(a+3b)(a+2b)}, \frac{(a+b)(a-b)}{(a+3b)(a+2b)(a-b)},$

$\frac{(a-3b)(a+3b)}{(a-b)(a+2b)(a+3b)}$ **15** $\frac{(a-3b)^2}{(a-4b)(a+4b)(a-3b)}, \frac{(a+4b)^2}{(a-4b)(a+4b)(a-3b)}, \frac{(a-4b)^2}{(a-4b)(a+4b)(a-3b)}$

17 $\frac{3}{8}$ **18** $\frac{7}{9}$ **19** $\frac{37}{36}$ **21** $\frac{b^2c - ac^2 + a^2b}{abc}$ **22** $\frac{6y^2w - 2xw^2 + 3x^2y}{6xyw}$ **23** $\frac{15t^2 - 6y^2 + 10x^2}{30txy}$

25 $\frac{y^2}{x(x+y)}$ **26** $\frac{-2x^2 + 9y^2}{3(x+2y)}$ **27** $\frac{-y(x+2y)}{x^2-y^2}$ **29** $\frac{19}{30}$ **30** $\frac{5}{3}$ **31** $\frac{1}{4}$ **33** $\frac{a^2 + 2b^2 + c^2}{abc}$

34 $\frac{12p^2 - 9q^2 + 2r^2}{6pqr}$ **35** $\frac{15t^2 + 20s^2 + 18r^2}{30rst}$ **37** $\frac{a^2-1}{a^2}$ **38** $\frac{3y-2x}{6y}$ **39** $\frac{x^2+y^2}{x^2y^2}$ **41** $\frac{a^2-b^2}{a(3a+b)}$

42 $\frac{2(x^2+y^2)}{x(x+2y)}$ **43** $\frac{2(x^2+5y^2)}{x(x-2y)}$ **45** $\frac{6xy}{(2x-y)(2x+y)}$ **46** $\frac{-22xy}{(x+3y)(x-3y)}$ **47** $\frac{y}{y-2x}$

49 $\frac{x^3+3y^3}{xy(x-3y)}$ **50** $\frac{9x^3-4y^3}{xy(3x-2y)}$ **51** $\frac{9x^3-8y^3}{2xy(2y-3x)}$ **53** $\frac{3xy}{(x+3y)(x-3y)}$ **54** $\frac{-2x^2+16y^2}{(x-4y)(x+4y)}$

55 $\frac{y}{y-x}$ **57** $\frac{x+2y}{x-3y}$ **58** $\frac{1}{x+4}$ **59** $\frac{x-2}{x+4}$ **61** $\frac{-2(4x-5)}{(x+1)(x-1)(2x-1)}$

62 $\frac{6x^2}{(x-3)(x+3)(2x-3)}$ **63** $\frac{9y^2}{(x+y)(2x-y)(x-2y)}$ **65** $\frac{2(r^2-6s^2)}{(r+s)(r-s)(r+2s)}$ **66** $\frac{r+3s}{(r+s)(r-s)}$

67 $\frac{a+b}{a-b}$

EXERCISE 4.3

1 4 **2** $\frac{18}{7}$ **3** $\frac{7}{5}$ **5** $\frac{15}{16}$ **6** $\frac{7}{18}$ **7** 21 **9** a **10** $c+3$ **11** $\frac{a}{2a-1}$ **13** $\frac{9-x^2}{5}$

14 $\frac{x}{3}$ **15** $\frac{xy}{6}$ **17** $\frac{x-y}{x+y}$ **18** $\frac{y-x}{y+2x}$ **19** $\frac{(x+y)(x-y+5)}{(x-y)(x+y+3)}$ **21** $\frac{x+5}{x+2}$ **22** $\frac{-1}{x+1}$

23 $\frac{2}{x-1}$ **25** $\frac{-1}{x+1}$ **26** $\frac{-2(x+1)}{x+3}$ **27** $\frac{-1}{2x+3}$

EXERCISE 4.4

1 $2a^2x$ **2** x^2 **3** $b-a$ **4** $(a+b)^2$ **5** $\dfrac{3}{2xy^2}$ **6** $x(x+1)$ **7** $\dfrac{x-3y}{x-4y}$ **8** $\dfrac{4x}{y}$

9 $\dfrac{(x-y)(x+2y)}{(2x+y)(3x-y)}$ **10** $\dfrac{x+2y}{x}$ **11** $\dfrac{x+y}{x+2y}$ **12** $\dfrac{(x+2y)(x+3y)(2x+y)}{(x-y)(x+3y)(2x+y)}, \dfrac{(2x-y)(x-y)(2x+y)}{(x-y)(x+3y)(2x+y)},$

$\dfrac{(4x+y)(x-y)(x+3y)}{(x-y)(x+3y)(2x+y)}$ **13** $\dfrac{(a+2b)(a-2b)}{(a-3b)(a+3b)(a-2b)}, \dfrac{(a+3b)^2}{(a-2b)(a-3b)(a+3b)}, \dfrac{(a-3b)^2}{(a-2b)(a-3b)(a+3b)}$

14 $\dfrac{3}{x+3}$ **15** $\dfrac{-5x^3 - 18x^2y - 18xy^2 + 13y^3}{(2x+5y)(x+3y)(x-2y)}$ **16** $\dfrac{x^2 - 9xy - 3y^2}{(2x-y)(x+2y)(x-2y)}$ **17** $\tfrac{7}{8}$ **18** $\dfrac{x}{3}$

19 $\dfrac{3x-2}{x-1}$ **20** $\dfrac{-1}{2x+1}$ **21** $\tfrac{67}{29}$

EXERCISE 4.5

1 $\dfrac{3x^2y^2}{2z^2}$ **2** $\dfrac{x+3y}{3x+y}$ **3** $\dfrac{x+y}{2x+y}$ **4** 1 **5** $\dfrac{x-1}{x+6}$ **6** $\dfrac{y^2(x-2y)}{2x(2x+y)}$ **7** $\dfrac{5x^2 - 16xy - 15y^2}{3y(x-2y)}$

8 $\dfrac{6xy}{(2x-y)(2x+y)}$ **9** $\dfrac{2}{a-2}$ **10** $\dfrac{-1}{2a+3}$

EXERCISE 5.1

1 11.55 s, 50.82 s **2** $1288.86 **3** 125, 135, 145 **5** $403.23, $423.73 **6** 27.03, 30.03 **7** 27, 28
21 169π in² **22** 205.2 in² **23** 64 ft² **25** 784π in² **26** 980 in² **27** 288π yd³ **29** 288 ft³
30 6 ft² **31** 42π in² **33** 216 cm³ **34** 288 in³ **35** 135 **37** $\tfrac{140}{9}$ **38** .87 **39** $4\pi/3$
41 26.34π **42** 297π **43** 5.525

EXERCISE 5.2

1 Yes, there is exactly one second number for each first number. **2** No, there is more than one second number for some first number. **3** No, there is more than one second number for some first number.

5 $\{(1,1), (3,3), (5,5), (7,7)\}$, yes **6** $\{(f,b), (0,a), (0,l), (t,l)\}$, no **7** $\{(b,p), (a,a), (l,r), (l,k)\}$, no **9** yes
10 yes **11** yes **13** $-2, 1, 10$ **14** $12, -3, -8$ **15** $2, 3, 7$ **17** $-5, -3, 1, 5$ **18** $-7, -1,$
$2, 8$ **19** $4, 3, 1, 5$ **21** $\{(-1,-7), (0,-3), (1,1), (2,5)\}$, yes **22** $\{(-2,7), (-1,4), (0,1), (1,-2)\}$, yes
23 $\{(1,5), (10,4), (17,3), (22,2)\}$, yes **25** $\{(-2,-1), (0,3), (2,7), (3,9)\}$, yes **26** $\{(-2,9), (-1,7), (1,3),$
$(4,-3)\}$, yes **27** $\{(-3,13), (-1,3), (1,1), (3,7)\}$, yes **29** 2 **30** 3 **31** $2x + h + 2$ **33** $\{(-2,-3)\}$
34 $\{(-1,1)\}$ **35** $\{(-1,2), (-2,7)\}$

EXERCISE 5.4

1 3249 in^3 **2** $1.56 \times 10^3 \text{ cm}^3$ **5** Yes, there is exactly one second number for each first number.   **6** Yes, there is exactly one second number for each first number.   **7** No, there is more than one second number for some first number.   **8** yes   **9** no   **10** $-7, -3, 5$ **11** $4, 3, 2$ **12** $\{-8, -2, 7\}$ **13** $\{(-1,-5), (0,-1), (1,3), (3,11)\}$ **14** $\{(-3,6), (1,6), (4,27), (5,38)\}$ **15** 3   **16** $\{(1,0), (3,10)\}, \{(-1,6), (0,1), (1,0), (2,3), (3,10), (-1,-2), (0,-2), (2,4)\}$

EXERCISE 5.5

1 204   **3** Yes, there is exactly one second number for each first.   **4** No, there is more than one second number for some first number.   **5** Yes, there is exactly one second number for each first.   **6** Yes, there is exactly one second number for each first.   **7** $-11, -1, 4$ **8** $\{(-2,-2), (0,4), (1,7), (3,13)\}$.

EXERCISE 6.1

1 $12a^7$ **2** $6a^7$ **3** $4a^3$ **5** $8a^6b^9$ **6** $9a^8b^4$ **7** $64a^6$ **9** 2048   **10** 4096   **11** 4096   **13** 177,147   **14** 19,683   **15** 177,147   **17** 8   **18** 32   **19** 27   **21** 243   **22** 125   **23** 216   **25** 3125   **26** 49   **27** 16   **29** 16   **30** 19,683   **31** 2401   **33** 5184   **34** 65,536   **35** 1   **37** $\frac{9}{8}$ **38** $\frac{1}{15,625}$ **39** $\frac{2401}{256}$ **41** $\frac{8a^6b^9}{c^3}$ **42** $\frac{9a^2b^8}{c^4}$ **43** $\frac{a^{12}b^4}{c^{12}}$ **45** $6a^5b^3$ **46** $10a^5b^2$ **47** $15a^5b^4$ **49** $5bc^2$ **50** $\frac{3c^4}{d^2}$ **51** $\frac{3d^4}{b}$ **53** $2a^5b^4$ **54** $3a^5x^7y^3$ **55** $a^5b^5c^6$ **57** $\frac{4w}{27x}$ **58** $\frac{81w^{12}}{32x^{10}y}$ **59** $\frac{2y^6}{9w^{10}x^2}$ **61** w^3y **62** $\frac{27x^3y^4}{w^5}$ **63** $\frac{4a}{3b^4}$ **65** x^2y **66** $\frac{y^{11}}{w^5}$ **67** $\frac{x^6y}{w^7}$ **69** $\frac{81x^{20}}{16y^4}$ **70** $\frac{8}{27x^3y^{12}}$ **71** $\frac{64x^4}{9}$

EXERCISE 6.2

1 $\frac{1}{a^2}$ **2** $\frac{3}{a^4}$ **3** $\frac{1}{4a^2}$ **5** $\frac{a+b}{ab}$ **6** $\frac{9}{a^2b^4}$ **7** $\frac{-1}{3a^2b}$ **9** $\frac{1}{8}$ **10** $\frac{1}{4}$ **11** $\frac{1}{25}$ **13** $\frac{1}{216}$ **14** $\frac{1}{125}$ **15** 3   **17** 2   **18** 729   **19** $\frac{1}{49}$ **21** 4   **22** $\frac{81}{16}$ **23** $\frac{3}{4}$ **25** $\frac{2}{3}$ **26** $\frac{25}{4}$ **27** 324   **29** a^3b^{-2} **30** $4a^2b^{-4}$ **31** a^{-1} **33** $5a^7b^3c^2d^{-1}$ **34** $x^6y^5w^2a^3b^{-2}$ **35** $4a^5b^2c^4fg^{-1}$ **37** $\frac{2}{a^3b}$ **38** $\frac{3n^2}{m}$ **39** $\frac{b^3}{5a^2}$ **41** a^2 **42** b^4 **43** $\frac{1}{c^2}$ **45** $\frac{a^2}{c^2}$ **46** $\frac{a^2b^2}{c}$ **47** $\frac{b^3}{a^3c^5}$ **49** $\frac{9xy^2}{2w^5}$ **50** $\frac{6a^2y^3}{b^3}$

51 $\dfrac{9r^5s^4t^8}{8}$ 53 $\dfrac{a^2}{b^8}$ 54 $\dfrac{z^{24}}{w^6}$ 55 $\dfrac{x^{21}y^{21}}{27z^{15}}$ 57 $\dfrac{b^4}{81a^{20}}$ 58 $\dfrac{1}{2ab^3}$ 59 $\dfrac{9b^6}{a^4}$ 61 $\dfrac{3b-4a}{a^2b^2}$

62 $\dfrac{5x+y}{-x^2y^2}$ 63 $\dfrac{x-y}{xy}$

EXERCISE 6.3

1 $6a^2b^8$ 2 $3a^2b^3$ 3 $2x^{3/2}y^2$ 5 $\dfrac{2b^{1/2}}{a^{1/2}}$ 6 $\dfrac{2a^{2/3}}{b^{1/3}}$ 7 $\sqrt{a^2-b^2}$ 9 4 10 2 11 0.5

13 4 14 27 15 4 17 8 18 4 19 0.1 21 $2a^3$ 22 $2a^2$ 23 $3a^2$ 25 $5xy^2$

26 $3x^2y^3$ 27 $3y^2$ 29 $\dfrac{2x}{3y^2}$ 30 $\dfrac{3x^2}{2y^3}$ 31 $\dfrac{6x^2}{7y^5}$ 33 $\sqrt[3]{x^2}$ 34 $\sqrt[5]{y^4}$ 35 $\sqrt[5]{a^2}$ 37 $\dfrac{1}{\sqrt[3]{a^2}}$

38 $\dfrac{1}{\sqrt[7]{b^3}}$ 39 $\dfrac{1}{\sqrt{c}}$ 41 $\sqrt[5]{x^3}$ 42 $\sqrt[3]{x}$ 43 $\sqrt[7]{y^3}$ 45 $4\sqrt[3]{ab^2}$ 46 $27\sqrt[4]{a^2b}$ 47 $3\sqrt{x^3y}$

49 $\sqrt[12]{x^8y^3}$ 50 $\sqrt[6]{x^2y^3}$ 51 $\sqrt[18]{x^6y^{15}w^4}$ 53 $12y$ 54 $\dfrac{10a^{1/3}}{b^{2/5}}$ 55 $12x^{1/7}y$ 57 $\dfrac{2a^{2/5}}{b^{1/3}c^2}$ 58 $\dfrac{6xy}{5w}$

59 $\dfrac{x^{3/2}w^2}{6y}$ 61 $\dfrac{3x^2}{4y^{1/5}}$ 62 $\dfrac{7}{9y^2w^{2/7}}$ 63 $\dfrac{1}{x^3y^2}$

EXERCISE 6.4

1 6 2 11 3 9 5 3 6 2 7 5 9 $2\sqrt{7}$ 10 $3\sqrt{5}$ 11 $2\sqrt[3]{3}$ 13 $3\sqrt[3]{2}$ 14 $9\sqrt{2}$

15 $3\sqrt[4]{3}$ 17 9 18 32 19 12 21 6 22 8 23 4 25 30 26 60 27 12 29 $\tfrac{3}{2}$

30 $\tfrac{4}{3}$ 31 $\tfrac{3}{2}$ 33 $\dfrac{\sqrt{21}}{7}$ 34 $\dfrac{\sqrt{77}}{11}$ 35 $\dfrac{\sqrt{10}}{8}$ 37 $\dfrac{\sqrt[3]{6}}{3}$ 38 $\dfrac{\sqrt[3]{10}}{2}$ 39 $\dfrac{\sqrt[4]{306}}{6}$ 41 $3ab^2$

42 $4a^3b$ 43 $2a^3b^2$ 45 $2a^2b\sqrt{2b}$ 46 $5a^3b^2\sqrt{3}$ 47 $3xy\sqrt[3]{2x}$ 49 $x^3yw\sqrt[3]{w}$ 50 $x^2y^2w^2\sqrt[3]{y^2w}$

51 $xy^2w\sqrt[6]{xy^2w^3}$ 53 $6x^3y^2$ 54 $6x^5y$ 55 $15x^2y^4\sqrt{y}$ 57 $4x^3y^2w^2\sqrt[3]{xy^2}$ 58 $2x^2yw^2\sqrt[5]{y^3}$

59 $3x^2y^2w^2\sqrt[4]{yw^2}$ 61 $5a^2b^2$ 62 $\dfrac{b^2}{3ac}$ 63 $\dfrac{5a^2}{3b^2c}$ 65 $\dfrac{3b^3c\sqrt{a}}{4a}$ 66 $\dfrac{7aw\sqrt{abw}}{6b}$ 67 $\dfrac{2b\sqrt{bc}}{3a^2}$

69 $\dfrac{a\sqrt[4]{24ab^2c^2}}{3bc}$ 70 $\dfrac{2ab\sqrt[5]{2}}{3c}$ 71 $\dfrac{3bc\sqrt[3]{b^2a}}{4a}$

EXERCISE 6.5

1 $\sqrt{3}$ 2 $\sqrt{3}$ 3 $\sqrt{5}$ 5 $\sqrt[10]{a^4}$ 6 $\sqrt[8]{a^6}$ 7 $\sqrt[9]{a^3}$ 9 $\sqrt[6]{a^3},\sqrt[6]{a^2}$ 10 $\sqrt[10]{a^5},\sqrt[10]{a^2}$

11 $\sqrt[6]{a^2},\sqrt[6]{a}$ 13 $\sqrt[30]{5^{15}a^{15}b^{15}},\sqrt[30]{2^{10}a^{10}b^{20}},\sqrt[30]{3^6a^{18}b^{12}}$ 14 $\sqrt[12]{3^6a^6b^6},\sqrt[12]{3^4a^8b^4},\sqrt[12]{3^3a^9b^3}$

15 $\sqrt[60]{a^{40}b^{20}},\sqrt[60]{a^{15}b^{30}},\sqrt[60]{a^{48}b^{36}}$ 17 $6\sqrt{3}+3\sqrt{2}+\sqrt{30}$ 18 $4\sqrt{5}+15\sqrt{2}+\sqrt{30}$ 19 $32+11\sqrt{2}$

21 $\sqrt{3} - 1$ **22** $\sqrt{5} + 1$ **23** $\dfrac{\sqrt{7} - 5}{6}$ **25** $4 - \sqrt{15}$ **26** $\dfrac{8 + 5\sqrt{14}}{-11}$ **27** $\dfrac{31 + 7\sqrt{35}}{13}$ **29** $\sqrt{2}$

30 0 **31** $6\sqrt{5}$ **33** $2\sqrt{2} - \sqrt{3}$ **34** $12\sqrt{2} + 4\sqrt{5}$ **35** $5\sqrt{6} + 2\sqrt{3}$ **37** $\dfrac{11(9\sqrt{2} - 32\sqrt{3})}{24}$

38 $\dfrac{7\sqrt{6} - 7\sqrt{3}}{3}$ **39** $\dfrac{9\sqrt{10} + 5\sqrt{5}}{-5}$ **41** $2 + 2\sqrt[3]{2} + 2\sqrt{2}$ **42** $15\sqrt{3} + 9\sqrt[3]{3}$ **43** $10\sqrt[3]{2} - \sqrt{2}$

45 $\sqrt{2a}$ **46** $\sqrt{2a}$ **47** $\sqrt[3]{3ab^2}$ **49** $\sqrt[4]{4a^2b^2}$ **50** $\sqrt[9]{64a^6b^3}$ **51** $\sqrt[6]{25a^2b^4}$ **53** $\dfrac{2a + 5\sqrt{ab} + 2b}{4a - b}$

54 $\dfrac{3a + 5\sqrt{ab} - 2b}{a - 4b}$ **55** $\dfrac{3a + 2\sqrt{ab} - 8b}{9a - 16b}$ **57** $\dfrac{3a + 2\sqrt{a} - 4\sqrt{b}}{a - 4}$

58 $\dfrac{-2a + 12\sqrt{a} - 3\sqrt{ab} - 3\sqrt{b}}{9 - a}$ **59** $\dfrac{4(2\sqrt{a} + \sqrt{b})}{4 - a}$ **61** $\dfrac{2(a + b)}{a - b}$ **62** $\dfrac{3(a + b)(2a - 5\sqrt{ab} + 2b)}{(a - 4b)(4a - b)}$

63 $\dfrac{-10a^2 + \sqrt{ab}(46b - 21a) + 14ab + 21b^2}{(4a - 9b)(a - b)}$

EXERCISE 6.6

1 $8 + 6i$ **2** $10 - 4i$ **3** $6 + 6i$ **5** $5 + 5i$ **6** $9 - 5i$ **7** $2 + 10i$ **9** $3 + 4i$ **10** $-1 + 47i$

11 $31 + 29i$ **13** $-29 - 29i$ **14** $-61i$ **15** $1 + 31i$ **17** $\dfrac{26 + 7i}{25}$ **18** $\dfrac{60 + 11i}{61}$ **19** $-i$

21 $\dfrac{1 + 23i}{10}$ **22** $\dfrac{-24 - 7i}{25}$ **23** $\dfrac{-33 + 4i}{13}$ **25** $\dfrac{-34 - 12i}{325}$ **26** $2i$ **27** $\dfrac{546 - 128i}{425}$

29 $\dfrac{-713 + 1271i}{2210}$ **30** 0 **31** $\dfrac{-152 - 36i}{1525}$

EXERCISE 6.7

1 256 **2** 729 **3** 3125 **4** 2401 **5** $11{,}664$ **6** $6{,}561$ **7** $\tfrac{64}{81}$ **8** $46{,}656$ **9** $6a^5b^3c$

10 $\dfrac{9b^4}{4a^2}$ **11** $5a$ **12** $\dfrac{1}{2bc}$ **13** $\tfrac{1}{9}$ **14** $\tfrac{1}{125}$ **15** $\dfrac{4x^2y}{3w^3}$ **16** $\dfrac{a^2}{b^{10}}$ **17** $\dfrac{3y + 4x}{xy}$ **18** 4 **19** 2

20 0.2 **21** 0.3 **22** 8 **23** 25 **24** $7x^2y$ **25** $6xy^3$ **26** $\dfrac{2x^2}{3y^{10}}$ **27** $\dfrac{7x^2y^4}{4w}$ **28** $\dfrac{4xy^{1/2}}{3y}$

29 $\dfrac{32}{a^3y^4}$ **30** $4y^2$ **31** $6a^2b$ **32** $\sqrt[9]{27x^6}$ **33** $\sqrt{3}$ **34** $5\sqrt{5} + 2\sqrt{2}$ **35** $\dfrac{36 - 17\sqrt{6}}{73}$

36 $\dfrac{-1 + \sqrt{35}}{17}$ **37** $\dfrac{2(a^2 + 6b)}{a^2 - 9b}$ **38** $47 - i$ **39** $\dfrac{27 - 8i}{13}$ **40** 0 **41** 2

EXERCISE 6.8

1 6561 **2** 81 **3** $531{,}441$ **4** 81 **5** $3a^5b^4$ **6** $27a^{15}b^{12}$ **7** $\dfrac{a^5}{2b^3c^4}$ **8** $\dfrac{a^{12}}{b^{12}}$ **9** 9 **10** $\tfrac{1}{27}$

11 $\sqrt{9a^2b^3}$ **12** $5\sqrt{2} - 4\sqrt{5}$ **13** $\dfrac{4\sqrt{15} - 11}{17}$ **14** $\dfrac{-19 - 25i}{29}$ **15** $\dfrac{-6(1 + 2i)}{5}$

EXERCISE 7.1

1 {3,4} **2** {4,5,9} **3** {Arizona, New Mexico} **5** {Eisenhower} **6** {Neil Armstrong} **7** {white, red} **9** {1,5} **10** {1,5} **11** {5,−5} **13** identity **14** conditional equation **15** identity **17** conditional equation **18** conditional equation **19** identity **29** equivalent **30** equivalent **31** not **33** not **34** not **35** equivalent **37** not **38** not **39** equivalent

EXERCISE 7.2

1 4 **2** $\dfrac{w}{y}$ **3** 3 **5** 2 **6** −4 **7** −3 **9** 2.5 **10** 3 **11** −2 **13** 2 **14** 3 **15** −1 **17** 1 **18** −2 **19** −5 **21** $-\dfrac{1}{26}$ **22** $-\dfrac{9}{2}$ **23** $-\dfrac{13}{4}$ **25** 6 **26** −4 **27** 4 **29** ∅ **30** −6 **31** 8 **33** 4 **34** 3 **35** −2 **37** 3 **38** −5 **39** −1 **41** 3 **42** 4 **43** 2 **45** a^2 **46** $2ab$ **47** $\dfrac{b^2}{a}$

EXERCISE 7.3

1 $x > 2$ **2** $x > 3$ **3** $x \geq -1$ **5** $x < 3$ **6** $x \leq -4$ **7** $x < 1$ **9** $x \geq 2$ **10** $x > 4$ **11** $x > 3$ **13** $x < -3$ **14** $x < -3$ **15** $x \leq 4$ **17** $x < -3$ **18** $x \leq 1$ **19** $x < 1$ **21** $x \geq -1$ **22** $x > 3$ **23** $x > 3$ **25** $x > 10$ **26** $x < 21$ **27** $x \leq 6$ **29** $x > 6$ **30** $x > 9$ **31** $x < 36$

EXERCISE 7.4

1 41, 26 **2** 12 **3** 16, 24 **5** 60 cents **6** 44 years, 22 years **7** 78 days **9** 38 in first, 36 in second **10** 210 mi/h **11** 336 mi **13** 60 mi/h **14** 59 mi/h, 66 mi/h **15** 75 min **17** 50 mL **18** 7 **19** 8 at $5.00, 6 at $8.50 **21** First, $1.30, second, $1.10 **22** 3 h **23** 30 days **25** 3 h **26** 4 h **27** 2.5 h

EXERCISE 7.5

1 {Truman, Johnson, Nixon} **2** 30, 60, 90 **3** identity **4** conditional equation **7** yes **8** no **9** no **10** 2 **11** 3 **12** ab **13** 12 **14** $x > 3$ **15** $x \geq 4$ **16** $x > -3$ **17** $x \leq 15$ **18** 35, 14 **19** $8750, $9600

EXERCISE 7.6

1 {Spiro Agnew, Gerald Ford} **2** 11, 13, 19 **3** no **4** no **5** 15, 21 **6** 0 **7** 4 **8** a^2b **9** $x > 5$ **10** $x \geq 2$ **11** $x \geq 15$

EXERCISE 8.1

1 4 **2** -3 **3** 5 **5** -1 **6** 10 **7** -1 **9** 3 **10** 9 **11** 12 **13** 5 **14** -2 **15** 7 **17** 8 **18** 4 **19** 10 **21** -3 **22** 1 **23** 3 **25** 1 **26** 0 **27** 8

EXERCISE 8.2

1 $\{x|x < -\frac{2}{3}\} \cup \{x|x > 2\}$ **2** $\{x|x < -3\} \cup \{x|x > 0.8\}$ **3** $\{x|x < -2.5\} \cup \{x|x > 3\}$ **5** $\{x|x < -4\} \cup \{x|x \geq \frac{5}{3}\}$ **6** $\{x|x \leq -\frac{5}{3}\} \cup \{x|x > 1\}$ **7** $\{x|x < -1\} \cup \{x|x \geq \frac{9}{4}\}$ **9** $\{x|x < \frac{1}{3}\} \cup \{x|x \geq \frac{11}{2}\}$ **10** $\{x|x > -\frac{2}{7}\} \cup \{x|x \leq -3.5\}$ **11** $\{x|x < -\frac{9}{4}\} \cup \{x|x \geq -0.8\}$ **13** $-5 < x < 1.5$ **14** $-4 < x < \frac{8}{3}$ **15** $-\frac{10}{3} < x < 6$ **17** $-\frac{7}{2} \leq x < \frac{2}{7}$ **18** $-\frac{9}{5} \leq x < \frac{5}{9}$ **19** $-\frac{8}{3} \leq x < \frac{7}{2}$ **21** $-\frac{11}{4} \leq x < -\frac{1}{2}$ **22** $-\frac{7}{2} < x \leq -\frac{4}{7}$ **23** $\frac{1}{2} < x \leq \frac{9}{5}$

EXERCISE 8.3

1 2 **2** 12 **3** 5 **4** 4 **5** 4 **6** 9 **7** 2 **8** ∅ **9** 6 **10** 3 **11** $\{x|x < -1\} \cup \{x|x > 3\}$ **12** $-2 < x < 1$ **13** $-\frac{11}{2} < x < 1$ **14** $\{x|x < -2\} \cup \{x|x > \frac{10}{3}\}$ **15** $-\frac{1}{3} \leq x < \frac{5}{2}$ **16** $\{x|x < -\frac{1}{2}\} \cup \{x|x \geq \frac{13}{4}\}$ **17** $\{x|x \leq \frac{1}{3}\} \cup \{x|x > 4\}$ **18** $-\frac{6}{5} < x \leq 2$

EXERCISE 8.4

1 8 **2** 5 **3** 11 **4** 7 **5** 5 **6** $-\frac{2}{3} < x < \frac{9}{2}$ **7** $\{x|x < -\frac{1}{2}\} \cup \{x|x > \frac{17}{5}\}$ **8** $\{x|x \leq -\frac{7}{4}\} \cup \{x|x > -\frac{1}{3}\}$ **9** $\frac{3}{2} < x \leq \frac{19}{6}$

EXERCISE 9.1

1 $\frac{2}{7}$ **2** $\frac{2}{3}$ **3** $\frac{7}{8}$ **5** $\frac{3}{8}$ **6** 3 **7** 3 **9** $\frac{x}{y}$ **10** $a - b$ **11** $x^2 - xy + y^2$ **13** $64:1$ **14** $9:4$ **15** 12 and 16 **17** $3:1$ **18** 6 and 10 **19** 20, 15, and 30 **21** 0.0889 **22** $\frac{5}{9}$ **23** 7.84

EXERCISE 9.2

1 12 **2** 12 **3** 3 **5** 15 **6** $\frac{1}{3}$ **7** ± 15 **9** ± 5 **10** ± 6 **11** 4 **13** $x = 6, y = 8$ **14** $x = -2, y = -3$ **15** $x = 3, y = 6$ **17** $x = 12, y = 8$ **18** $x = 10, y = 3$ **19** $x = \pm 4$ **21** 154 in² **22** 4 **23** 32 mi **25** 2.94 L **26** Smith $333.33, Brown $466.67 **27** $2\frac{2}{7}$ m

532 | Answers

EXERCISE 9.3

1 $a = kb$ **2** $a = kb^2$ **3** $z = kxy$ **5** $a = k/b$ **6** $y = k/x^2$ **7** $z = k/\sqrt{w}$ **9** $a = kbc/de$ **10** $y = kz^2/(v + w)$ **11** $b = k/w$ **13** 30 **14** 24 **15** 4 **17** $y = 64/3, b = -5/2, a = -29/4$ **18** $y = 1, x = \pm 3$ **19** $p = 2, q = \sqrt{5}, r = 16$ **21** 30.2 cal **22** 9248 lb **23** 144.9 ft **25** 110.25 mi **26** 32.011 in **27** 8 lb

EXERCISE 9.4

1 $\frac{12}{13}$ **2** $\frac{3}{35}$ **3** $\frac{7}{25}$ **4** $\frac{10}{9}$ **5** $\frac{a+b}{a+2b}$ **6** $\frac{4}{25}$ **7** 24 and 36 **8** 15, 20, and 25 **9** $1,445,000, $2,167,500, and $2,890,000 **10** 1 **11** $\frac{39}{4}$ **12** 8 **13** 15 **14** 2 V **15** 11.35 **16** 1.25 dyn **17** 146.25 poundals **18** $\frac{16}{375}$ ohms **19** 980 lb **20** 7.5 lb **21** 7.99×10^{-4} dyn **22** 30 eggs

EXERCISE 9.5

1 $\frac{20}{3}$ **2** 3 cookies per boy **3** 111 g **4** 20 **5** $x = 21, y = 14$ **6** 4 sacks **7** 16 in **8** 100 **9** 136,125 ft-lb **10** 24 A

EXERCISE 10.1

1 (4,2) **2** (3,−2) **3** (−3,1) **5** (3,5) **6** (2,−5) **7** (1,−3) **9** (6,3) **10** (−2,7) **11** (5,8) **13** (4,1) **14** (2,5) **15** (6,−5) **17** (2,8) **18** (5,−7) **19** (7,5) **33** inconsistent **34** dependent **35** independent **37** independent **38** dependent **39** inconsistent

EXERCISE 10.2

1 (3,−1) **2** (2,−5) **3** (−4,3) **5** (3,$\frac{1}{2}$) **6** (5,−$\frac{2}{3}$) **7** ($\frac{3}{5}$,1) **9** ($\frac{4}{5}$,−$\frac{1}{5}$) **10** ($\frac{5}{4}$,−$\frac{1}{4}$) **11** (1,$\frac{1}{4}$) **13** (2,−3) **14** (2,2) **15** ($\frac{1}{2}$,$\frac{3}{2}$) **17** (5,7) **18** (8,6) **19** (9,−7) **21** (3,−2) **22** (6,1) **23** ($\frac{1}{4}$,−2) **25** (−5,4) **26** (3,−$\frac{1}{5}$) **27** ($\frac{1}{3}$,$\frac{5}{2}$) **29** (−$\frac{5}{6}$,$\frac{3}{2}$) **30** (8,−9) **31** (1,−$\frac{1}{3}$) **33** ($\frac{3}{7}$,−$\frac{4}{7}$) **34** (−1,2) **35** ($\frac{1}{4}$,$\frac{3}{4}$) **37** (3,4) **38** (5,4) **39** (2,4) **41** (3,4) **42** (−2,3) **43** (2,4)

EXERCISE 10.3

1 sucker, 20 cents; apple, 30 cents **2** 2 qt whipping cream, $\frac{1}{2}$ qt half-and-half **3** 5 piano, 3 organ **5** 10 one-bedroom, 7 two-bedroom **6** 165 mi at 55 mi/h, 110 mi at 40 mi/h **7** 1880 ft² with one coat,

P. 118
121-122
138
139
142
147
148
249
250
254

500 ft² with two coats   **9** $1 pennant, $2 corsage   **10** 18, shorter hike; 17, longer hike   **11** $130, $160   **13** $2\frac{1}{2}$ h, second; $1\frac{1}{2}$ h, third   **14** nineteen 18-year-olds, ten 19-year-olds, three 20-year-olds   **15** 4 at $4.60, 6 at $4.80, 2 at $5.80

EXERCISE 10.4

1 7, 5, 14, 20, −6   **2** −4, 2, −12, −14, 2   **3** 4, −2, 24, 6, 18   **5** 22   **6** −5   **7** −7   **9** 9   **10** 10   **11** −11   **17** $D = -7, D_x = -7, D_y = -49, x = 1, y = 7$ **18** $D = -7, D_x = 7, D_y = -42, x = -1, y = 6$ **19** $D = 53, D_x = -159, D_y = 212, x = -3, y = 4$ **21** (5,−2)   **22** (14,18)   **23** inconsistent   **25** $(\frac{5}{2},-\frac{7}{2})$ **26** $(-\frac{1}{3},-1)$ **27** $(\frac{2}{3},0)$ **29** $(-\frac{1}{6},-\frac{1}{6})$ **30** $(\frac{9}{8},\frac{3}{4})$ **31** $(\frac{1}{7},\frac{2}{7})$

EXERCISE 10.5

1 1, −1, 2   **2** 4, −2, 1   **3** −3, 4, 2   **5** $\frac{1}{2}$, 2, 1   **6** 2, $-\frac{1}{3}$, −1   **7** −3, 1, $\frac{3}{2}$ **9** −17   **10** 6   **11** 254   **17** $\frac{1}{3}, \frac{1}{6}, -1$ **18** $\frac{3}{4}, \frac{5}{8}, 2$ **19** 2, −1, 3

EXERCISE 10.6

1 (3,5)   **2** (2,2)   **3** (1,0)   **4** (3,5)   **5** dependent   **6** inconsistent   **7** inconsistent   **8** dependent   **9** (4,−3)   **10** $(\frac{1}{3},\frac{2}{5})$ **11** $(-\frac{1}{2},\frac{2}{3})$ **12** (2,4)   **13** $(-\frac{1}{2},\frac{4}{3})$ **14** $(\frac{1}{4},-6)$ **19** (1,4)   **20** $(1,\frac{1}{4})$ **21** $(-\frac{2}{3},5,\frac{1}{2})$ **22** $(1,\frac{1}{3},-3)$ **23** $(4,-\frac{5}{6},\frac{1}{5})$

EXERCISE 10.7

2 $b = 18$ **3** (4,3)   **4** (−2,5)   **5** $(\frac{1}{3},-\frac{2}{5})$ **6** $(\frac{1}{4},\frac{3}{2})$ **7** (1,1,1)   **8** $(0,\frac{2}{3},\frac{1}{2})$

EXERCISE 11.1

1 0, −5   **2** 0, 7   **3** 0, −21   **5** $0, \frac{7}{4}$ **6** $0, \frac{100}{3}$ **7** $0, -\frac{8}{7}$ **9** $0, \frac{2}{3}$ **10** $0, -\frac{5}{2}$ **11** $0, \frac{6}{7}$ **13** 2, −2   **14** 7, −7   **15** $\frac{9}{2}, -\frac{9}{2}$ **17** $3\sqrt{3}/5, -3\sqrt{3}/5$ **18** $4\sqrt{2}/3, -4\sqrt{2}/3$ **19** $\sqrt{5}/4, -\sqrt{5}/4$ **21** $2i, -2i$ **22** $12i, -12i$ **23** $8i/3, -8i/3$ **25** $i\sqrt{3}, -i\sqrt{3}$ **26** $i\sqrt{11}/2, -i\sqrt{11}/2$ **27** $i\sqrt{5}/9, -i\sqrt{5}/9$ **29** 1, 2   **30** 2, 3   **31** 2, −3   **33** $\frac{1}{2}, 3$ **34** $\frac{1}{3}, 2$ **35** $-\frac{1}{2}, 2$ **37** $-\frac{1}{5}, 5$ **38** $\frac{2}{3}, 4$ **39** $\frac{2}{5}, 3$ **41** $5, -\frac{2}{7}$ **42** $3, -\frac{3}{4}$ **43** $2, -\frac{2}{3}$ **45** $-2, \frac{3}{5}$ **46** $-1, \frac{2}{3}$ **47** $\frac{2}{3}, \frac{3}{2}$

EXERCISE 11.2

1 6, −2   **2** 3, −9   **3** $1, \frac{1}{3}$ **5** 2, 1   **6** 3, 2   **7** 5, 3   **9** 1, −2   **10** −4, −2   **11** 2, −5   **13** $2, \frac{1}{2}$ **14** $3, \frac{2}{3}$ **15** $5, \frac{3}{4}$ **17** $\frac{2}{5}, \frac{5}{3}$ **18** $\frac{3}{5}, \frac{5}{4}$ **19** $\frac{2}{7}, \frac{7}{2}$ **21** $-\frac{3}{5}, -\frac{2}{3}$ **22** $-\frac{6}{5}, \frac{5}{6}$ **23** $\frac{5}{3}, -\frac{1}{2}$

25 $1 \pm \sqrt{2}$ **26** $2 \pm \sqrt{3}$ **27** $3 \pm \sqrt{5}$ **29** $4 \pm 2i$ **30** $3 \pm i$ **31** $5 \pm 3i$ **33** $2.414, -0.414$
34 $4.414, 1.586$ **35** $-2.268, -5.732$ **37** $0.366, -1.366$ **38** $1.079, -0.412$ **39** $2.823, 0.177$

EXERCISE 11.3

1 $a = 1, b = 5, c = -1$ **2** $a = 2, b = -5, c = -2$ **3** $a = -3, b = -5, c = -8$ **5** $3, 1$ **6** $2, 5$ **7** $4, 2$ **9** $3, \tfrac{2}{5}$ **10** $2, \tfrac{1}{3}$ **11** $2, \tfrac{3}{7}$ **13** $\tfrac{2}{3}, \tfrac{3}{4}$ **14** $\tfrac{3}{4}, \tfrac{4}{5}$ **15** $\tfrac{4}{5}, \tfrac{3}{7}$ **17** $-\tfrac{2}{5}, \tfrac{3}{4}$ **18** $-\tfrac{3}{4}, \tfrac{2}{3}$ **19** $-\tfrac{2}{3}, -\tfrac{1}{2}$ **21** $2 \pm \sqrt{5}$ **22** $3 \pm \sqrt{2}$ **23** $\tfrac{1}{2}(2 \pm \sqrt{3})$ **25** $2 \pm 3i$ **26** $3 \pm 2i$ **27** $\tfrac{1}{2}(3 \pm i)$ **29** $\tfrac{1}{2}(1 \pm i\sqrt{3})$ **30** $\tfrac{1}{3}(2 \pm i\sqrt{5})$ **31** $\tfrac{1}{3}(3 \pm i\sqrt{2})$ **33** $2, -\tfrac{2}{5}$ **34** $4, -\tfrac{2}{5}$ **35** $\tfrac{1}{3}, \tfrac{2}{3}$ **37** $-3 \pm \sqrt{7}$ **38** $\tfrac{1}{2}(17 \pm 3\sqrt{37})$ **39** $\tfrac{1}{4}(-5 \pm i\sqrt{43})$ **41** $3.732, 0.268$ **42** $4.414, 1.586$ **43** $7.236, 2.764$ **45** $2.207, 0.793$ **46** $1.244, 0.089$ **47** $1.411, 0.089$

EXERCISE 11.4

1 2 **2** 1 **3** no solution   **5** $-2, -3$ **6** 0 **7** 5 **9** 3 **10** no solution   **11** 3 **13** 2 **14** 4 **15** -2 **17** $-3, 2$ **18** -2 **19** $4, 0$ **21** 2 **22** $\tfrac{5}{16}$ **23** 2 **25** 2 **26** 3 **27** $3, -1$ **29** 1 **30** 0 **31** 2 **33** 2 **34** 0 **35** 1 **37** 2 **38** 3 **39** -1

EXERCISE 11.5

1 $\dfrac{2s}{a+l}$ **2** $\dfrac{2s}{n} - l$ **3** $\dfrac{ar}{a-r}$ **5** $\sqrt{\dfrac{2h}{g}}$ **6** $\sqrt{\dfrac{A}{\pi}}$ **7** $\dfrac{p^2 g}{2}$ **9** $\tfrac{1}{32}(v \pm \sqrt{v^2 - 64y})$ **10** $\sqrt{\dfrac{v^2 - 2g}{h}}$ **11** $\dfrac{2c\sqrt{2}}{3} \approx 0.94c$ **13** $\sqrt{\dfrac{4A}{nc}}$ **14** $\sqrt{\dfrac{360A}{\pi\theta}}$ **15** $\sqrt{\dfrac{360A}{\pi\theta - 180s}}$ **17** $-2, -1, 1, 2$ **18** $-2, 2, -4i, 4i$ **19** $-\tfrac{1}{2}, \tfrac{1}{2}, -\tfrac{1}{3}, \tfrac{1}{3}$ **21** $1, 2$ **22** $2, 3$ **23** $2, -2, \tfrac{3}{2}, -\tfrac{3}{2}$ **25** $2, -2, 3, -3$ **26** $2, -2, 3, -3$ **27** $1, -1, 2, 4$ **29** $-3, -\tfrac{3}{5}$ **30** $6, -1$ **31** $-10, 4$ **33** $c = 15$ **34** $c = 41$ **35** $c = 9.487$ **37** $b = 20$ **38** $b = 24$ **39** $a = 9.798$ **41** 12 and 16 cm   **42** 15 and 36 m   **43** $10\sqrt{162} = 127.279$ ft   **45** 5 **46** $\sqrt{40} = 6.325$ **47** $\sqrt{52} = 7.211$

EXERCISE 11.6

1 $\{x | 1 < x < 4\}$ **2** $\{x | -1 < x < 4\}$ **3** $\{x | -2 < x < 6\}$ **5** $\{x | x < -1\} \cup \{x | x > 2\}$ **6** $\{x | x < 3\} \cup \{x | x > 8\}$ **7** $\{x | x < -1\} \cup \{x | x > 5\}$ **9** $\{x | -2 < x < \tfrac{5}{2}\}$ **10** $\{x | -\tfrac{1}{3} < x < \tfrac{7}{2}\}$ **11** $\{x | x < -\tfrac{2}{3}\} \cup \{x | x > \tfrac{3}{4}\}$ **13** $\{x | -3 < x < 7\}$ **14** $\{x | -5 < x < -1\}$ **15** $\{x | x < 3\} \cup \{x | x > 10\}$ **17** $\{x | x < -\tfrac{5}{6}\} \cup \{x | x > 2\}$ **18** $\{x | x < -5\} \cup \{x | x > \tfrac{3}{4}\}$ **19** $\{x | -\tfrac{1}{4} < x < 3\}$ **21** $\{x | -\tfrac{1}{3} < x < \tfrac{2}{3}\}$ **22** $\{x | -\tfrac{7}{4} < x < \tfrac{2}{3}\}$ **23** $\{x | x < \tfrac{1}{3}\} \cup \{x | x > \tfrac{5}{2}\}$ **25** $\{x | -1 < x < 2\}$ **26** $\{x | -3 < x < 4\}$ **27** $\{x | x < 4\} \cup \{x | x > 8\}$

EXERCISE 11.7

1 $0, 7$ **2** $0, -\frac{5}{2}$ **3** $\pm\sqrt{7}$ **4** $\pm\frac{i}{2}\sqrt{10}$ **5** $\pm\frac{1}{3}\sqrt{14}$ **6** $5, -3$ **7** $7, -\frac{3}{2}$ **8** $\frac{1}{4}, -\frac{9}{2}$ **9** $7, -4$ **10** $\frac{3}{2}, -2$ **11** $\frac{6}{5}, -\frac{4}{3}$ **12** $\frac{7}{4}, -\frac{5}{3}$ **13** $-\frac{7}{2}, \frac{5}{3}$ **14** $\frac{9}{4}, -\frac{5}{2}$ **15** $4, -2$ **16** $5, -4$ **17** $\frac{2}{3}, -\frac{5}{3}$ **18** $1, \frac{2}{3}$ **19** $\frac{3}{2}, \frac{1}{3}$ **20** $1 \pm \sqrt{5}$ **21** $5 \pm i$ **22** $3 \pm 4i$ **23** $5, -6$ **24** $7, -3$ **25** $\frac{7}{3}, -\frac{5}{2}$ **26** $\frac{5}{3}, -\frac{4}{3}$ **27** $2 \pm \sqrt{3}$ **28** $5 \pm \sqrt{7}$ **29** $-3 \pm i$ **30** $4 \pm 2i$ **31** $\frac{1}{3}(3 \pm 5i)$ **32** $\frac{1}{5}(2 \pm i\sqrt{5})$ **33** $2b/c, -b/2c$ **34** $r + t, r - 3t$ **35** $1.098, -4.098$ **36** $0.380, -4.380$ **37** $3.886, -0.386$ **38** $1.148, -0.348$ **39** $\frac{1}{2}(1 \pm 3i)$ **40** $\frac{1}{3}(2 \pm \sqrt{58})$ **41** $1, \frac{1}{2}$ **42** 3 **43** $1, 5$ **44** $3, -1$ **45** $\frac{3}{2}, 1$ **46** 5 **47** $2, -2, \frac{1}{2}, -\frac{1}{2}$ **48** $2, -2, 1, -1$ **49** $c = 50$ **50** $c = \sqrt{89} = 9.434$ **51** $b = 24$ **52** $a = \sqrt{72} = 8.485$ **57** $\{x | x < -\frac{1}{3}\} \cup \{x | x > 5\}$ **58** $\{x | x < \frac{5}{9}\} \cup \{x | x > \frac{9}{2}\}$ **59** $\{x | -3 < x < \frac{5}{3}\}$ **60** $\{x | -\frac{10}{3} < x < -\frac{1}{4}\}$

EXERCISE 11.8

1 $\pm 6i/5$ **2** $\pm\frac{6}{5}$ **3** $0, \frac{36}{25}$ **4** $3, -6$ **5** $-\frac{5}{2}, \frac{2}{5}$ **6** $1 \pm \sqrt{3}$ **7** $\frac{1}{2}(-1 \pm 3i)$ **8** $\frac{1}{2}(2 \pm \sqrt{7})$ **9** $\frac{1}{3}(1 \pm 4i)$ **10** 9 **11** 6 **12** $-5, -4, 1, 2$ **13** $2, 0$ **14** $\{x | -\frac{1}{3} < x < \frac{7}{2}\}$ **15** $\{x | x < 3\} \cup \{x | x > 10\}$ **17** all points on the line $x + 2y - 10 = 0$

EXERCISE 12.1

1 3 **2** 3 **3** 5 **5** exponent, logarithm **6** logarithm, exponent **7** $\frac{1}{2}, 121$ **9** $2, 8$ **10** $5, 8$ **11** $\log_4 32 = \frac{5}{2}$ **13** $\log_2 64 = 6$ **14** $\log_6 36 = 2$ **15** $\log_6 \frac{1}{216} = -3$ **17** $\log_{64} 4 = \frac{1}{3}$ **18** $\log_{36} 6 = \frac{1}{2}$ **19** $\log_9 27 = \frac{3}{2}$ **21** $7^2 = 49$ **22** $49^{1/2} = 7$ **23** $5^{-4} = \frac{1}{625}$ **25** $5^1 = 5$ **26** $3^0 = 1$ **27** $8^{5/3} = 32$ **29** 4 **30** $\frac{2}{3}$ **31** $\frac{3}{2}$ **33** 2 **34** 25 **35** 625 **37** 81 **38** $\frac{1}{81}$ **39** $\frac{1}{36}$ **41** $\log_5 10$ **42** $\log_8 12$ **43** 1.7

EXERCISE 12.2

1 $4, 3, 7$ **2** $9, 6, 3$ **3** $4, 2, 8$ **5** -1 **6** $\frac{2}{3}$ **7** 1 **9** $\log_5 48$ **10** $\log_4 5$ **11** $\log_4 1296$ **13** $\log_{10} 2xy$ **14** $\log_9 \frac{8x}{y}$ **15** $\log_8 \sqrt[3]{xy^2}$ **17** 13 **18** 9 **19** 3 **21** 0 **22** 3 **23** 2 **25** $\frac{8}{3}$ **26** $\frac{10}{3}$ **27** 5 **29** $3x$ **30** $x + 2y$ **31** $3x + y$ **33** $y + z + 1$ **34** $y + 1 - 2x - z$ **35** $y + \frac{z}{2}$

EXERCISE 12.3

1 a, b, d **2** a, b, c **3** b, c, d **5** a, c, d **6** a, b, c **7** a, c, d **9** 4.673×10^3 **10** 4.73×10^1 **11** 8.92×10^{-1} **13** 8.07×10^2 **14** 4.106×10^3 **15** 5.07×10^{-2} **17** 8.50×10^2 **18** 8.5×10^2 **19** 3.60×10^3 **21** 1 **22** 3 **23** 2 **25** 4 **26** 3 **27** 5 **29** -1 **30** -3 **31** -6 **33** 1.30 **34** $9.48 - 10$ **35** $8.30 - 10$ **37** 1.78 **38** 3.78 **39** $8.82 - 10$ **41** 1.60 **42** 2.96 **43** 3.90 **45** 0.15 **46** $8.70 - 10$ **47** 0.36

EXERCISE 12.4

1 2.8621 **2** 1.9196 **3** $9.5877 - 10$ **5** $8.6911 - 10$ **6** $7.1399 - 10$ **7** 1.4914 **9** $8.7482 - 10$ **10** 0.6990 **11** 0.9031 **13** $8.8451 - 10$ **14** $6.9542 - 10$ **15** $7.3010 - 10$ **17** 3.0000 **18** 5.0000 **19** 3.5145 **21** 5.7875 **22** 2.9350 **23** 1.8704 **25** $9.1399 - 10$ **26** $8.3945 - 10$ **27** $7.6243 - 10$ **29** 4.5798 **30** 5.4771 **31** $9.5391 - 10$ **33** $4.6990 - 10$ **34** 2.1038 **35** 2.6830 **37** $9.8932 - 10$ **38** $8.1430 - 10$ **39** $6.6866 - 10$ **41** 0.7868 **42** $9.9106 - 10$ **43** 2.9196 **45** $7.7955 - 10$ **46** $6.7138 - 10$ **47** 2.2117 **49** 6.8936 **50** $9.7141 - 10$ **51** $7.3302 - 10$ **53** 3.8397 **54** 1.9098 **55** 2.9693 **57** 2.7234 **58** 3.5778 **59** $9.2702 - 10$

EXERCISE 12.5

1 14.6 **2** 5920 **3** 34,100 **5** 303 **6** 7630 **7** 4.15 **9** .176 **10** .00622 **11** .000397 **13** .0468 **14** .914 **15** .00419 **17** 2.664 **18** 37.35 **19** 548.3 **21** .6863 **22** .002363 **23** .01672 **25** 55.05 **26** 6.045 **27** 3062 **29** .7816 **30** .004285 **31** .0002178

EXERCISE 12.6

1 60.7 **2** 3.59 **3** .416 **5** 1.72 **6** .0887 **7** 17.8 **9** 1.92 **10** 0.451 **11** 0.0793 **13** 7.47 **14** 0.395 **15** 0.237 **17** 21.7 **18** 0.937 **19** 0.480 **21** 322 **22** 3.12 **23** 3.41 **25** 47.2 **26** 0.458 **27** 1.29 **29** 147.1 **30** 0.07742 **31** 220.2 **33** 2.731 **34** 0.9621 **35** 0.4546 **37** 22.39 **38** 0.4023 **39** 0.05356 **41** 2.81 **42** 0.356 **43** 2.07 **45** 1.06 **46** .957 **47** 2.03 **49** $(-1.10)10^3$ **50** .929 **51** 10.4 **53** 128 **54** 1.13 **55** 1.21

EXERCISE 12.7

1 4 **2** $\frac{3}{2}$ **3** exponent, logarithm **4** logarithm, exponent **5** 4, 7 **6** $\log_7 5$ **7** $\log_{1/4} \frac{1}{8} = \frac{3}{2}$ **8** $(\frac{1}{2})^{-4} = 16$ **9** 125 **10** $\frac{1}{27}$ **11** $\frac{2}{3}$ **12** $\frac{3}{2}$ **13** 27 **14** 16 **15** $\log_4 19$ **16** $\log_5 4$

17 $\log_9 4$ **18** $\log_8 72$ **19** $\log_5 \sqrt{2}$ **20** $\log_3 (8\sqrt{2})$ **21** 1.92 **22** 0.23 **23** 0.67 **24** 2.04 **25** 2,.6739 **26** −3,.4871 **27** −2,.0934 **28** 8,.1107 **29** −1,.9045 **30** −1,.4921 **31** 15 **32** 35 **33** 224 **34** .6187 **35** 1.3288 **36** 8.1694 − 10 **37** 3.9250 **38** 206 **39** 254 **40** 1.71 **41** .372 **42** .971 **43** 1.54 **44** .459 **45** .0713 **46** 733.1 **47** 23.83 **48** 6.0449 **49** .9464 **50** 2.65 **51** 2.70 **52** 2.716 **53** 35.8 **54** .410 **55** 17.0 **56** 2.67 **57** 4.15 **58** 48.1

EXERCISE 12.8

1 $\frac{1}{8}$ **2** 27 **3** $-\frac{3}{2}$ **4** $-\frac{1}{2}$ **5** 4 **6** 32 **7** 10 **8** 3 **9** 10 **10** 2 **11** 0 **12** 1.9360 **13** 4.6096 **14** 7.8451 − 10 **15** 8.1163 − 10 **16** 215 **17** 55.3 **18** 0.9135 **19** 105 **20** .00634 **21** 1.27 **22** .320 **23** .367 **24** 3.87 **25** .3453

EXERCISE 13.1

5 $\frac{3}{5}, \frac{4}{5}, \frac{3}{4}, \frac{4}{3}, \frac{5}{4}, \frac{5}{3}$ **6** $\frac{5}{13}, \frac{12}{13}, \frac{5}{12}, \frac{12}{5}, \frac{13}{12}, \frac{13}{5}$ **7** $\frac{7}{25}, \frac{24}{25}, \frac{7}{24}, \frac{24}{7}, \frac{25}{24}, \frac{25}{7}$ **9** $\frac{5}{13}, \frac{12}{13}, \frac{5}{12}, \frac{12}{5}, \frac{13}{12}, \frac{13}{5}$ **10** $\frac{7}{25}, \frac{24}{25}, \frac{7}{24}, \frac{24}{7}, \frac{25}{24}, \frac{25}{7}$ **11** $\frac{8}{17}, \frac{15}{17}, \frac{8}{15}, \frac{15}{8}, \frac{17}{15}, \frac{17}{8}$ **13** 1.9 **14** 19 **15** 88 **17** 0.5640 **18** 0.8675 **19** 0.8744 **21** 0.2952 **22** 1.6003 **23** 0.0963 **25** 0.3175 **26** 0.1512 **27** 0.9560 **29** 0.9255 **30** 0.4929 **31** 0.7339 **33** 5°50′ **34** 19°10′ **35** 36°40′ **37** 51°20′ **38** 64°40′ **39** 71°30′ **41** 32°21′ **42** 20°11′ **43** 37°56′ **45** 56°18′ **46** 65°17′ **47** 73°52′

EXERCISE 13.2

1 $a = 272, b = 504, B = 61°40′$ **2** $a = 5.86, b = 4.31, B = 36°20′$ **3** $b = 74.6, a = 36.1, A = 25°50′$ **5** $b = 178, c = 286, A = 51°30′$ **6** $A = 37°40′, b = 149, c = 188$ **7** $B = 34°20′, a = 473, c = 573$ **9** $B = 22°10′, b = 2.29, c = 6.06$ **10** $B = 57°30′, b = 177, c = 210$ **11** $A = 28°40′, a = 39.3, c = 81.8$ **13** $A = 60°20′, B = 29°40′, a = 632$ **14** $B = 32°50′, A = 57°10′, a = 7.77$ **15** $A = 43°40′, B = 46°20′, b = 66.0$ **17** $A = 31°33′, B = 58°27′, c = 40.77$ **18** $A = 49°13′, B = 40°47′, c = 95.34$ **19** $A = 61°35′, B = 28°25′, c = 100.3$ **21** 1.43×10^3 ft **22** 202 ft, 21 ft **23** 118 ft/s **25** 190 mi/h **26** 150 ft **27** 421 mi/h, 87°

EXERCISE 13.3

1 0.9744 **2** −0.1392 **3** −0.6494 **5** −0.9013 **6** −0.8491 **7** −1.6534 **9** −6.4971 **10** −1.1257 **11** 0.7919 **13** $C = 74°10′, b = 319, c = 332$ **14** $B = 79°10′, b = 20.2, c = 17.7$

15 $A = 65°50'$, $a = 68.8$, $c = 53.6$ **17** $B = 45°50'$, $b = 568$, $c = 486$ **18** $C = 46°20'$, $b = 4.13$, $c = 3.03$ **19** $C = 108°50'$, $a = 63.5$, $b = 33.0$ **21** $A = 59°7'$, $b = 3721$, $c = 4628$ **22** $A = 25°1'$, $a = 4041$, $b = 6570$ **23** $B = 42°40'$, $a = 5661$, $c = 3851$ **25** two solutions **26** one solution **27** no solution **29** $B = 68°30'$, no solution **30** $B = 13°20'$, $A = 132°20'$, $a = 620$ **31** $A = 50°20'$, $B = 93°20'$, $b = 315$, $A' = 129°40'$, $B' = 14°$, $b' = 76.4$ **33** 75.9 km **34** 176 m **35** 39 **37** 456, 204, 660 mi **38** 13° **39** 2 min

EXERCISE 13.4

1 11 **2** 70 **3** 54 **5** 46 **6** 86 **7** 100 **9** 356 **10** 527 **11** 529 **13** 43° **14** 96° **15** 63° **17** 148°40' **18** 67°50' **19** 99°40' **21** 42°0' **22** 52°30' **23** 37°50' **25** 1 h 9 min **26** 1300 m **27** 101° between the first and second legs of the flight

EXERCISE 13.5

1 $\frac{8}{17}, \frac{15}{17}, \frac{8}{15}, \frac{15}{8}, \frac{17}{15}, \frac{17}{8}$ **2** 0.5200 **3** 3.2041 **4** 0.7967 **5** 0.1503 **6** -0.9474 **7** 0.9158 **8** 16°30' or 163°30' **9** 71°10' **10** $B = 48°30'$, $a = 272$, $c = 410$ **11** $B = 17°50'$, $b = 236$, $a = 732$ **12** $A = 32°30'$, $B = 57°30'$, $c = 508$ **13** $C = 64°20'$, $b = 370$, $c = 342$ **14** $B = 58°10'$, $C = 43°30'$, $c = 233$ **15** 57° **16** 226 **17** 218 m **18** 549 mi

EXERCISE 13.6

1 $A = 60°30'$, $b = 137$, $c = 279$ **2** $B = 51°20'$, $b = 334$, $c = 427$ **3** $A = 42°40'$, $B = 47°20'$, $b = 576$ **4** $C = 76°50'$, $a = 310$, $b = 376$ **5** $c = 62$, $A = 42°$, $B = 36°$ **6** $C = 109°$

Index

Abscissa, 179
Absolute value, 66
Accuracy, 43
Addition:
  axioms of, 64, 65
  cancellation theorem for, 69
  of decimal fractions, 3
  of fractions, 27, 143
  law of signs for, 66
  of monomials, 67
  of polynomials, 67
  of radicals, 220
Additive inverse, 3
Algebraic solution, 403
Ambiguous case, 485
Angle:
  complementary, 463
  cosecant of, 463
  cosine of, 462
  cotangent of, 462

Angle:
  definition of, 461
  of depression, 475
  of elevation, 475
  measure of, 462
  secant of, 463
  sine of, 462
  supplementary, 483
  tangent of, 462
  vertex of, 461
Antilog, 439
Approximate numbers:
  addition, 42
  products of, 41
Area, lateral, 165
    total surface, 165
Arithmetic, fundamental
    theorem of, 11
Associative axiom:
  of addition, 2, 65

Associative axiom:
  of multiplication, 10, 75
Axes, coordinate, 179
Axiom:
  associative: for addition, 2, 65
    for multiplication, 75
  closure: for addition, 65
    for multiplication, 75
  commutative: for addition, 65
    for multiplication, 75
  distributive, 10, 65, 75
  reflexive, 68
  symmetric, 68
  transitivity, 68

Base:
  of exponential term, 191
  of logarithm, 419

Base:
  numbers to various, 93
Binary numerals, 93–96
Binary operation, 65
Binomial(s):
  definition of, 64
  product of two, 107
  square of, 108
Briggsian logarithms, 429

Calculators, 49, 463
Cancellation theorem:
  for addition, 69
  for multiplication, 76
Carrying, 3
Characteristic:
  definition of, 430
  method for obtaining, 430
Charts, 163
Circle, 165
Closure axiom:
  for addition, 65
  for multiplication, 75
Coefficient, 64
Combined variation, 301
Combining similar terms, 64
Common factors, 115
Common logarithm, 429
Commutative axiom:
  for addition, 2, 65
  for multiplication, 10
Complementary angles, 463
Completing the square, 375–376
Complex fractions, 151
Complex numbers, 227
Computation, logarithmic, 443–444
Conditional equation, 240
Cone, right circular, 165
Constant:
  definition of, 175
  of variation, 301
Coordinate axes, 179
Coordinate system, rectangular, 179
Coordinates of point, 179
Cosecant, definition of, 463
Cosine(s):
  definition of, 462
  law of, 493
Cotangent, definition of, 462

Cramer's rule, 339, 347
Cubes, sum or difference of two, 119

Decimal fraction(s):
  addition of, 3
  conversion of, 20
  division of, 13
  multiplication of, 11
  repeating, 21
  subtraction of, 3
Decimal point, 2
  locating the, 11, 13
  reference position of, 429
Degree, 462
Denominate numbers:
  accuracy of, 43
  computation involving, 43
  conversion of, 44
  definition of, 43
Denominator(s), 12, 19, 64, 133
  least common, 20
  lowest common, 143
  rationalizing, 213, 220
Dependent equations, 318
Dependent variable, 176
Determinants:
  definition of, 339
  of second order, 339
  of third order, 345–347
    minor of element of, 346
    in terms of minors, 346
  use of, in solving equations, 339, 347
Difference, 3, 64
  of two cubes, 119
  of two squares, 119
Digits, 2
  significant, 43, 429
Direct variation, 301
Direction of a line, 476
Distance, directed, 179
Distributive law, 10, 65, 75
Dividend, 12, 85
Division, 12–14
  of decimal fractions, 13
  of fractions, 31
  law of exponents for, 86
  law of signs for, 85
  of monomials, 85–87
  of polynomials, 87

Division:
  of radicals, 213
  zero in, 14
Divisor, 12, 85
Domain, 175

Element:
  of determinant, 346
  of set, 50
Elimination:
  by addition or subtraction, 323, 345
  by substitution, 324, 345
Empty set, 51
Equality, axioms of, 68
Equation(s):
  conditional, 240
  definition of, 240
  dependent, 318
  equivalent, 241, 315
  fractional, 275–288, 382
  inconsistent, 318
  independent, 318
  linear, 247, 315
  quadratic, 367–418
  in quadratic form, 395
  radical, 389
  rationalizing the, 390
  root of, 240
  simultaneous, 315–366
  solution set of, 240
Expansion:
  of a determinant of third order, 345
  of second order, 339
  in terms of minors, 345
Exponent(s):
  definition of, 77, 419
  fractional, 207
  integral, 191
  law of: for division, 86
    for multiplication, 77
  negative, 199
Exponential form, 420
Expression, 64
Extremes, 295

Factors, common, 115
  definition of, 9, 20, 64
  of difference of two squares, 119

Index

Factors, common: of quadratic trinomials, 123, 124
   of sum or difference of two cubes, 119
Formulas, solution of, 395
Fraction(s):
   addition of, 27, 143
   algebraic, 133–162
   complex, 151
   conversion of, 19
   definition of, 64
   denominator of, 12, 19, 64, 133
   division of, 31
   equal, 133
   fundamental principle of, 19, 134
   generating, 21
   members of, 12
   multiplication of, 31, 86, 135
   numerator of, 12, 19, 64, 133
   product of, 19
   quotient of two, 32
   reduction to lowest terms, 135
   signs of, 134–135
Fractional equations, 275–288, 382
Fractional exponents, 207
Fractional inequalities, 279–282
Function:
   definition of, 175
   domain of, 175
   graph of, 179
   notation for, 176
   range of, 175
   value, 176
Fundamental operations of algebra, 64

Generating fraction, 21
Geometry, numerical, 164
Graph:
   of an equation in two variables, 316
   of a function and a relation, 179–182
Graphical representation of positive integers, 1–2
Graphical solutions, 317
Grouping, symbols of, 78

Hindu-Arabic number system, 1
Hypotenuse, 164, 462, 473

Identity, 240
Imaginary numbers, 227
Inconsistent equations, 318
Independent equations, 318
Independent variable, 176
Index of radical, 207
Inequality:
   definition of, 253
   equivalent, 253
   fractional, 279
Integers, positive, 1
Intercepts, 316
Interpolation, 434, 439
Intersection of sets, 51
Inverse variation, 301
Irreducible, 115
Isolating a radical, 390

Joint variation, 301

Law(s):
   of cosines, 493–495
   of exponents, 86
   of signs: for addition, 66
      for division, 85
      for multiplication, 76
   of sines, use of, 484–486
Least common denominator, 20
Linear equation(s):
   graph of, 316
   systems of: in one variable, 247
      in three variables, 345
      in two variables, 315
Linear inequality, 253
Literal part, 64
Logarithm:
   base of, 419
   Briggsian, 429
   characteristic of, 430
   common, 429

Logarithm:
   definition of, 419
   mantissa of, 430
   properties of, 425
Logarithmic computation, 443–444
Logarithmic form, 420
Logarithmic table, 433–435
Lowest common denominator, 143

Mantissa:
   definition of, 430
   method for obtaining, 433
Means, 295
Metric system, 42, 43
Minor of a determinant, 346
Minuend, 3
Monomial(s):
   addition of, 67
   definition of, 64
   division involving, 85–87
   multiplication of, 77
   product of, 77
   quotient of two, 85–87
Multinomial, 64
Multiplication:
   associative axiom for, 10
   cancellation theorem for, 76
   closure axiom for, 75
   commutative axiom for, 10
   of decimal fractions, 11
   of fractions, 31, 86, 135
   identity element for, 10, 75
   inverse for, 75
   law of exponents for, 76
   law of signs for, 76
   of monomials, 77
   of polynomials, 219
   of radicals, 213
   table, 9
   by zero, 10

Natural number, 19
Negative of a sum, 69
Negative exponents, 199
Notation, scientific, 41, 429
Null set, 51
Number(s):
   complex, 227
   composite, 11

Number(s):
  denominate, 43
  imaginary, 227
  mixed, 21
  prime, 11, 20
  rational, 19
  real, 64
  reciprocal of, 75
  root of, 207
  square root of, 207
Number system:
  to bases other than 10, 93–96
  Hindu-Arabic, 1
  rational, 19
  real, 64
Numerator, 12, 19, 64, 133
Numerical geometry, 164
Numerical trigonometry, 461–508

Oblique triangle, 484
Open sentence:
  definition of, 239
  truth set for, 240
Operation(s):
  binary, 65
  four fundamental, 64
Ordered pairs of numbers, 175
Ordinate, 179
Origin, 179

Parallelogram, 164
Parentheses, removal of, 69
Partial product, 11
Percentage, 35
Perfect squares, 124, 375
Place value, 2
Plotting point, 179
Polygon, regular, 164
Polynomial(s):
  addition of, 67
  definition of, 64
  multiplication of, 219
  product of, 219
  quotient of two, 87
  square of, 111
Positive integers, exponents, 191
Power:
  of a number, 191
  of power, 192
  of product, 192

Power:
  of quotient, 192
Prime number, 11, 20
Principal root, 207
Prism, 165
Product(s), 9
  of approximate numbers, 41
  of fractions, 19, 136
  of monomials, 77
  of sum and difference of two numbers, 111
  of two binomials, 107
  of two polynomials, 77
  of two powers, 77, 191
  of two radicals, 213
Properties of logarithms, 425
Proportion, 295–300
Pyramid, 165
Pythagorean theorem, 395

Quadrants, 179
Quadratic equation:
  definition of, 367
  solution of: by completing the square, 375–376
    by factoring, 367–369
    by use of formula, 381
Quadratic formula, 381
Quadratic inequalities, 403–404
Quadratic trinomial:
  definition of, 123
  factors of, 123, 124
Quadrilaterals, 164
Quotient:
  of two fractions, 31
  of two polynomials, 87
  of two powers, 191
  of two radicals, 213

Radical(s):
  addition of, 220
  changing order of, 219
  equations, 389
  index of, 207
  product of, 213
  quotient of two, 213
  simplification of, 213
Radicand, 207
Range, 175
Ratio, 289
  trigonometric, 462

Rational numbers, 19
Rationalizing denominators, 213, 220
Ray, 461
Real numbers:
  definition of, 64
  set of, 64
Reciprocal, 75
Rectangle, 164
Rectangular coordinate system, 179
Reduction to lowest terms, 135
Reference position for decimal point, 41, 429
Relations, 175
  graph of, 179–182
Remainder, 88
Removal of parentheses, 69
Repetend, 21
Replacement set, 240
Right triangle, solution of, 473
Root:
  of equation, 240
  of number, 207
  principal, 207
Rounding off numbers, 41

Scientific notation, 41, 429
Secant, definition of, 463
Sentence, open, 239
Set(s):
  complement of, 51
  definition of, 50
  element of, 50
  empty, 51
  equality of, 50
  intersection of, 51
  notation for, 50
  null, 51
  replacement, 52
  solution, 194
  truth, 240
  union of, 51
  universal, 51
Side:
  adjacent, 462
  opposite, 462
Significant digits, 43, 429
Similar terms, 64
Simplification:
  of complex fraction, 151

Simplification:
  of exponential expressions, 208
  of radicals, 213
Simultaneous equations, 315–366
  solution pair, 315
  solution set, 240
Sine(s):
  definition of, 462
  law of, 484
Solids, 165
Solution pair, 315
  of quadratic equations (*see* Quadratic equation)
Solution set, 240
  of stated problems, 257
Sphere, 165
Square:
  completing, 375–376
  of a polynomial, 111
Subset, 50
Substitution, elimination by, 324, 345
Subtraction, 3, 64
Subtrahend, 3
Sum, 2
Symbols:
  of grouping, 78
  of inequality, 253
Symmetric axiom of equality, 68

Systems of linear equations, 315–366
Systems of numbers to bases other than 10, 93–96

Tables:
  conversion factors, 43
  of logarithms, use of, 433–435
  of natural functions, use of, 463–465
  of weights and measures, 44
Tangent, 462, 463, 501
Term, 64
Terms, similar, 64
Transitivity axiom of equality, 68
Transposing, 247
Trapezoid, 164
Triangle(s), 164
  oblique, 484
  right, 164, 473
  similar, 296
Trigonometric ratios:
  of acute angle, 462
  of obtuse angle, 483
Trigonometry, numerical, 461–508
Trinomials:
  definition of, 64
  quadratic, 123

Trinomials:
  that are perfect squares, 124
Truth set, 240

Union of sets, 51
Universal set, 51

Variable:
  definition of, 175, 239
  dependent, 176
  independent, 176
  replacement set for, 240
Variation, 301–306
  combined, 301
  constant of, 301
  direct, 301
  inverse, 301
  joint, 301
Venn diagram, 51
Vertex, 461
Vertical lines and graphs, 182

Weights and measures, tables of, 44

Zero:
  definition of, 3, 65
  in division, 14
  as exponent, 87
  as factor, 78
  in multiplication, 10